APPLIED MINERALOGY

A QUANTITATIVE APPROACH

Dedication

DIOLCH O GALON I THELMA
AM
EU AMYNEDD
AC AM
EU CHYMORTH

APPLIED MINERALOGY

A QUANTITATIVE APPROACH

Meurig P. Jones

Mineral Resources Engineering Department
Imperial College, London

Graham & Trotman

A member of the Kluwer Academic Publishers Group

First published in 1987 by

Graham and Trotman Ltd.
Sterling House
66 Wilton Road
London SW1V 1DE
UK

Graham and Trotman
Kluwer Academic Publishers Group
101 Philip Drive
Assinippi Park
Norwell, MA 02061
USA

British Library Cataloguing in Publication Data

Jones, Meurig P.
    Applied mineralogy: a quantitative approach.
    1. Mineralogy, Determinative
    I. Title
    549′.1   QE367.2

ISBN 0 86010 510 5 (hardback)
ISBN 0 86010 511 3 (paperback)

LCCCN 86-083175

Typeset in Great Britain by
Acorn Bookwork, Salisbury, Wiltshire

Contents

Chapter 4   Mineral identification   46

Chapter 5   The polarising microscope in applied mineralogy   58

Chapter 9   X-rays, electron beams and miscellaneous methods of mineralogical analysis

Chapter 10   The role of mineralogy in mineral processing

Preface

This book is based on the contents of a number of undergraduate and post-graduate courses given by the author at the Royal School of Mines, Imperial College, London. The material has also formed the basis of seminars carried out in Chile, China, Czechoslovakia, Indonesia, Portugal, Thailand, Turkey, the USA and the United Kingdom, on "Mineralogy Applied to the Mineral Industry". This kind of study is sometimes referred to as "process mineralogy".

The primary purpose of this book is to show students of mineral engineering and practising engineers (who are seldom specialist mineralogists) how to obtain and use quantitative information on the mineralogical characteristics which affect the treatment processes and uses of rocks, ores and mineral products. The book will also be useful to students and specialists working in the fields of geology, petrology, extractive metallurgy, mining and chemical engineering. It differs from most other books on mineralogy in that the approach is essentially practical throughout and that every effort is made to cover methods of analysis that produce quantitative, numerical mineralogical results which can then be used for process design and process control purposes.

The book is not directed specifically at the experienced crystallographer–mineralogist, nor does it set out to describe in detail the theoretical bases of crystallography, crystal growth, crystal chemistry or crystal optics – these subjects are already adequately dealt with in a number of excellent textbooks. Instead, close attention is paid to the methods that are now available for collecting the detailed numerical data on mineralogical characteristics demanded by the mineral production industry and by those industries that use mineral-based raw materials.

The manual skills needed to make accurate analyses of mineralogical features such as mineral proportions and grain-size distributions can only be acquired by constant practice. The book therefore gives details of a number of laboratory exercises which have been designed to improve the student's skill and to develop his or her confidence in the results obtained.

In addition, the book describes a selection of mechanical and electronic devices that are now used to collect the necessary mineralogical data. It also provides information that will allow the engineer to select the most appropriate measuring system for a particular analytical function. Although the analytical procedures and the measuring devices described are directed primarily towards the investigation of minerals and mineral-based materials, they can also be used to study many other types of materials: for example, metals, alloys, slags, cements, concrete, archaeological artefacts and so on.

Each chapter provides some background information on its particular topic: this is followed by brief details of the basic principles that lie behind the methods or the devices concerned. Where appropriate, detailed practical instructions are given for measuring the various mineralogical characteristics, and examples are provided of the ways in which the measured information is used. Most of these examples are drawn from the mineral industry.

The Appendices give practical exercises and tables of mineral characteristics which will be of value to the mineralogist: tables giving some of the physical properties and the chemical compositions of the minerals most likely to be encountered by the mineral engineer in the course of his or her work; a list of the elements in alphabetical order with their chemical symbols, atomic numbers and relative atomic masses; a table of the elements in order of increasing atomic numbers; a selection of minerals in order of their mean atomic numbers, and their subsequent arrangement according to their elemental concentrations. The tabulation of mineral characteristics forms the basis of a simple mineral identification procedure which can be used to identify small mineral grains: this procedure needs only modest skills and simple equipment of the kind that can be found in most field laboratories and in mineral processing plants.

Chapter 1

Introduction to applied mineralogy

1. General background

1.1. The need for minerals

We use vast quantities of minerals and mineral-based products in our industrialised society and we could not possibly maintain our way of life without a large and regular supply of many kinds of these minerals. In fact, virtually everything we use has strong mineral connections.

Sometimes a mineral is used by us more or less in its natural form because it has some specific, valuable property – for example, diamond is often cherished because of its great beauty and is also used for cutting rocks and metals because of its extreme hardness; various clay minerals are used because they have remarkable properties of plasticity and absorption; certain kinds of mica are used because of their very special thermal resistance, and so on.

On other occasions, a mineral may have no intrinsic worth but may contain a chemical component of great value. Thus, the mineral chalcopyrite ($CuFeS_2$) consists of about 34% copper and the mineral is collected in order to recover this valuable metal. Much of the copper thus obtained is made, for example, into electrical components used in "consumer durable" goods. Similarly, some hematite (Fe_2O_3) is used in its mineral form as a source of red pigment; today it is also widely used as a source of the iron metal which is the major component in the steels em-ployed in buildings and bridges and in all forms of vehicles.

There is hardly any aspect of our life that does not ultimately depend on a regular supply of a variety of minerals. The production of crops relies heavily on phosphate- and potash-based fertilisers as well as on organometallic insecticides; crop harvesting requires complex metallic machinery which is fuelled by fossil fuels; transportation of the products from farm to consumer requires roads, railways, trucks, ships, silos, refrigerators and so forth. The food is eventually cooked in metal or ceramic utensils, on metal stoves and is eaten from china plates. Without minerals we would have no schools, hospitals or concert halls; we would not be able to indulge in hobbies such as photography or fibreglass canoeing; we would be forced to use paper money for currency; transportation would be primitive; and long-range communication would have to rely on smoke signals!

The total amount of minerals produced in a country is often an excellent measure of its prosperity. Most of the developed countries, such as the USA or the UK, are major producers and major consumers of minerals; the developing countries (such as Chile and Mexico) are major mineral producers but their consumption of minerals is only modest, whilst both mineral production and consumption are very restricted in less developed countries such as Niger or Malawi.

New uses are being found for some of the well-known minerals which have been items of trade for centuries. For example, zircon (ZrSiO$_4$) has long been valued as a semi-precious mineral used in jewellery; nowadays it is also being used as the raw material for zirconia (ZrO$_2$) high-temperature refractory materials; it is also in demand as a source of the element hafnium which is used in nuclear reactors (the formula of natural zircon is more correctly (Zr,Hf)SiO$_4$ and, although the amount of hafnium in ordinary zircon is usually very small, zircons from some localities show a hafnium content of up to 5%). "New" minerals are also being found for new needs; for example, columbite (Fe,Mn)(Nb,Ta)$_2$O$_6$, which until a few years ago was a valueless component of many tin-bearing alluvial deposits, is now mined as a source of niobium for use in corrosion-resistant steel alloys and superconducting tin alloys.

Only rarely does a mineral cease, with time, to have any commercial value, although this can happen when a mineral's function is taken over by another material or when a traditional function ceases to exist. Thus, oil shale was, at one time, a major source of petroleum but it has been almost completely replaced (for the time being, at least) by natural liquid petroleum; flint is no longer used for making axes or arrow-heads and it is unlikely to come back into vogue!

Some of the minerals which are produced in very large quantities, such as fuels and the metal ores, are listed in Table 1.1. Some of the more exotic minerals and their common uses are given in Table 1.2. In order to obtain these quantities of saleable minerals the world's mines must produce at least 10^{10} tonnes of untreated rock each year, and this huge annual production is increasing (although less rapidly nowadays than in recent decades). This increase in production is needed in part to meet the demands of a steadily increasing world population and also to meet the progressive demands of society for more and more structures,

TABLE 1.1
Annual production figures and major uses of the commoner minerals in 1982

Mineral	Millions of tonnes	Major use
Coal	3962	Power raising
Asbestos	4	Insulation
Barite	7	Drilling fluids
Bentonite and fuller's earth	8	Fillers
Bauxite	79	Aluminium production
Chromium-bearing concentrates	8	Production of chromium
Copper ores (estimated) (metal production 8.2 × 10^6 t)	820	Electrical conductor
Feldspar	4	Ceramics
Fluorspar (fluorite)	4	Metallurgical uses
Gold ores (estimated) (metal production 1300 t)	130	Jewellery
Gypsum	65	Plasterboard
Iron concentrates	783	Production of iron
Kaolin	16	Fillers; ceramics
Magnesite	11	Refractories
Manganese ore	23	
Crude petroleum	2700	Transportation
Phosphate rock	123	Fertilisers
Platinum ores (estimated) (metal production 200 t)	200	Catalysts, jewellery
Potash	26	Fertilisers
Salt	170	De-icing; chemical base
Pyrite (and sulphur)	50	Sulphuric acid
Talc	7	Filler
Tin ores (estimated) (metal production 213 000 t)	213	Tin plate
Titanium-bearing concentrates	6	Paints, paper
Zinc ore (estimated) (metal production 7 × 10^6 t)	70	Castings
Total (excluding petroleum)	6789 million tonnes	
Cement (clay, limestone, gypsum)	12 000 million tonnes	

Source: *World Mineral Statistics*, 1978–1982. British Geological Survey.

TABLE 1.2
Annual production figures and major uses of some less common mineral-based materials

Material	Production (thousands of tonnes)	Common uses
Antimony	50	Alloying constituent
Arsenic	20	Insecticide
Cobalt	21	Metal plating; alloying
Diatomite	1500	Insulator; abrasive, filler
Graphite	535	Refractory; lubricant; paint
Lithium-bearing minerals	5	Pyrotechnics
Mercury	6	Detonators; fungicides
Molybdenum	91	Lubricants
Nickel	623	Metal plating; alloying
Rare-earth minerals	47	Television tube phosphors
Sillimanite	437	Refractory
Tungsten	46	Alloying; light bulbs
Vanadium	33	Alloying
Zircon	724	Zirconium metal; refractory

roads and material goods of all kinds. This tonnage of mineral material works out, on average, at more than 5 tonnes per head of world population per year, although in the developed economies the amount used per head averages closer to 50 tonnes per year.

1.2. Sources of minerals

What are the sources of these large amounts of minerals? Almost all of our minerals are obtained directly from the earth's crust, but a small, steadily increasing, proportion is, and will be, obtained from secondary sources, i.e. products will be obtained from recycled mineral-based materials such as metal and glass that have already been used at least once before. This is a well-established practice and has been in use at least since the beginning of the Bronze Age, when old tools were re-melted to make new ones. That part of the crust which is accessible to us as a source of minerals is perhaps 5000 m deep; this accessible crustal layer covers virtually all of the land surface of the earth as well as much of the shallower seas. Although the total mass of material in the crust is enormous (of the order of 10^{19} tonnes), only a very small proportion of this mass can be exploited for our benefit. The *average* crustal composition (based on thousands of chemical analyses) is shown in Table 1.3, which shows that the elements oxygen, silicon, aluminium and iron account for nearly 90% of the total, whilst the economically important elements like copper, tin, zinc, lead and gold occur in only minute proportions. Fortunately, the rocks of the earth's crust are not of uniform composition – some rocks con-

TABLE 1.3
Average chemical composition of the earth's crust (to mineable depth)

Element	Crustal average
Major elements (per cent):	
Oxygen	46.60
Silicon	27.72
Aluminium	8.13
Iron	5.00
Magnesium	2.09
Calcium	3.63
Sodium	2.83
Potassium	2.59
Titanium	0.44
Phosphorus	0.10
Manganese	0.09
Selected minor elements (those likely to be of economic value) (parts per million):	
Carbon	200
Fluorine	625
Sulphur	260
Chlorine	130
Vanadium	135
Chromium	100
Cobalt	25
Nickel	75
Copper	55
Zinc	70
Strontium	375
Zirconium	165
Niobium	20
Molybdenum	1.5
Silver	0.07
Tin	2
Tungsten	1.5
Platinum	0.01
Gold	0.004
Mercury	0.08
Lead	13
Uranium	1.8

tain even less of the rare elements than the crustal averages given in the table, but happily others contain unusually high, and extremely valuable, concentrations of elements and minerals. These concentrations are sometimes hundreds of times richer than their crustal averages. Where the minerals are potentially of economic worth, these natural concentrations are called *mineral deposits* and it is from these quite rare crustal features that the minerals of commerce are obtained. Any mineral deposit from which it is possible to produce *metallic* products is called an *ore* (see section 2.5 below.)

1.3. Saleability of minerals

In earlier times it was comparatively easy to find mineral deposits from which saleable products could be directly extracted from the earth's crust by using selective, manual mining methods. One of the oldest examples of selective mining was the extraction of flint pebbles from the workings at Grimes Graves in Norfolk, where the flints were recovered from deep shafts whilst the accompanying worthless chalk was left in the mine.

Nowadays the output from a mining operation generally consists of a complex mixture of worthless and valuable minerals and this mixture has no immediate or readily apparent value. Table 1.4 compares the qualities of some of the mineral products that can be sold on the open market with the qualities of the raw materials which are produced during typical mining operations. It shows that the desired mineral often occurs within an ore in vanishingly small proportions; further-

more, such minerals often occur as minute grains which are disseminated throughout the enclosing rock in a highly irregular manner (see Fig. 1.1 in colour section). Such minerals are extremely difficult to recover in saleable form.

1.4. The role of the mineral engineer

The main task of the mineral engineer is to convert the seemingly worthless material produced by the mining engineer into valuable, saleable products. Figure 1.2 shows the sequence of procedures usually undertaken to meet the needs of the customer for mineral products. It also shows the important place occupied by the mineral engineer in this sequence. It can be seen that a suitable mineral deposit must, first, be *located* by the geologist, with valuable assistance from the geochemist and the geophysicist; the deposit is *tested* by the drilling engineer; the working mine, where the deposit is *severed* from the earth's crust, is supervised by the mining engineer; the *treatment* plant, in which the mined material is converted into a saleable raw material, is designed and operated by the mineral engineer; and finally the factory, where the raw material is converted into *consumer products*, may be in the charge of a metallurgist or a ceramicist. Some mineral-based waste materials are *recycled* – the reprocessing of this waste is undertaken by the mineral engineer.

As will be demonstrated in this book, mineralogical information is a prerequisite of all the mineral engineer's work. He or she must have large amounts of mineralogical data to

TABLE 1.4
Qualities of some typical materials: as mined, and as sold

Material	Quality as mined	Usual saleable quantity	Typical customer
Gravel	Wide size range	Closely graded sizes	Civil engineer
Iron ore	20–55% iron occurring as a number of iron-rich minerals: wide size range	Greater than 55% iron – in coarse fragments	Metallurgist
Tin ore	0.01–1% tin in the form of cassiterite (nominally SnO_2)	Up to 75% tin	Metallurgist
Gold ore	5–15 parts per million (by weight)	Essentially pure gold	Metallurgist, jeweller
Coal	20–90% combustible matter in a wide size range	About 90% combustible matter in graded sizes	Power stations, metallurgist, householder
China clay	20–25% of mixed clay materials	About 100% of a specific clay mineral	Ceramicist, paper technologist
Diamond	1 part by weight of diamond in every 1–10 million parts of rock	Pure diamond – if possible in its original (natural) size	Jeweller, toolmaker

Fig. 1.1 Typical ore specimen.

a

b

1 cm

Fig. 4.5 Calcite and scheelite: (a) in ordinary light; (b) in ultraviolet light.
The scheelite shows a marked blue fluorescence.

Fig. 6.2 Photomicrograph illustrating the complex nature of a fine-grained specimen containing many sulphide minerals (taken with vertically reflected white light). The darker yellow particles are chalcopyrite; the lighter yellow particles are pyrite; the galena is white and the grey particles are siliceous gangue minerals. Most particles appear to be completely liberated.

Fig. 10.2 Photograph of beach sand. Virtually every grain is liberated.

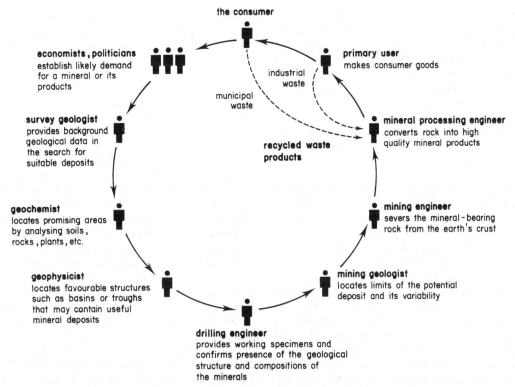

Fig. 1.2 Location and exploitation of mineral resources.

delineate the boundaries of a mineral deposit and to outline the variations of quality that occur within that deposit. Quantitative mineralogical information is also essential when designing, commissioning, testing and operating a mineral treatment plant. Finally, long-term mineralogical monitoring of old waste dumps must be carried out to ensure that they remain safe.

2. Definitions

2.1. What is mineralogy?

Mineralogy is the scientific study of minerals: it involves the study of their:

internal structures and compositions;
physical and chemical properties;
classification;
mode of formation; and
occurrences and associations.

Originally, the science of mineralogy was limited to the study of materials from the earth's crust: nowadays, it also includes the study of any debris, such as meteorites and tektites, that falls on to the earth from outer space. Recently, mineralogy has been expanded to include the study of lunar

specimens and it will, eventually, be expanded still further to take in the study of planetary materials, when they become available.

The terms "applied mineralogy" and "industrial mineralogy" are used in this book to describe the practical study of those minerals which are used for industrial or commercial purposes (this field of study is also sometimes called "process mineralogy"). These industrial minerals are in the main derived from the earth's crust, but artificially made (synthetic) mineral-like materials are also usually included in the term "applied mineralogy". Figure 1.2 shows the role of process mineralogy in ore deposit development, and the recycling of waste products.

2.2. What is a mineral?

It is extraordinarily difficult to provide a sufficiently precise, yet concise, definition of the term *mineral* that will satisfy every mineralogist. Historically, a mineral was at one time taken to be any stone from which metal could be produced. In current everyday language the term is used to describe a natural material that is clearly of neither animal nor vegetable origin. The mineralogist, on the other hand, usually defines a mineral as "a naturally occurring, homogeneous, solid substance which is generally of inorganic origin: it usually has a well-defined crystalline structure and a chemical composition that lies between well-defined limits". This definition is necessarily long and cumbersome because it tries to encompass all the materials which are generally accepted by the mineralogist as being "mineral". Even so, as we shall see, it still does not cover all eventualities and it is important to study the definition carefully in order to understand the meaning of the term "mineral".

2.2.1. A mineral is a naturally occurring substance

This requirement seems to rule out any mineral-like materials made by an industrial process or made in a laboratory – even though these materials may be indistinguishable from "natural" minerals. Thus the crystalline, mineral-like *phases* that occur in metallurgical slags, or those which are formed in the brickwork lining of a smelting furnace, do not, strictly speaking, fit the definition of a mineral. Nor do the artificially made crystalline phases which are found in coal ash, or in cement and concrete. In practice, however, the term "mineral" is nowadays almost always extended to include these materials, along with synthetic gems and semi-precious "minerals" that have been made in the laboratory. In addition, the large quantities of magnetic iron oxide (magnetite) made during the roasting of some iron ores, as well as the silicate minerals made in ceramic products, are included within the applied mineralogist's field of study.

2.2.2. A mineral is homogeneous

This means that a mineral must consist of only a single phase. Homogeneity is sometimes difficult to prove since a "mineral" must be homogeneous at very high magnification – it is not enough that it appears to be uniform to the naked eye. It must certainly be homogeneous at the limit of optical resolution (about the micrometre level); it should also, if possible, be homogeneous at the limit of electron microscopy (the nanometre range); but, clearly, most minerals cannot be homogeneous at the atomic scale.

Fragments of some rocks, such as fine-grained basalts, may initially appear to be homogeneous and can therefore be mistaken for minerals, but even a low-power microscope will show that they are composed of many different minerals in the form of minute grains. Similarly, the mineral particles found in river gravels and in beach sands may appear to be homogeneous to the naked eye but in the electron probe microanalyser (Chapter 9, section 9) a particle may turn out to be a complex mixture of finely divided minerals.

Coal is a complex mixture of materials – some truly mineral, others clearly of organic origin. Consequently, coal does not, strictly speaking, qualify to be called a mineral.

2.2.3. A mineral is solid

This suggests that there can be no liquid minerals; however, water and naturally occurring liquid mercury are usually regarded as minerals.

2.2.4. A mineral is inorganic

This part of the definition rules out coaly materials: it should also exclude much calcium carbonate (calcite) in limestones and calcium phosphate minerals in phosphate deposits, since it can be clearly shown that these materials are organic in origin (they are the remains of plants, shells and bones, respectively). In some cases, however, it may be very difficult to determine whether a material has, in fact, been formed as the result of organic process, or is the product of inorganic action. For example, is the graphite found in some gold ores formed by an inorganic process or by the metamorphic alteration of an organic material similar to coal? Much of the calcite and aragonite ($CaCO_3$) in a shelly limestone must have been formed by organic agencies; but other calcite grains in the same rock will have been formed by inorganic processes of solution and crystallisation. These difficulties, which are associated with establishing a material's origin, are often avoided by accepting that calcite, aragonite, graphite and calcium phosphate are minerals – however they may have been formed!

2.2.5. Minerals are crystalline

Most minerals have well-defined crystal structures, i.e. they consist of well-ordered arrangements of atoms. It is probable that *all* minerals possess some measure of crystallinity, although the degree of crystallinity may be small and is sometimes difficult to determine. However, any regularity of internal structure is reflected in many of the physical properties of a mineral. Thus, where the crystal structure, although regular, varies in different directions then the mineral will be *anisotropic* and many of its properties will vary with crystal orientation. For example, optical properties, such as refractive index and reflectivity, will vary along different crystallographic axes (see Chapter 5). Similarly, the chemical reactivity and the hardness of an anisotropic mineral will vary on different crystal planes. As will be shown later, these variations can be measured and used to derive information about the crystal structure of a mineral.

Some minerals which have exactly the same

TABLE 1.5
Some commonly occurring polymorphic minerals

Mineral	Composition	Crystal structure	Hardness (Mohs' scale)	Specific gravity
Calcite	$CaCO_3$	Hexagonal	3	2.71
Aragonite	$CaCO_3$	Orthorhombic	3.5–4	2.94
Quartz	SiO_2	Hexagonal	7	2.65
Tridymite	SiO_2	Orthorhombic	7	2.27
Cristobalite	SiO_2	Tetragonal	6–7	2.33
Coesite	SiO_2	Monoclinic	7.5	3.01
Rutile	TiO_2	Tetragonal	6–6.5	4.2
Anatase	TiO_2	Tetragonal	5.5–6	3.9
Brookite	TiO_2	Orthorhombic	5.5–6	3.9–4.1
Graphite	C	Hexagonal	1–2	2.23
Diamond	C	Isometric	10	3.51
Pyrite	FeS_2	Isometric	6–6.5	5.02
Marcasite	FeS_2	Orthorhombic	6–6.5	4.89

chemical compositions nevertheless crystallise to produce more than one structural form. This phenomenon is known as *polymorphism* and the different mineral forms that exist with a single chemical composition are known as *polymorphs*. Table 1.5 lists some of the commoner polymorphic minerals and shows how some of the important properties of these polymorphs can differ markedly from one to another.

There are some minerals which show only very poor evidence of crystallinity: these minerals lack any marked degree of ordered atomic arrangement and are called *amorphous* minerals. The most obviously amorphous minerals are the liquids, water and mercury. Other mineral-like materials such as opal, a solid silica gel ($SiO_2.nH_2O$), show very poorly developed crystallinity. These amorphous materials are externally of irregular shape and can absorb (or adsorb) a wide variety of chemical elements from an aqueous environment. They can, therefore, show a wide range both of chemical composition and of physical properties.

Some radioactive minerals also lack a well-developed crystalline structure. These minerals were originally formed as well-crystallised solids but their crystal structures have since been almost totally destroyed by α-particle radiation derived from their radioactive components, the elements uranium and thorium. These minerals are called *metamict* minerals; they tend to retain their external shapes during the process of "metamictisation" and they become, in effect, *pseudomorphs* of the original well-crystallised minerals. Their lack of internal atomic regularity is shown by their anomalous optical properties (for example, they may be isotropic where they

should be anisotropic) and by their poor X-ray diffraction patterns (see Chapter 9, section 6). When a metamict mineral is annealed (i.e. kept at a high temperature for a few hours) an ordered, crystal structure may be produced.

2.2.6. The chemical composition of a mineral lies within well-defined limits

Very few, if any, minerals have fixed and definite chemical compositions. Diamond (C), galena (PbS) and quartz (SiO_2) are three minerals whose compositions are nearly constant, but diamond often contains variable amounts of nitrogen, galena almost invariably contains small amounts of silver, and quartz commonly contains trace amounts of titanium and iron.

Most minerals show a wide range of composition and even the concentrations of the major elements in these minerals may vary widely. This compositional range is not random, however; it is controlled by well-defined laws of chemistry and by the rules which govern the development of atomic structures. For example, the plagioclase feldspar (commonly written *felspar*) group of minerals forms a complete, isomorphous, solid-solution series ranging in composition from $NaAlSi_3O_8$ (albite) to $CaAlSi_2O_8$ (anorthite). In this series the amount of aluminium (Al^{3+}) varies along with the proportions of Ca^{2+} and Na^+ so as to maintain electrical neutrality within the minerals under all circumstances.

The effects of compositional variability are important when one is trying to identify an unknown mineral because the minerals in an isomorphous series can also show a wide range of physical properties. For example, the densities of the minerals in the columbite–tantalite series

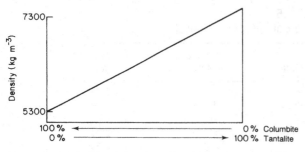

Fig. 1.3 Variation of density with change in composition in an isomorphous mineral series (columbite–tantalite: $(Fe,Mn)(TaNb)_2O_6$).

$(Fe,Mn)(Nb,Ta)_2O_6$ vary linearly from $5300\ kg\ m^{-3}$ for columbite $(Fe,Mn)(Nb)_2O_6$, to $7300\ kg\ m^{-3}$ for tantalite $(Fe,Mn)(Ta)_2O_6$ (Fig. 1.3).

There is a large number of *solid-solution* (iso-morphous) series similar to, but often much more complex than, the columbite–tantalite series, but it does not always follow that there is a simple, readily discernible relationship between the compositions of these minerals and their physical properties.

Some minerals which exist as true single-phase minerals at elevated temperatures may *exsolve* (precipitate) into two distinct phases at lower temperatures. At ambient temperature, either the two exsolved phases form comparatively large contiguous masses or the minor component occurs as very fine grains widely dispersed throughout the larger, host component. These very fine-grained precipitates are often preferentially aligned along certain crystallographic axes of the host and are difficult to see without using high-magnification microscopes (see Fig. 1.4). The combined pairs of exsolved phases often show quite unexpected physical properties. For example, the single-phase mineral ilmenite $(FeTiO_3)$ containing some excess *dissolved* iron is only moderately magnetic, but when that excess iron is exsolved from the ilmenite it may form a second phase of thin parallel lamellae of mag-

netite, and the combined ilmenite–magnetite particle then becomes appreciably magnetic.

Small amounts of fine-grained impurities (rather than exsolved phases) may also be very difficult to see but they may, nevertheless, greatly affect the properties of the host minerals. For example, calcite $(CaCO_3)$ is normally a colourless or white mineral but small amounts of finely disseminated carbon particles can make it look distinctly blue-grey in colour. Again, thin, almost invisible coatings can greatly alter the surface-chemical properties of minerals – for instance, quartz particles from alluvial deposits are frequently coated with iron oxide and in a froth-flotation process this coating can make the quartz behave like particles of iron oxide. Surface films formed by oxidation of the underlying minerals may also be difficult to see but they can also markedly alter the surface properties of minerals – for example, many sulphide minerals quickly "tarnish" with the formation of sulphate or thio-sulphate films, and these films affect flotation properties of the minerals and their rates of re-action during industrial leaching processes.

A detailed knowledge of the chemical composition of a mineral is thus not always sufficient information on which to base an identification of that mineral – the same chemical constituents can be arranged in different ways to form different minerals (see Table 1.5). For example, both diamond and graphite consist of crystalline carbon. In graphite, the carbon atoms form an arrangement of hexagonal plates: these plates are only very loosely held together so that they easily split apart and the mineral is very soft: these plates form an excellent lubricating medium between two moving metal faces. Diamond, on the other hand, forms such a strong, compact structure that the mineral is the hardest known substance and it can be used for cutting any other material. Many other polymorphous minerals are known (good examples are pyrite and marcasite, FeS_2; calcite and aragonite, $CaCO_3$; anatase, brookite and rutile, TiO_2).

2.3. Rocks

Rocks are naturally occurring coherent aggregates of mineral grains. Sometimes, these grains are so firmly locked together that the rocks are hard, and resistant, like granite or dolerite. At other times, the individual grains are only loosely cemented together, as in a friable sandstone, which can be disintegrated by hand.

Some rock types, such as sandstone and marble, consist almost entirely of a single mineral (i.e.

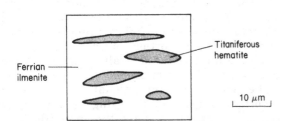

Fig. 1.4 Complex particle consisting of iron rich (ferrian) ilmenite and exsolved, orientated titanium rich (titanifer-ous) hematite.

TABLE 1.6

Modal (i.e. mineralogical) analysis of a complex ore from Derbyshire (this material has been worked for the recovery of barite, fluorite, smithsonite and galena)

Mineral	Weight (per cent)
Barite ($BaSO_4$)	26.3
Calcite ($CaCO_3$)	15.5
Nacrite (clay)	13.6
Quartz (SiO_2)	12.2
Fluorite (CaF_2)	8.7
Hematite (Fe_2O_3)	10.7
Smithsonite ($ZnCO_3$)	7.9
Witherite ($BaCO_3$)	3.0
Galena (PbS)	1.2
Cerussite ($PbCO_3$)	0.6
Magnesite ($MgCO_3$)	0.1
Kurnakite (Mn_2O_3)	0.2
	100.0

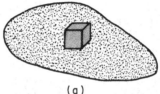

Fig. 1.5 (a) A grain of galena in an irregularly-shaped particle containing galena and pyrite; (b) liberated particle consisting only of a single galena grain.

quartz and calcite respectively); others are very complex and can contain ten or more different minerals. For instance, a lead ore from Derbyshire contains calcite ($CaCO_3$), fluorite (CaF_2), galena (PbS), cerussite ($PbCO_3$), sphalerite (ZnS), chalcopyrite ($CuFeS_2$), barite ($BaSO_4$), nacrite ($Al_2Si_2O_5(OH)_4$), quartz (SiO_2), hematite (Fe_2O_3), smithsonite ($ZnCO_3$), witherite ($BaCO_3$), magnesite ($MgCO_3$), kurnakite (Mn_2O_3) – see Table 1.6.

2.4. Mineral deposits

As we saw in section 1.2 above, any unusually rich, naturally occurring concentration of a particular mineral (or element) is called a *mineral deposit*. Although the proportion of that mineral (or that element) is, by definition, unusually high compared to the average crustal composition, the mineral (or element) may nevertheless only account for a very small fraction of the mass of that deposit. For example, a copper-bearing deposit rarely contains more than 2 or 3%, by weight, of copper-bearing minerals and a potentially valuable diamond deposit need contain only 1 part of diamond to a million parts of rock.

2.5. Ores

An *ore* is a mineral deposit from which it is possible to produce a metal by using technological methods that already exist or can reasonably be inferred. There are tin ores, gold ores, copper ores and so forth, but by convention there are no "clay

ores" or "gypsum ores". Many mineral deposits are not ores at present, but they may become ores when technological developments allow them to be mined and treated successfully.

2.6. Mineral grains and mineral particles

A mineral *grain* is a feature that consists of only one type of mineral. Mineral grains may occur as free particles (see below) or they may be embedded in a rocky matrix.

A mineral *particle* is a discrete fragment of rock that can contain any number of different minerals. When a particle contains only *one* mineral that particle is also a "liberated" grain (Fig. 1.5).

3. Mineral nomenclature

The method of naming minerals is both illogical and highly unsatisfactory. The names of the very common minerals – minerals that have been known and have been used for ages – are well established and often date back to antiquity. For example, "quartz" is derived from the German and alludes to the abundance of this mineral where ore veins cross: the name "feldspar" is also derived from German and refers to the wide occurrence of this mineral in nature. "Zircon" is an old word which has been derived from the Persian for "golden colour", while "diamond" is derived from the Greek and appropriately means "invincible".

Almost all the minerals that have been named more recently have names ending in *-ite*. Some of these modern names provide clues to the mineral composition; for instance, stibnite (Sb_2S_3) contains antimony (or stibium), and cuprite (Cu_2O) is a copper-bearing mineral (the Latin word for copper is *cuprum*). Minerals are often named after the localities where they were first found or where they are found in abundance – for example, ilmenite ($FeTiO_3$) occurs in the Ilmen Mountains in Russia, and strontianite ($SrCO_3$) was named after the Strontian area in Scotland (the element strontium was later named because it was found in the mineral strontianite!). Minerals are sometimes named after some obvious property: thus, celestite ($SrSO_4$) is named after its "celestial-blue" colour, and the name of barite ($BaSO_4$) which is a dense, white mineral, is derived from a Greek word meaning "heavy". Today, newly found minerals are frequently named after a famous (or, sometimes, not-so-famous) person; for example, the new lunar mineral *armalcolite* was named after *Arm*strong, *Al*drin and *Col*lins, the "lunarnauts" who first brought back rock specimens that contained the mineral.

All names for new minerals must now be submitted to an international group of eminent mineralogists who judge whether the names are appropriate – but new names are still not required to follow any logical system of nomenclature.

4. Mineral classification

In 1976 there were over 3500 named minerals (P. Emburey, *New Mineral Names 1926–1976*). New minerals are being added to this list every year. These are mainly minerals that have been discovered through the use of new analytical techniques such as electron probe microanalysis (described in Chapter 9, section 9). Some "miner-

Classification	Structure	Example
Nesosilicates	Independent tetrahedra * △	Forsterite $Mg_2(SiO_4)$
Sorosilicate	Two tetrahedra sharing a single oxygen ▷◁	Hemimorphite $Zn_4(Si_2O_7)(OH)_2 \cdot H_2O$
Cyclosilicate (Ring silicate)	Closed rings of tetrahedra : each tetrahedron shares two oxygens	Beryl $Be_3Al_2(Si_6O_{18})$
Inosilicate (Chain silicate)	Chains of tetrahedra	Anthophyllite $Mg_7(Si_4O_{11})_2(OH)_2$
Phyllosilicate (Sheet silicate)		Kaolinite $Al_4(Si_2O_5)_2(OH)_8$
Tektosilicate	3-D framework of tetrahedra	Quartz , SiO_2

*Silicate tetrahedra are made up of relatively large oxygen ions at the corners and small silicon ions at the centres

SiO_4 ○ Oxygen   ● Silicon

Fig. 1.6 Examples of structures formed from silicate tetrahedra. (After Berry *et al.*, 1983.)

als" are also removed from the list when modern methods of study have shown that a substance that had previously been considered as a single mineral consisted, in fact, of a mixture of minerals.

Such a large number of different minerals must, for convenience, be classified into smaller groups, and a single group should, as far as possible, contain all related mineral species. The most fundamental characteristics of any mineral are its crystal structure (or lack of structure) and its chemical composition. Consequently, these characteristics are always used as the major classification features. In the past it was often much easier to determine the composition of a mineral than it was to establish its crystal structure. Therefore, early mineralogists adopted a method of mineral classification which is based on a mineral's dominant anion content. This method is still employed; the minerals are initially classified into major categories such as native elements, sulphides, sulpho-salts, oxides, halides, carbonates, borates, sulphates, phosphates, tungstates, silicates and so forth. This categorisation is very useful since minerals that fall into any single category all show distinct family features and, furthermore, often tend to occur together in nature. For example, the "native elements" category includes gold, silver, copper, lead, mercury, the platinoid-group elements and iron; they are all very dense, they are all soft and malleable, and all are opaque and have high reflectivity values. Furthermore, the least chemically reactive of these minerals often occur together in

alluvial deposits. Where necessary, the anion-based major categories are subdivided according to the crystal structures of the minerals. Thus the silicate minerals, which make up about a quarter of all known minerals, are subdivided into smaller groups which are based on the manner in which the basic silicate building blocks – the structural SiO_4 tetrahedra – are grouped or are linked together into sheets, chains and so on (Fig. 1.6).

5. Mineral identification

As mentioned above, there are a few thousand different known mineral species. Many of the individual minerals have wide ranges both of chemical composition and physical properties and, consequently, it is not surprising that it is often extremely difficult to identify an unknown mineral specimen unequivocally. These already difficult problems of identification are greatly accentuated if the unknown mineral only occurs as very small grains in a hard rock and is, therefore, very difficult to obtain in a pure form. Further problems arise when the mineral occurs in minute quantities within a rock and is therefore, difficult to collect in large enough amounts for traditional methods of examination and analysis. Chapter 4 provides more details of these problems and how they can be overcome. For a simple determinative scheme which can be used for identifying small mineral grains, see Appendix 2.

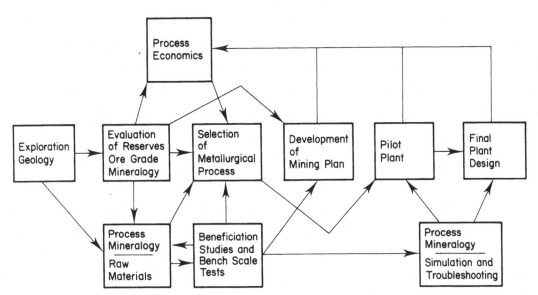

Fig. 1.7 Various inter-related stages in mineral deposit development requiring a wide range of information (after Shapiro).

6. What information does the mineral engineer need?

It often happens that a mineral deposit can only be exploited satisfactorily when the mineral engineer in charge has a clear, detailed knowledge of the nature and the distribution of the minerals in that deposit (Fig. 1.7). He or she needs information regarding the identities, compositions, sizes and proportions of the various *minerals* in the deposit: detailed information is also needed about the sizes, shapes, compositions and textures of the *particles* which are produced during a treatment operation. These details are constantly changing within a working mineral treatment plant and, in order to achieve effective control of such an operation, the engineer must have a steady stream of quantitative mineralogical data from large numbers of routine test specimens. One of the primary aims of this book is to show how such information can be collected and used.

7. Conclusion

It is difficult to provide a simple definition of a mineral but, in the mineral industry, most natural materials that are clearly not of animal or vegetable origin are taken to be minerals, including man-made, mineral-like substances.

There are many thousands of minerals and, for convenience, they are classified, first, on the basis of their chemical components and then by crystal structures.

Minerals of commerce are required in vast quantities at low cost. There are plentiful resources of all minerals in the earth's crust but only those resources which occur in comparatively high-quality concentrations are economically attractive. Even so, a great deal of quantitative mineralogical information is always needed before a mineral deposit can be profitably worked, and the provision of this information often requires the services of skilled mineralogists.

Chapter 2

Sampling of mineralogical materials

1. Introduction

Mineralogical information is required for many purposes: geologists need it during field surveys and during mineral exploration campaigns; mining engineers need it whilst they are evaluating potential ore bodies and also when they are planning their mining schedules; mineral processing engineers use it when they are designing mineral treatment processes and when they are operating these processes; environmental engineers require it to ensure the safe disposal of waste mineral products; ceramicists seek it when designing and manufacturing their products; concrete technologists produce artificial minerals and want mineralogical information concerning the quality of their products. How can this mineralogical information be obtained?

Clearly, it is not possible to examine in detail the total mass of any bulk material that is under study; for example, it is entirely impractical to measure the mineralogical characteristics of a *complete* ore body, which may be a mass of rock weighing over 10^8 tonnes. Similarly, it is not feasible even to attempt to measure the mineralogical characteristics of *all* of the material passing through a treatment plant (which can be treating up to 10^5 tonnes per day). Instead, the geologist, the mining engineer and the mineral engineer must rely on the information that they obtain by measuring only very small, selected portions of the large amounts of material involved in the deposit, mine or mill. These small portions

of material taken from the much larger entities are called *samples*, and the procedure used to produce these samples is known as *sampling*. The purpose of sampling is to provide small portions of a much larger entity from which the character of that entity can be inferred.

Sampling is the single most important factor in any analytical procedure. Good sampling is sometimes an expensive exercise, but bad sampling can be disastrous – as the following cautionary tale shows. Early in this century a local landowner carried out some drilling to evaluate the coal seams that were thought to occur beneath his estate near Selby, in Yorkshire. The results of the borings were very disappointing: they showed that, although coal seams did exist in the area, they were very thin and, consequently, were of no practical value. In the early 1960s the National Coal Board carried out further exploratory drilling in the same area – in fact, one of their new bore-holes was sited in the same field as one of the original ones. Cores from the new holes showed the presence of five separate coal seams, each of which was of workable thickness. The work done by the National Coal Board (now British Coal) demonstrated clearly that the amount of coal recovered from the coal seams during the first drilling programme must have been hopelessly inadequate, i.e. the earlier programme had not produced adequate samples of the coal seams. Additional exploration drilling by British Coal in the area has since proved the existence of a major coalfield with reserves totalling about 2×10^9

tonnes of readily available coal. At the time of writing, coal is already being produced in large quantities from this field and, in the next few years, production will build up to the designed output of 10^7 tonnes per year. At this rate, the coalfield can continue to work for 200 more years! In fact, it is one of the largest deep coal-mining projects that has ever been undertaken – and all this is taking place in an area where poor sampling procedures had produced unreliable information and had failed to detect any workable deposits.

This example shows how important it is to ensure that the specimens that are produced as a result of a sampling procedure accurately represent the material that *should* be sampled. It was vitally important, in the Selby investigation, to ensure that the coal seams, at least, were properly sampled; and, of course, it was almost as important from the mining point of view that the rock strata between the coal seams were correctly sampled too. Similar problems may arise when a weathered rock outcrop is sampled by a geologist in order to infer the mineralogical characteristics of a large mass of underlying unweathered rock: in this instance, it is important that the sample does not consist only of the weathered, easily obtained, outer skin of that outcrop. Instead, the sample should consist of the unweathered inner material which may be difficult to break (and therefore is not always collected).

Similarly, if the output from a treatment process is being sampled then the sample must contain *all* the components of that output; for example, it is only too easy to forget about, or ignore, the very finest-grained particles – small particles which are carried in suspension in the large volumes of water which often exist within mineral treatment plants. These "fines" are often decanted away along with the excess water collected during the sampling operation and any information they contain is lost.

2. Definitions

Before proceeding further it would be well to define our terms. The *target population* is the population about which information is required; the *sampled population* is the population (or mass of material) from which a sample is drawn. It is up to the engineer to ensure that these two populations are the same. A *sample* is any portion taken from a statistical population and whose properties are studied to gain information about that population; a *representative sample* is a small

portion which has, on average, the same values for certain properties as the bulk material from which it was derived. Such a representative sample is often assembled from a number of smaller portions (samples) taken from the population. For example, a representative sample may be the whole of a drill core, or it may be made up of a number of portions taken from such a core. The *average* result obtained from many representative samples provides a good estimate of the result that would have been obtained had all the bulk material been measured.

The *sampling frame* is the set of instructions for collecting the samples. For example, the sampling frame may be the drilling grid used to position a group of boreholes, or it may be the counting grid used during an image analysis procedure (discussed in Chapter 6). This frame need not be a regular grid, but it frequently is.

In *simple random sampling*, arrangements must be made so that all possible values of the property which is being investigated have equal chances of being collected in a sample.

In *systematic sampling* the samples are taken from the total population according to a fixed, cyclic procedure. This procedure must not match any cyclic pattern in the sampled material; for example, a mill product can be periodically sampled but the sampling period must be arranged so as not to coincide with any periodic surge in the mill output.

In *stratified sampling* the total population is divided into convenient sub-populations. Thus, a population of particles having a very wide size range is first screened into a convenient number of sub-populations (called *strata*). Each stratum is sampled and measured separately and conclusions can then be inferred about the total overall population. In *stratified random sampling*, each stratum is randomly sampled.

3. Producing a sample

The initial sample taken in the field or in a plant is often of considerable weight and bulk, and it must then be subsampled to provide a manageable amount of material for measurement. During the subsampling stages it is important that the features of interest are retained, in unaltered form, in the subsample. Thus, if the aim of the sampling procedure and of the subsequent analysis is to determine the grain-size distribution of a specified mineral, then the grains of this mineral must *not* be broken by the procedures which are used to produce the sample (see below).

It is obviously important that the mineralogical feature that *should* be measured is, in fact, the one that actually *is* measured. For example, there is no purpose in measuring the moisture content of a sample that has been allowed to dry out if the object of the exercise is to determine the natural water content of that specimen.

Naturally occurring materials such as alluvial deposits, soils and beach sands, and industrially produced materials such as crushed rocks, are often difficult to sample because of wide variations in the sizes, shapes, masses and compositions of the various particles that they contain. In these circumstances, it is often advantageous to remove (or, at least, reduce) some of these variations before carrying out the sampling procedure. Thus, particles of roughly the same size can be produced by *screening* and these screened fractions are much easier to sample than the unsized material (see the description of stratified sampling above). Samples of the size fractions can then be analysed and, if necessary, these samples can be re-mixed in the appropriate proportions and used for further test work.

There are even some occasions (although, admittedly, rare) when it is advantageous to take selected, *typical* portions, rather than to take representative samples of a bulk material. For example, if the aim of a mineralogical study is to examine the nature of the native gold grains within a gold ore then it may help the investigator if he selects ore fragments which are known (perhaps by inspection or from previous analysis) to contain unusually high proportions of that gold. Fragments which are known *not* to contain gold, e.g. fragments of the enclosing country rock rather than of the gold-bearing vein, need not be collected for this kind of study since they cannot provide any specific information about the nature of the gold grains.

The cost of analysing any material is related, in part, to the total amount of material that has to be collected and examined – and, in general, the smaller the amount, the smaller the cost. So there are good financial, as well as practical, reasons for limiting the amount of material that has to be prepared and analysed. The only way to obtain the correct, minimum amount of material is by good sampling practice.

The procedures that must be employed during a sampling exercise vary with the nature and with the amount of material being sampled, and also with the kind of information required and the accuracy being sought. Consequently, the provision of a good representative sample always requires some preliminary knowledge of the material being sampled; it also demands some experience in the application of sampling principles, and a detailed knowledge of the type of information that is needed.

4. Information required

The geologist and the mineral engineer use the collected samples to derive some, or all, of the following mineralogical information:

(a) the bulk chemical composition of a sample;
(b) the names of the minerals that occur in the sample;
(c) the proportions of each of these minerals;
(d) the chemical compositions of the various mineral species;
(e) the size distributions of the mineral *grains* and/or the size distributions of broken mineral *particles*;
(f) the spatial inter-relationships between the mineral grains – i.e. the texture – of an "unbroken" rock or ore;
(g) the sizes, shapes, compositions and textures of the particles that occur in particulate specimens; and
(h) the manner in which all these features change from one part of an ore body to another, and how they change in the products from the various processes in a treatment plant.

It is very important to realise that the accuracy of the results which are obtained during any mineralogical investigation cannot be better than the accuracy of the sampling procedures used to produce the analysed samples. The errors introduced into the final results by using poor, biased sampling techniques *cannot* be offset by carrying out statistical manipulations of those results.

However, it must be acknowledged that it can be very difficult to obtain good samples from many kinds of rocks and mineralogical materials – particularly, as we shall see, when the mineralogical features that we wish to measure are large (e.g. the mineral grains in coarse-grained rocks), or when the proportion of a desired feature within a material is vanishingly small (such as the gold grains which occur in the "tailings", or discard, product from a gold recovery plant). In addition to these difficulties, the sampling procedures that we are forced to use often involve a number of sampling stages before we can arrive at a suitable weight of material for the final analysis. Statistical errors are inevitably introduced during each of these sampling stages, but these can be calculated. The variances of these errors are cumulative, i.e.

$$\sigma^2_{\text{total}} = \Sigma\sigma^2_1 + \sigma^2_2 + \ldots + \sigma^2_n$$

Consequently, the results obtained from the final, measured samples will often show significant (but unavoidable) errors. The mineralogist must accept that such errors do occur, and must not falsely claim that the final results are better than they really are. For example, even where a method that was used during a sampling sequence can be shown to have cumulative statistical errors in excess of 10% relative, it is not unusual to see the results quoted to two or even three significant figures. If the investigation demands such high accuracy, then the *appropriate* (often very large) samples must be taken.

5. Primary sampling

The first, or primary, stage of sampling is often carried out by the geologist in the field. The vast mass of a potential ore body is sampled by drilling, pitting (i.e. by sinking small-diameter, shallow pits) or cutting trenches through the ore deposit.

5.1. Core-drilling

Hard, consolidated rocks are generally sampled by a method called *core-drilling*. This method is especially useful when the rocks are concealed, at great depth, beneath the earth's surface. Core-drilling employs an annular cutting edge (or *bit*) attached to a vertical, rotating metal pipe (the *drill string*). This cutting edge is armoured with a hard material such as tungsten carbide or, more usually, diamond. A cylinder of more or less undamaged rock is collected inside the hollow bit (Fig. 2.1) and this solid rock cylinder, when withdrawn from the hole, forms the geologist's *primary sample*. One of the great advantages of collecting undamaged solid cores is that they retain the spatial relationships of the mineralogical features within the sampled rock.

It is not always easy to produce such complete samples, however. Important parts of a core can be lost if the ore is fragile and easily fragmented, and the use of inappropriate drilling procedures can give rise to other losses. When core losses do occur (for whatever reason) then the information obtained by the examination of the recovered core must be treated with caution. In fact, great efforts must be made to ensure that *all* (or, at worst, very nearly all) of the core *is* recovered from a drill hole, because the geologist has no information

(a)

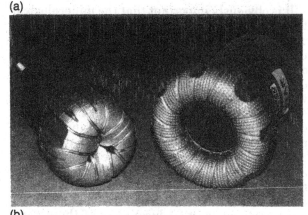

(b)

Fig. 2.1 (a) Rotary drill bits – the left-hand type breaks the rock into fragments which are forced up around the outside of the bit by water pumped down the drillpipe and through the nozzles at the centre; the right-hand type is a coring bit, which cuts an annular drill hole and the resultant cylinder of undamaged rock passes up the centre. (b) Drill cores – these range in diameter from a few centimetres to 20 cm or larger.

concerning the nature of the lost portions of rock (cf. the poor-quality, initial drilling for coal in Selby, described in section 1 above).

It is, of course, possible (but unlikely) that an incomplete core can still be representative of the rock mass being drilled – provided that the inadequacies of the drilling operation are not related to the mineralogical characteristics of the rock, i.e. if only small, *random* portions of the core have been lost then the recovered core may still adequately represent the sampled rock.

On the other hand, it is always likely that some core will be lost when drilling through soft, unconsolidated or porous strata. In these instances, the lost portions of core are almost always different in character from the adjacent, recovered core. Since only the recovered core can be analysed, the results will almost certainly be biased and, unhappily, there is no way of knowing which way the bias applies, i.e. whether the recovered core underestimates or overestimates some quality of the bulk material.

5.2. Percussion drilling

Soft, unconsolidated and porous rocks are sampled by using either drills that rely on a hammering action or drills having a rotating action but which use "roller" bits instead of coring bits (see Fig. 2.1). In both instances, the rock at the bottom of a drill hole is purposely broken into small fragments. These fragments are carried to the surface by a strong, circulating flow of water. The rock fragments recovered from such a drill hole are often quite adequately representative of some bulk characteristics of the rocks surrounding the hole. Thus, the overall, bulk chemical analysis of the broken cuttings of rock from different levels in the drill hole will mirror the compositional variations in the solid rock – provided there has been no loss of selected components into pores or fissures in the walls of the bore-hole. The recovered rock fragments, however, do *not* retain their spatial relationships, nor do they necessarily retain the natural grain sizes of the various minerals. These samples are therefore inevitably of less value to the mineralogist than good, complete cores obtained by diamond core-drilling.

5.3. Pitting and trenching

Pitting and trenching methods are often used when soft, unconsolidated, near-surface materials are sampled down to about 20 m from the surface. If needed, all the material that has been removed from a pit or a trench can be used as a single, bulk primary sample. With materials obtained from pits and trenches, of course, the relationships between the various structures will have been lost. However, features in the walls of the pit (or the trench) will retain their spatial relationships and can be viewed and sampled separately, if necessary.

Further details of the various sampling methods used in the field are outside the scope of this book, but these details can be found in many excellent books that deal with mineral prospecting and the evaluation of ore bodies.

5.4. Treatment plants

Large numbers of routine primary samples are taken by the mineral engineer from a number of locations within a mineral treatment plant. Many of these samples are taken from fast flowing streams of "pulp" (a mixture of water and solid particles). The best way to take a sample from this kind of material is to make a "cut" across the stream at some point where the pulp is in free fall. This cut (see Fig. 2.2) must be made at right angles to the direction of flow of the stream and either the complete stream must be diverted for a short time into a sample container, or the container must be made to cut across the stream at a uniform speed so that all parts of the stream are sampled for the same length of time. There is a very appropriate adage used by mineral process-

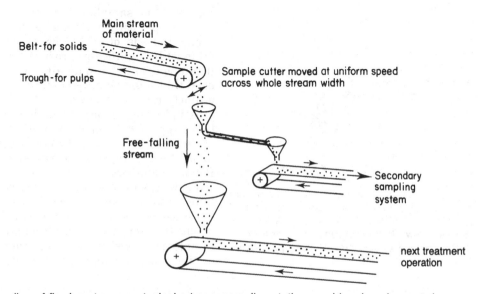

Fig. 2.2 Sampling of flowing streams – typical primary sampling station used in mineral processing operations.

ing engineers which states that "one should sample the *whole* of a flowing stream of pulp for part of the time rather than sample *part* of that stream for the whole time", i.e. it is not good practice to obtain a sample by continuously cutting just one part of a flowing stream with a permanently sited stationary cutting device (Fig. 2.3), because one part of such a stream is unlikely to be representative of the whole.

This adage is based on the observation that mineral particles inevitably segregate according to size, shape and mass at every possible opportunity – especially when attempts are made to produce uniform mixtures! Large particles tend to separate from small particles; flat particles and round particles react differently; high-density and low-density particles tend to follow different paths. (These effects are not unexpected; after all, if different particles did *not* tend to react in these ways the physical separations of minerals that take place in treatment plants and in laboratories would be impossible.) Consequently, the procedures used for sampling mineralogical materials tend to rely on controlling this inevitable segregation; i.e. since segregation by size, shape and density is known to occur and is virtually impossible to prevent, why not use the segregation effects to improve sampling efficiency?

5.5. Commercial products

In many instances, and especially where commercially valuable materials such as mineral concentrates are being bought and sold, there are well-established and legally binding sampling procedures which must be strictly adhered to. The appropriate British Standard specification and the procedures laid down by the American Society for Testing Materials are quite unequivocal about the details of particle size, sample mass and sampling methods that *must* be used in these circumstances. Whilst such precisely defined sampling conditions need not necessarily arrive at the "true" values within the sampled material, they do at least tend to produce results that are strictly *comparable* from one day to the next or from one laboratory to another.

6. Secondary sampling

The primary sample taken from an ore body or from an operating plant may weigh many tens (or even hundreds) of kilograms. This mass of material must be significantly reduced in weight before it can be examined in detail. The amount of material that must be taken to form a good sample is a function of particle size.

Sampling efficiency is, fundamentally, a function of particle *numbers*: the greater the number of particles in the primary sample, the easier it is to produce a representative secondary sample. Particle numbers can, of course, be increased by reducing the particle size (but see below for exceptions to this rule): a linear size reduction of 50% increases the number of particles by a factor of about 8. Consequently, although a secondary sample may consist of only one-eighth of the mass of the primary sample, it will still contain the

(a)

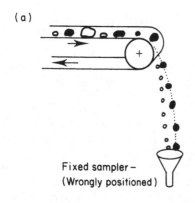

Fixed sampler –
(Wrongly positioned)

(b)   small, light   large, heavy   small, light

(c)

pulp in →

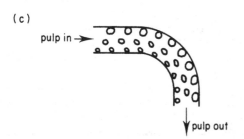

↓ pulp out

Fig. 2.3 (a) Illustration of segregation effect that occurs when mineral particles of different sizes, shapes and densities are allowed to fall freely in their natural trajectories. The diagram also shows (b) how segregation tends to occur across a troughed conveyor belt or (c) within a pipe carrying pump.

same (very large) *number* of particles – provided the particle size has been reduced by a half (say, from 2 cm to 1 cm).

However, it is not always easy to provide an adequate *number* of suitable particles to make a good sample. Consider, for example, a cubic metre of an imaginary gold ore that contains $10\,\mathrm{g\,t^{-1}}$ of native gold; when the relative densities of the various components are taken into account this is roughly equivalent to 1 part per million by volume. In the unlikely event that all the gold in that cubic metre of rock occurs as a single, small cube (Fig. 2.4) then that cube will be 1 cm in size. ($1\,\mathrm{m^3} = 10^6\,\mathrm{cm^3}$, and 1 part per million of $10^6\,\mathrm{cm^3} = 1\,\mathrm{cm^3}$.) If the large block of ore were crushed, this cube of gold would deform rather than break into smaller fragments, since gold is malleable. Consequently, the *only* way to obtain a good estimate of the gold content of the original block of ore is to analyse it all! Thus, if the block of ore is broken in half then one half will contain no gold whilst the other half will contain $20\,\mathrm{g\,t^{-1}}$ of gold. However, the gold fragments in a real gold ore are more likely to be about $10\,\mu\mathrm{m}$ in size (rather than 1 cm) and we can see that a cubic metre of an ore that contains $10\,\mathrm{g\,t^{-1}}$ of native gold will then contain 10^9 gold grains; i.e.

$1\,\mathrm{m^3} = 10^{18}\,\mu\mathrm{m^3}$;

1 grain of gold is $10^3\,\mu\mathrm{m^3}$ in volume;

1 part per million (by volume) $= 10^{-6}$;

number of grains $= 10^{18} \times 10^{-6}/10^3 = 10^9$.

If these gold grains are *uniformly* distributed throughout the ore then the average volume of ore associated with each gold grain is $1\,\mathrm{mm^3}$ and this weighs only about 3 mg; i.e.

$1\,\mathrm{m^3} = 10^9\,\mathrm{mm^3}$; number of gold grains $= 10^9$;

∴ volume of ore associated with each gold grain $= 10^9/10^9\,\mathrm{mm^3} = 1\,\mathrm{mm^3}$.

In practice, the gold in such an ore is most unlikely to be uniformly distributed and a hundred times (or even a thousand times) this minimum amount of 3 mg would be needed to form a reasonable sample. (If, on the other hand, the gold grains were $100\,\mu\mathrm{m}$ cubes then the *average* quantity of rock associated with each grain would be 3 g and a reasonable sample weight would be between 300 g and 3 kg.)

If the primary sample contains particles of a wide size range (say 1 mm to 10 cm, for example) then the numbers of the larger particles may be quite small and sampling can become very inaccurate. The sampling accuracy can be much improved by the use of stratified sampling, mentioned above: the primary sample is screened into several suitable size fractions, and each size fraction can then be sampled separately. The materials from the different size fractions can be studied separately or they can be recombined in the appropriate proportions to become a re-formed sample of the original bulk material (Fig. 2.5).

The problems of sampling have been dealt with in great detail by Gy (1982), and the sampling theory devised by him is often used to calculate the mass of specimen needed to provide a required degree of accuracy. This theory is especially useful for determining the amount of sample needed to establish global values such as the proportions of various elements in a material. The theory takes into account the particle size, the minerals that are present, the particle shape and the degree of liberation of the minerals. Gy's basic equation is:

$$\frac{ML}{L-M} = \frac{Cd^3}{\sigma^2}$$

where M is the minimum mass required to provide an adequate sample (in grams),

L is the gross mass (g) of the material being sampled,

C is a constant ($\mathrm{g\,cm^{-3}}$) for the material,

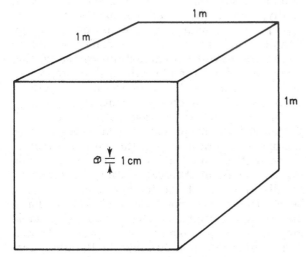

Fig. 2.4 Illustration of low volumetric concentration. The small speck at the centre of the cube is equivalent in size to a $1\,\mathrm{cm^3}$ cube. The volume of the larger cube is $(10^2)^3\,\mathrm{cm^3} = 10^6\,\mathrm{cm^3}$, i.e. the speck represents a concentration of 1 part per million.

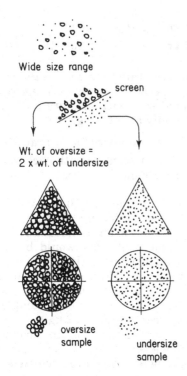

Fig. 2.5 Stratified sampling. A material of wide size range is divided into two or more size fractions. Each fraction is weighed and then sampled individually; each sample may then be analysed and the separate results combined by calculation. Alternatively, the samples can be physically recombined in the appropriate ratios to produce a single representative sample of the original material.

d (cm) is the size of the largest fragment in the sampled material, and

σ is the statistical error that can be accepted in the analytical result (usually the assay, or elemental composition of the sample).

Since L is usually large in relation to M, the above equation generally becomes

$$M = \frac{Cd^3}{\sigma^2}$$

The sampling constant C takes into account the mineral content and degree of liberation of the sample:

$$C = fglm$$

where f is a shape factor (normally taken to be 0.5, except for gold ores for which $f = 0.2$),

g is a factor which depends on the particle size range (g varies from 0.25 to 1.0),

l is a liberation factor which varies from 0 (completely homogeneous material) to 1.0 for completely heterogeneous material, and

m is a mineralogical composition factor, calculated from

$$m = \frac{1-a}{a}[(1-a)r + at]$$

where r is the mean density of the valuable mineral,

t is the mean density of the gangue, and

a is the fractional average mineral content of the material being sampled.

Gy's equation shows the advantage of sampling material after it has been finely ground, since the value d^3 dominates that equation. For many mineralogical purposes, however, it may be unwise (or even completely unacceptable) to grind the material that has to be sampled.

6.1. Secondary sampling techniques

Ordinarily it is the geologist who is involved in collecting the primary samples from rock outcrops, from working mines, and from exploratory drill holes. The mineral engineer, on the other hand, is responsible for collecting the bulk samples from operating plants and from mineral treatment test programmes. It is the mineralogist who is responsible for subsampling the primary samples, preparing the working specimens and, finally, making the actual measurements.

A good secondary sample is produced by sampling the primary sample and, if the sampling procedures have been carried out correctly, this small secondary sample will still be representative of the original bulk material in important features such as elemental concentrations and mineral proportions. However, it need not (and, probably, does not) provide a good representative sample of the grain sizes of the various minerals in the original material, nor is it likely to provide a good sample of the rock texture. A secondary sample may have to be further reduced in bulk before it becomes the "working" sample for a mineralogical analysis. This "working" sample may be made up into a number of prepared specimens which are then suitable for the analytical procedures. These measuring procedures may, of course, themselves further sample the prepared specimens (see Image Analysis in Chapters 6 and 7).

6.1.1. Coning and quartering
As shown above it is almost always necessary for

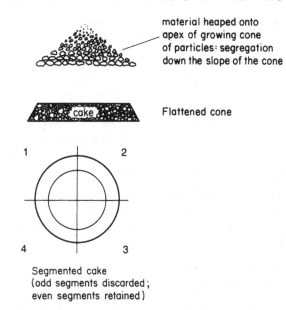

material heaped onto apex of growing cone of particles: segregation down the slope of the cone

Flattened cone

1   2

4   3

Segmented cake
(odd segments discarded;
even segments retained)

Fig. 2.6 "Coning and quartering" method of representative sampling.

the mineralogist to reduce the weight of a primary sample before it can be conveniently measured. One of the most suitable ways of doing this with coarse material (up to a few centimetres in size) is by the coning and quartering method (Fig. 2.6). In this method the material is first heaped into an accurately shaped, conical pile. This is best done by repeatedly placing small amounts of the material on to the apex of a growing pile of mineral particles. A properly prepared cone will show a well-marked radial segregation of mineral particles – the larger, rounder, heavier particles will tend to collect around the periphery of the cone, and the smaller, flatter, lighter particles will tend to congregate nearer the apex.

As we have seen, this segregation is inevitable and, as a result, equal-sized radial segments of the cone will tend to contain similar amounts of all the different kinds of particles. The finished cone is flattened into a circular "cake" a few centimetres thick by firmly pressing a flat board on to its apex. The "cake" is marked off into four equal segments: one pair of *opposite* segments is discarded (or stored for further tests) and the other two segments become the new, reduced-weight sample. Although the procedure is called "coning and quartering" it is really a coning and *halving* operation since the mass of the new, retained sample should be almost exactly half the mass of the original sample. If necessary, this halving operation can be repeated on the retained sample; it is, however, often advisable to reduce the particle *size* between successive reductions in

sample *mass* (see above). In some instances care must be taken at this stage to ensure that important characteristics like mineral grain sizes are not changed by any reduction of the particle size.

6.1.2. Nine-point sampling

Another useful method of sampling is called the *nine-point method*. A flattened cake of material (produced in the same manner as in the cone and quartering method) is marked into eight equal segments. Roughly equal radially oriented portions are taken from each segment with a specially shaped scoop. A ninth portion is taken from the centre of the cake. Further small groups of nine increments can be taken to obtain a close approximation to a pre-determined total mass of sample; e.g. it may be necessary to use equal masses of sample in a group of tests (Fig. 2.7). This sampling procedure is especially useful for sampling fine-grained dry materials.

6.1.3. Riffling

Another commonly used sampling procedure – one which is normally restricted to materials of small particle size (less than a few millimetres in diameter) – uses the riffle-type sample splitter. This is a metal box divided into an even number (usually eight or sixteen) of compartments of equal width. These compartments alternately slope in opposite directions (see Fig. 2.8). The material to be sampled is spread into a layer of uniform thickness on a special scoop which is the same width as the riffler. The material is then poured steadily through the riffler and two

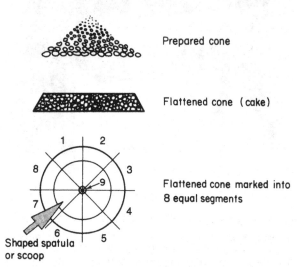

Prepared cone

Flattened cone (cake)

Flattened cone marked into 8 equal segments

1 | 2
8 | 3
| 9
7 | 4
6 | 5
Shaped spatula or scoop

Fig. 2.7 "Nine-point" sampling method. Eight radial portions are taken from the marked segments; a ninth portion is taken from the centre.

(a)

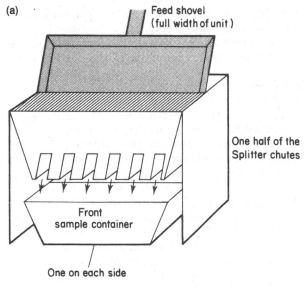

(b)

Fig. 2.8 (a) Diagram of a riffle sampler: the feed shovel contains a uniform thickness of mineral particles, and these particles are poured at a steady speed through the riffler. Half of the feed reports into the front sample container; the other half reports to the back sample container.
(b) Photograph of Gilson sample splitter (Courtesy of Christison Scientific Equipment Ltd., Gateshead).

"equal" halves are collected in boxes placed on either side of the unit. One half, say the one reporting to the front of the riffler, is discarded or stored for other purposes; the other half becomes the working sample. The retained half can be further reduced in mass by again being riffled, but the material reporting to the *front* of the riffler should then be kept whilst the other portion is discarded. This alternate discarding of the front and back fractions is a useful precaution which counters the effects of any slight variations in the width of the riffle spaces – such variations would tend to allow more than half the specimen to collect on one side of the riffle and, in this way, would tend to produce biased samples (see Practical no 1).

6.1.4. General comments

As mentioned above, it is not always easy to ensure that the features that *should* be measured are the features that actually *are* measured. It is important therefore that the desired information is not altered (or, worse still, destroyed) during the various sampling and preparation stages. If, for example, the object of an analysis is to determine the natural (unaltered) grain sizes of the mineral in an ore then every care must be taken to ensure that these mineral grains are not broken during the sampling procedure. In these circumstances it may not be advisable to follow blindly the usual sampling procedure of "size reduction followed by reduction in mass" described above. It is a useful practical rule that if the *grain* size of the material is to be established, then the *particle* size should not be reduced to less than about ten times the size of the largest grain expected to occur in the specimen. Thus, if the *largest* grain of cassiterite expected in a tin ore is 100 μm, then the particles of the ore can be reduced in size to about 1000 μm without much danger of breaking the cassiterite grains.

However, the sizes of the mineral grains in a specimen are seldom known with any degree of accuracy before the sampling is carried out (in fact, it may well be that the object of the analysis is to determine this value) and it is therefore well to err on the cautious side and to limit the amount of particle size reduction allowed during these sampling procedures. Similarly, if an analysis is being carried out to determine the textural characteristics of a rock, it is essential to examine rock fragments that are large enough to preserve this texture: this may mean that the rock specimens should, in some instances, be many square centimetres in area. Consequently, it is often necessary to carry out some preliminary measurements on a specimen to establish the approximate grain size distribution in order to determine (1) the most appropriate sampling procedure, and (2) the number of measurements that will be needed to achieve a required accuracy (see below).

The preparation of a good representative specimen which is ready for analysis can be a difficult operation and it often involves several distinct sampling stages. The sampling accuracy aimed at for any stage of the sampling procedure should match the accuracies that have been attained in all the earlier sampling stages, i.e. there is little point in using a highly accurate secondary sampling procedure if the primary sampling has been badly carried out. Nor is there much value in making highly accurate measurements on poorly representative specimens.

7. Measuring the sampled material

7.1. Specimen preparation

Once a suitable (and, therefore, small) weight of sample has been obtained it must be made up into a suitable specimen (or specimens) for analysis. In applied industrial mineralogical work, this often means that the sample is made into a number of polished sections which can then be examined by one of several analysing techniques (see below). Great care is necessary to ensure that the unbiased working sample is properly represented in this final polished specimen; for example, care must be taken to ensure that certain components are not plucked out of the specimen by the polishing procedure, and the prepared *section* must be an unbiased, random section through the specimen. Details of suitable specimen preparation procedures are given elsewhere (Practical no 9).

7.2. Mineralogical measurement of prepared specimens

Once a sample has been properly taken and a suitable polished specimen has been correctly prepared then the specimen can be measured by one of a number of available techniques.

7.2.1. Area measurement
Area-measuring systems can, if necessary, measure the *whole* of a specimen surface and such an analysis ensures that no sampling errors are introduced during the measuring procedure. In practice, the area-measuring method is more likely to be used to *sample* the specimen surface

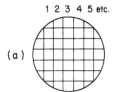

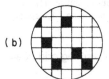

Fig. 2.9 Area analysis of a prepared specimen: (a) complete analysis; (b) surface sampling.

by measuring a relatively small number of regularly spaced, but unconnected, areas on that surface. These areas are generally positioned at the intersections of a regular grid pattern (see Fig. 2.9) spaced uniformly across the whole surface of the specimen. Care must, of course, be taken to ensure that the grid pattern does not coincide with some textural pattern in the specimen (for example, the measured areas must not be parallel to any schistosity or banding in rocks and ores – if they are, the results obtained will show a bias. It is also worth remembering that, from a statistical point of view, it is better to examine a small number of such areas on each of a large number of related specimens than to examine a large number of similar areas on a single specimen (Fig. 2.10). Further details on area-measuring procedures are given in Chapter 6, section 5.1.

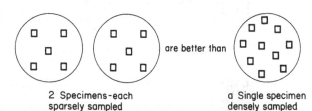

Fig. 2.10 Improved sampling obtained by examining a small number of areas on a large number of specimens (rather than examining the same number of areas on a single specimen).

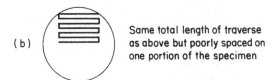

(a) Good spacing of sample *lines* covering the whole specimen surface with a widely-spaced grid

(b) Same total length of traverse as above but poorly spaced on one portion of the specimen

Fig. 2.11 Linear measuring grid spacing.

7.2.2. Line measurement and point counting

If the specimen is to be analysed by a line-measuring system or by a point-counting method (see Chapter 6, sections 5.2 and 5.3) then, of necessity, the specimen surface must itself be sampled (since there is no possibility of measuring the whole of the surface by these methods). So, once again, the measurements are carried out along (or at the intersections) of a suitable grid. This grid must be sufficiently diffuse so that a specimen surface can be sampled in a reasonable time, but it must not be so diffuse that the number of observations becomes too small to represent that surface. Again, in order to avoid bias, the orientation of the grid must not be parallel to any textural feature in the specimen and, as before, it is better to examine a number of specimens using a diffuse measuring grid than it is to examine a small part of a single specimen in great detail (Figs. 2.11 and 2.12). Further details on line measuring and point counting procedures are given in Chapter 6, sections 5.2 and 5.3.

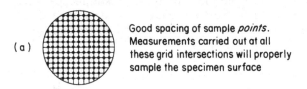

(a) Good spacing of sample *points*. Measurements carried out at all these grid intersections will properly sample the specimen surface

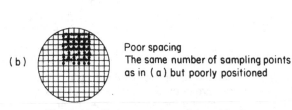

(b) Poor spacing
The same number of sampling points as in (a) but poorly positioned

Fig. 2.12 Point-counting grid positioning.

8. Sources of error during measurement

There are many sources of error that can be introduced by a mineralogist who is attempting, by any of the above methods, to make a quantitative mineralogical analysis of a prepared specimen. These errors can be classified as (1) operational errors, and (2) statistical errors.

8.1. Operational errors

The *operational errors* include simple mistakes such as the misidentification of minerals, the miscounting of the various features, and the misreading of scales (as described in greater detail in Chapter 6, section 6). These operational errors are almost impossible to quantify except by carrying out repeated checks on the precision and on the reproducibility of the results which are obtained. All these errors can, however, be significantly reduced by careful attention to detail.

8.2. Statistical errors

These can only be kept within acceptable limits by making a sufficiently large number of unbiased measurements. The binomial probability distribution is often used by mineralogists to calculate the number of measurements needed to obtain a good estimate of the proportion of a specified mineral in a specimen.

8.2.1. The binomial distribution

The standard deviation of a set of measured values is given by:

$$\sigma = \sqrt{\frac{pq}{N}}$$

where σ is the standard deviation, p is the proportion of the selected mineral, $q = 1 - p$, and N is the number of observations. The acceptable error is usually taken as the 2σ level, and the formula for calculating the absolute error (e) is:

$$e = 2\sigma\sqrt{\frac{pq}{N}}$$

N, the number of measurements needed, is then given by $4pq/e^2$, but if E is the *relative* error on p (i.e. $E = e/P$) then

$$N = \frac{4q}{pE^2}$$

It is clear that the statistical errors are inversely proportional to the *number* of observations that are carried out. More details on the binomial distribution are given in Chapter 6, section 7.1.

8.2.2. The Poisson relationship

When p is very small (and where there is no possibility of increasing p by any physical or chemical concentration procedures – see Chapter 3) it is necessary to employ the Poisson relationship to determine the number of observations that must be made to achieve a required accuracy.

The Poisson relationship states that the probability of finding a rare mineral is:

$$p(r) = \frac{\lambda r}{r!}\, e^{-\lambda}$$

where p is the probability of locating a rare phase, n is the number of observations needed, $np = \lambda$ = average success rate, and r is the number of successes. This relationship should be used in preference to the binomial equation when n is greater than about 100, when p is less than 0.1, and np is less than 5. These conditions apply, for example, when trying to determine the proportions of the gold minerals in a gold ore; and the equation is especially appropriate for use when analysing the discard products from gold ore treatment plants.

8.3. Conclusion

Many important mineralogical decisions are based on measurements which have been carried out on small samples taken from large bodies of solid rock or from large masses of flowing mineral pulp. Consequently, these samples should be as representative of the bulk material as possible. At the same time, the samples should be as small as possible in order to reduce the costs which are incurred during collection, handling and preparation of specimens, and during the ensuing measurement procedures. It is often extremely difficult to determine the minimum amount of material needed to form a "good" mineralogical sample, however, and in view of the many uncertainties that exist when a sample is taken, it is always wise to take a generous amount of material.

Most of the samples which are to be analysed will have been produced by a *sequence* of sampling operations and each of these operations will have its own inherent sources of error. These errors must be minimised by using good sampling and specimen-handling procedures (since there is little purpose in making highly accurate measurements on poor samples – nor is there much

purpose in making poor measurements on good samples). The variances of the errors (not the errors themselves) introduced at each sampling stage are cumulative and, therefore, it is advantageous if similar degrees of care and attention are taken at each of these stages. The sampling procedure(s) (Fig. 2.13) must not change the feature(s) of interest and, although mineralogical materials are particularly difficult to sample, the results obtained can always be improved if the sampling methods are carried out carefully and with due regard to the special problems which are involved.

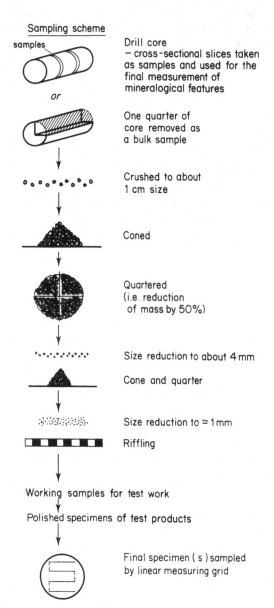

Fig. 2.13 Summary of sampling stages employed in the examination of a drill core specimen.

9. Summary

Mineralogical materials are difficult to sample because:

(1) the mass of material is often exceedingly large;
(2) this material often occurs in a wide range of particle sizes;
(3) these particles may vary widely in: (a) shape; (b) density; (c) mass; and (d) mineralogical composition;
(4) the target component(s) may only occur in very small proportions;
(5) it is usually necessary to establish a *distribution* of values rather than a single-global value.

The sampling procedures must be adapted to suit each type of material and preliminary data obtained from a sample must, if necessary, be used to refine the sampling method.

Bad sampling procedures may produce biased results and these biases cannot be offset by increasing the number of observations carried out on the samples.

The mineralogist and the mineral engineer require large amounts of swiftly produced, accurate mineralogical data. Good data can only be obtained from good samples: good samples are always expensive to produce, but poor sampling can have disastrous results.

Chapter 3

Fractionation of mineral particles

1. Introduction

There are many occasions when the mineralogist or the mineral engineer wishes to obtain clean, pure fractions of a selected mineral species. For instance, a few grams of clean, pure mineral may be needed for the determination of its chemical composition or its atomic structure (this particular need is less pressing nowadays than in the past because of the increasing availability of microanalytical techniques – see Chapter 9). The mineralogist may also have to collect a few grams of a radioactive mineral (such as thorite or some kinds of zircon) for radiometric age determination of a rock. It may also be necessary to isolate a large enough mass of pure mineral for the accurate determination of mineral properties such as density or magnetic permeability. In addition, the mineral engineer may find it necessary to concentrate the components from a mineral treatment process in order to carry out a detailed study of their characteristics and performances. Finally, the mineralogist often uses mineral fractionating procedures as part of a mineralogical analysis method where the aim is to determine mineral proportions, particle compositions and so on.

For most of these purposes it is very important to obtain high-quality products (for some purposes, the product should be absolutely pure). But it may not always be essential to achieve very high recovery values; i.e. as long as a product is sufficiently pure for the job in hand then it does not always matter if some of the mineral is lost during the separation procedures. (In the terminology of the mineral engineer, "grade is usually more important than recovery".)

2. Mineral liberation

When a mineral has to be separated from other minerals to produce a high-grade (nearly pure) product then some, at least, of the selected mineral must be fully *liberated* (or freed) from the accompanying minerals. Occasionally, as in beach sands, soils and alluvial deposits, the constituent minerals may have already been liberated by natural agencies. More commonly, the first stage in any laboratory separation procedure (as also in any mineral treatment plant) is to liberate the required mineral from its rocky matrix. This is normally done by grinding the rock to very fine particle sizes. *Complete* liberation is seldom possible by this method so, in practice, the rock is ground until 90–95% of the important mineral is liberated (Fig. 3.1).

Since good liberation is so important in the preparation of clean mineral fractions, and since laboratory-scale separations deal only with small amounts of material (a few tens of kilograms, at most) it is often possible to use expensive, unconventional liberation methods that are not economically viable on an industrial scale. Thus, it is possible to liberate the chemically inert minerals tourmaline, zircon and cassiterite from a quartz-rich pegmatite rock by dissolving the

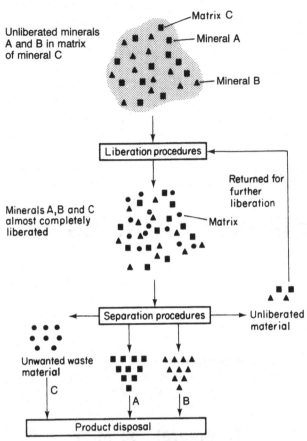

Unliberated minerals A and B in matrix of mineral C

Matrix C
Mineral A
Mineral B

Liberation procedures

Returned for further liberation

Minerals A,B and C almost completely liberated

Matrix

Separation procedures → Unliberated material

Unwanted waste material

C

A   B

Product disposal

Fig. 3.1 Schematic representation of a mineral recovery process (and the accompanying separation processes).

quartz in hydrofluoric acid. (This procedure should *on no account* be attempted without adequate safety equipment such as gloves, goggles and aprons; the correct polythene containers must be used for the acid and the dissolution must be carried out in a suitable fume cupboard.) Non-sulphide minerals such as cassiterite can be released from a pyrite-rich matrix by high-temperature oxidation of the pyrite to form a water-soluble sulphate; minerals locked in a carbonate (e.g. calcite) matrix can be released either by dissolving the calcite in hydrochloric acid or by calcining the rock to form calcium oxide ($CaCO_3$ + heat → CaO + CO_2), the calcium oxide can then be "slaked" (or hydrated) by the addition of water to give calcium hydroxide (CaO + H_2O → $Ca(OH)_2$); the calcium hydroxide is extremely fine-grained and can be decanted away from the insoluble minerals.

Selective solution methods and roasting methods such as those described above can provide almost 100% liberation of the insoluble or inert phases and, what may be even more valu-

able for the mineralogist, these phases are made available at their natural grain sizes.

3. Fractionating methods

Minerals which have been liberated, either by grinding or by chemical means, must usually be "sized" before they are separated from each other because the efficiencies of most separation methods are improved when closely sized fractions are used (Fig. 3.2).

3.1. Screening

The term "size" is difficult to define when used to describe irregularly shaped mineral particles. However, one commonly used method for "sizing" mineral particles is screening. A screen analysis categorises particles according to their ability to pass through an aperture of specified size and shape under a prescribed set of screening conditions. This method of sizing collects together all particles having roughly the same "second diameter". All the particles shown in Fig. 3.3 could be made to pass through a square aperture of size l and all the particles would be retained on a screen aperture of $(l + dl)$: they are, therefore, of the same screen "size" – even though they are clearly of different lengths and different volumes. Many screens are made of woven-wire "cloth" and the accuracy of the aperture size in such a cloth is limited by the tolerances demanded by the wire manufacturers and by the cloth weavers. Consequently, the smallest effective aperture for most purposes is about 50 μm in size (see Practical no 2). Nowadays, however, it is possible to obtain expensive "micro-mesh" screens: these are manufactured from a single sheet of metal to much higher accuracies than the woven-wire screens and can be made with apertures only a few micrometres in size. The "throughput" of a micro-mesh screen is, unfortunately, very small, and as a result, particles below about 50 μm are generally fractionated on the basis of their settling speeds in fluids: this procedure is called *classification*.

3.2. Classification

If uniformly shaped particles of the *same* mineral fall through a fluid then the larger particles will fall more rapidly than the smaller particles. If, on the other hand, a population of uniformly shaped monosize particles is made up of *different* miner-

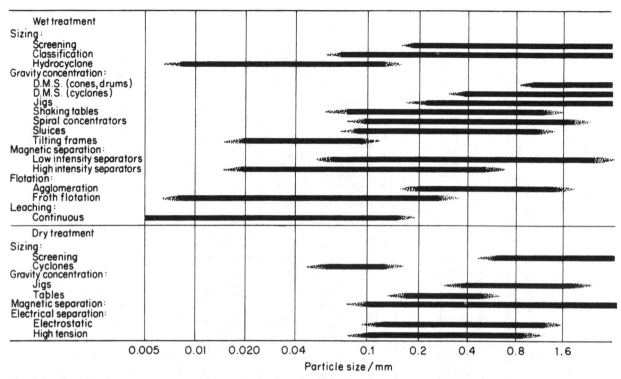

Fig. 3.2 Particle size ranges over which various mineral treatment processes work efficiently. The "fuzzy" zones at the top and bottom of the size ranges indicate losses of efficiency at these extreme values.

als then the densest particles will fall more rapidly through a fluid than the "light" particles.

The velocity attained by a small spherical particle in any fluid can be calculated from Stokes' Law, which states:

$$V_T = \frac{(D_s - D_f)gd^2}{18\eta}$$

where V_T = terminal velocity (for a sphere)
  D_s = density of the particles
  D_f = density of the enclosing fluid (this is usually water)
  d = particle diameter
  g = gravitational acceleration
  η = viscosity of the fluid

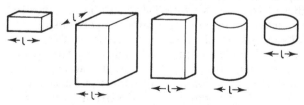

Fig. 3.3 Diagrammatic illustration showing how particles of the same square aperture screen size (l) can be of very different volumes.

The terminal velocity of a particle is thus affected by its size, its "effective density" $(D_s - D_f)$ and the viscosity of the medium. The terminal velocity, when substituted in the above equation, gives a value of d which can be used to characterise the particle "size": this "size" is referred to as the particle's *Stokesian diameter* or the *Stokes equivalent sphere diameter*.

Figure 3.4 is a nomogram which shows the variation in the settling speed of spheres of different diameters and different densities and this can be used as an easy way to determine settling times or particle diameters (see Practical no 2).

Despite the theoretical problems that are encountered in calculating accurate "size" values for irregularly shaped particles, it is nevertheless possible to use this equation as the basis of a mineral fractionating method for particles in the size range 5 to 75 μm. The Cyclo-sizer (Practical no 2) is a practical device which is widely used, especially for fractionating mono-mineralic particles into different size fractions. However, it is important to realise that when particles of different densities are fractionated in a Cyclo-sizer then the particles which collect in the same "size" category are not necessarily of even approximately the same mass. Table 3.1 shows the mass ratios between spherical particles of quartz and

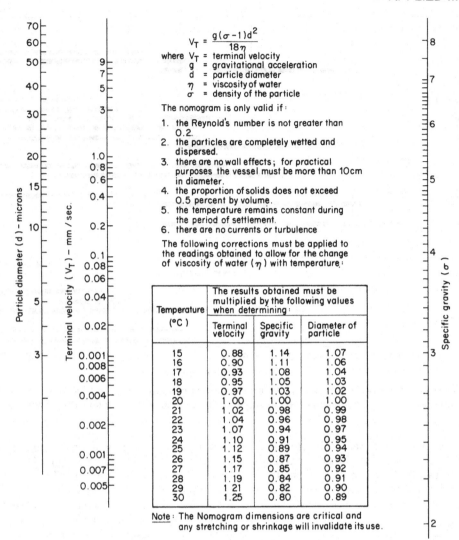

Fig. 3.4 Nomogram showing the variation in settling velocities of spheres of different diameters and different densities (e.g. a ruler laid from $V_T = 0.1$ to a specific gravity of 3.0 will show a particle diameter of 9.5 μm).

galena having the same terminal velocities. (The quartz and the galena particles are *nominally* of the same "size" but their masses differ considerably. In fact, the quartz particles are nearly three times as heavy as the equivalent galena particles!) These ratios show that it is

TABLE 3.1
Mass ratios of spherical particles of different densities which have the same terminal velocity

Classification fluid	Ratios of spherical diameters	Mass ratios	
		Quartz/galena	Quartz/sphalerite
Air	1:1	1.70:1	1.24:1
Water	1:1	2.84:1	1.67:1

most unwise to assume that particles which have the same Stokesian size are of equal mass – they seldom are.

Classification methods can also be used to fractionate composite particles. A composite mineral particle containing, say, the minerals quartz and galena, will have a terminal velocity that will depend, among other things, on the proportions of those two minerals (see Table 3.2); i.e. composite particles made up of the same proportions of the different mineral components will have the same terminal velocity and will collect in the same classification fraction. A series of fractionations would, in this instance, provide the basis for a liberation (or particle composition) analysis. Unfortunately, this relationship between particle composition and terminal velocity does *not* hold for particles that contain more than two minerals

TABLE 3.2

Variations in the size ratios and the mass ratios of composite particles which are of *equal terminal velocities in water* i.e. particles that would accumulate in the same density fractions (see Fig. 3.10).

% galena (or sphalerite) by mass in particle	Density fraction (kg m^{-3})	Size ratio (compared to pure quartz particle)	Mass ratio
Quartz + galena:			
0 (i.e. quartz)	2.60	1.00	1.00
10	2.78	0.95	0.91
20	2.99	0.90	0.83
30	3.23	0.85	0.76
40	3.52	0.80	0.68
50	3.86	0.75	0.62
60	4.28	0.70	0.56
70	4.79	0.65	0.51
80	5.45	0.60	0.45
90	6.31	0.55	0.40
100 (i.e. galena)	7.50	0.50	0.35
Quartz + sphalerite:			
0 (i.e. quartz)	2.60	1.00	1.00
10	2.69	0.97	0.95
20	2.80	0.94	0.90
30	2.91	0.92	0.86
40	3.02	0.89	0.82
50	3.15	0.86	0.78
60	3.29	0.84	0.74
70	3.44	0.81	0.70
80	3.61	0.78	0.67
90	3.80	0.76	0.63
100 (i.e. sphalerite)	4.00	0.73	0.60

and, since real mineralogical specimens are often multi-mineralic, the fractionating procedure can only rarely be used to determine mineral liberation.

3.3. Heavy liquids

Two particles of different densities can be separated by placing them in a fluid of intermediate density. One of the simplest examples is the separation of sand and sawdust in water (density 1000 kg m^{-3}); the sand (density 2650 kg m^{-3}) sinks and the sawdust (density about 900 kg m^{-3}) floats. Unfortunately, most minerals sink in water so that water is not a suitable separating medium and other "heavier" (denser) fluids must be used instead. These liquids are of four main types:

(1) organic compounds of the halogen elements, e.g. tribromomethane (bromoform);

(2) aqueous solutions of very dense salts, e.g. Clerici solution, a saturated aqueous solution of thallium(I) methanoate (formate) and thallium(I) propanedioate (malonate);

(3) aqueous suspensions of finely divided solids (rather than true liquids), e.g. quartz, magnetite, etc. in water;

(4) molten alloys, e.g. Wood's metal.

Many of the organic fluids are expensive and they are also toxic: they can be dangerous to use unless strict safety precautions are taken (see Practical no 5). Because of the potential dangers, great care must be taken to use as little as possible and every effort must be made to recover as much as possible of these liquids for re-use. The dense salts are also potentially dangerous and must be handled with great care. The solid suspensions are cheap and safe to use but their densities are difficult to control accurately and they are only really effective when used to separate coarse mineral particles. The molten salts and the molten alloys are difficult to use because the separations

TABLE 3.3

Commonly available heavy liquids

Name	Formula	Density (kg m^{-3})	Solvent	Main use
Trichloroethene	CHCl.CCl$_2$	1460	Ethanol	Coal separations
Tribromomethane (bromoform)	CHBr$_3$	2890	Ethanol or propanone	Silicates from "heavy" minerals
Tetrabromoethane	C$_2$H$_2$Br$_4$	2960	Ethanol	Silicates from ore minerals
Di-iodomethane (methylene iodide)	CH$_2$I$_2$	3200	Propanone	"Heavy" silicates e.g. ferromagnesian minerals from other silicates or from ore minerals
Clerici solution	Saturated solution of thallium(I) methanoate + thallium(I) propanedioate	4200	Water	Ore minerals e.g. separation of rutile/ilmenite

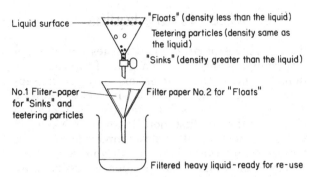

Fig. 3.5 Simple equipment for separating mineral particles in heavy liquids – batch operating system (for handling amounts up to 1 kg).

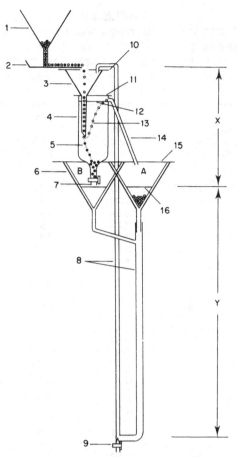

Fig. 3.6 Equipment for continuous separation of large quantities of mineral particles in heavy liquids. Details of the construction: (1) feed container; (2) feeder; (3) mixing funnel; (4) separating funnel; (5) heavy grains; (6) collecting funnels A and B; (7) pinchcock; (8) air-lift tubes; (9) pinchcock; (10) loose-fitting cover; (11) close-fitting cover; (12) liquid level in separating vessel; (13) light grains; (14) overflow carrying light grains; (15) loose-fitting cover on collecting funnels; (16) liquid level in collecting funnels.

must be carried out at slightly elevated temperatures and the convection currents which occur within the liquids tend to re-mix the heavy and the light fractions.

The true liquids (Table 3.3) have an advantage in that they can be diluted with suitable solvents and can be used to produce particle fractions that differ only very slightly in density – for example, low-density particles larger than about 100 μm in size can be separated into fractions that differ in densities by only about 100 kg m^{-3}. The heavy liquids can, in some instances, be used to carry out a liberation (or particle composition) analysis – (see section 3.3.7 of this chapter, below).

3.3.1. Batch separation in heavy liquids

Heavy-liquid separations of small quantities of minerals are usually carried out in batches in small glass funnels (Fig. 3.5). Larger amounts of material are treated in continuously operating separating devices (see below) which incorporate suitable pumps for recirculating the liquid (Fig. 3.6).

A batch separation is carried out by stirring the mineral grains into a heavy liquid of known density contained in a funnel fitted with a pinch clip (see Fig. 3.5). It is important that all the mineral grains are properly wetted by the liquid and that sufficient time is allowed for the separation to take place, i.e. time must be given for the dense particles to reach the bottom of the vessel. As shown in section 3.2 above, the rate of movement of a particle through a liquid depends on:

(1) the particle size (try shaking a half-full bottle of shampoo and note how the large air-bubbles trapped in the liquid rise much faster than the smaller bubbles do – this example is, of course, the reverse of the process that takes place when dense mineral particles are dispersed into a less-dense liquid: the large particles sink very swiftly but the finest particles may take many minutes to reach the bottom of the separating vessel);

(2) the density difference between particle and liquid: the greater the difference, the faster the rate of rise (or fall) of a particle;

(3) the viscosity of the liquid which, in part, depends on its temperature: an increase of temperature reduces viscosity (but may also reduce the density of the liquid).

Any particles that adhere to the sides of the separating vessel during a separation must be given the opportunity to rise (or fall): a glass rod can be used to free these particles. In the unusual circumstance where the density of a mineral particle exactly matches that of the liquid, the mineral will "teeter" – it will neither sink nor float but will remain suspended in the liquid.

After a separation the heavier, "sink" fraction is run out into a filter paper held in a lower funnel. The expensive heavy liquid is allowed to drain through the filter and is collected; it is then ready for immediate re-use. The "float" fraction is run into a second filter paper (both filter papers can be held in the same funnel – see Practical no 5) and the remaining liquid is also drained off for re-use. Any "teetering" particles can be either collected separately as a third density fraction, or included in one of the two main products: they are usually collected with the "sinks", but if a clean sink product is required then the teetering particles can easily be collected with the "floats".

All the fractions must be carefully washed with a suitable solvent to remove the adhering heavy liquid: these washings will contain appreciable amounts of valuable heavy liquid which must be recovered (see section 3.3.6 below).

3.3.2. Continuous separation in heavy liquids

Although many density separations are carried out in small "batches" of a few tens (or, at most, a few hundreds) of grams of material, there are occasions when it is necessary to treat much larger amounts of material: for example, when separating the light from the heavy minerals in a typical beach sand deposit it may be necessary to fractionate many kilograms of material in order to collect sufficient heavy minerals for test purposes.

A "continuous" separator (Fig. 3.6) can easily handle such amounts. It works on the same principle as the batch operation above. A mixture of heavy and light particles is continuously fed *into* a bath of liquid (this is done to ensure that all the particles are wetted by the liquid). The small amount of high-density particles falls to the bottom of the vessel and is removed at the conclusion of the separation procedure. The large amount of low-density particles floats and continually overflows into a funnel where the liquid is filtered off and returned to the separating vessel by a suitable pump. This can be an air lift pump, as illustrated in Fig. 3.6; more conveniently, it can be a peristaltic pump fitted between the drainage pipes from funnels A and B and the return pipe to the separating funnel.

3.3.3. Heavy-liquid separation by centrifuge

If the mineral particles are small – less than about 100 μm but greater than 10 μm – then the rate of separation of "heavies" from "lights" may

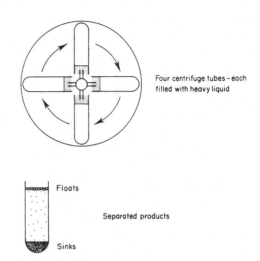

Fig. 3.7 Separation of mineral particles by heavy liquids in a centrifuge. Each of the four centrifuge tubes shown weighs very nearly the same; i.e. they should contain the same amount of liquid plus solid particles for proper balance.

be exceedingly slow. In fact, gentle convection currents within a liquid may prevent the very smallest particles from ever separating. In these circumstances the separation can be speeded up by using a centrifuge (Fig. 3.7). The particles are stirred into the appropriate liquid, which is held in a glass centrifuge tube. Two tubes (at least), placed diagonally, must be used to "balance" the centrifuge since an unbalanced, high-speed centrifuge can break up and become very dangerous. After centrifuging for a few minutes (the time depends on the speed of rotation and on the particle size) the "sinks" and the "floats" are poured into separate filter papers and are treated as in section 3.3.1.

It may be difficult to keep the floats and the sinks from re-mixing whilst they are being poured into their respective filter papers. However, with a little practice, it is possible to remove *all* the floats by gently rotating the tube whilst slowly pouring out the bulk of the liquid. The sinks and the remaining liquid can then be washed out with a jet of solvent from a wash-bottle.

3.3.4. Choosing a heavy liquid

The most commonly used liquids are listed in Table 3.3. These liquids can be diluted with the appropriate solvents to produce intermediate density values. Trichloroethene, tribromomethane and tetrabromoethane are much cheaper than the denser liquids di-iodemethane and Clerici solution. One of the former group of liquid is, therefore, used to produce preliminary heavy

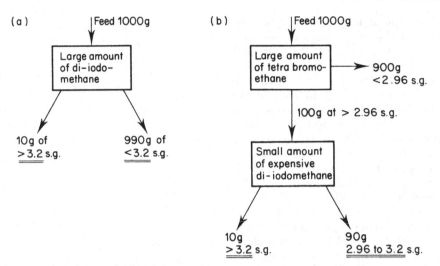

Fig. 3.8 Choosing a heavy liquid for separating large specimens. (a) This route requires the use of large amounts of expensive liquid to isolate the small proportion of heavy minerals; (b) the alternative route involves *two* separations; one involving the separation of a large specimen in comparatively cheap tetrabromoethane, and another requiring only small amounts of expensive di-iodomethane.

products when large amounts of material are being separated (see Fig. 3.8).

Trichloroethene is especially suitable for separating coal from sulphide minerals and from silicate minerals: it is often mixed with tetrabromoethane to obtain densities up to about 2000 kg m^{-3} during the study of coal specimens. The usual working range for tetrabromoethane lies between 2000 kg m^{-3} and 2960 kg m^{-3}: it is mainly used for separating quartz, feldspar and calcite (which have densities ranging from 2500 kg m^{-3} to about 2700 kg m^{-3}) from the metallic ore minerals which have densities well in excess of 3000 kg m^{-3}. Di-iodomethane is used in small amounts to separate mica and the heavier ferromagnesian silicates from the much denser ore minerals.

Clerici solution is used (sparingly) in separating, for example, the two major titanium-bearing minerals of commerce – rutile, 4200 kg m^{-3} and ilmenite 4700 kg m^{-3} – from each other. (Unlike other heavy liquids, the density of saturated Clerici solution increases with temperature, and it is possible, but *not* advisable for students, to use hot solutions with densities up to 4800 kg m^{-3}.)

3.3.5. Separations at very high densities
Liquid densities can sometimes be extended by using "special" fluids, but these fluids should not be used by unsupervised students (nor is their use generally recommended for others):

(1) Liquid mercury has a density of $13\,600 \text{ kg m}^{-3}$ and only the native metals and a few other minerals are denser; consequently, mercury can be used to float away virtually all other minerals from these native metals. Provided the separation is carried out swiftly, the metals will not amalgamate appreciably with the mercury, and can be recovered by passing the mercury through a chamois cloth. This method is particularly useful for concentrating gold and the platinoid elements from heavy-mineral concentrates.

(2) Suspensions of fine globules of mercury in tribromomethane can provide fluid densities ranging from about 3000 kg m^{-3} to 7000 kg m^{-3}, but these suspensions are difficult to use and are not recommended.

(3) The chlorides of zinc and lead become molten at moderate temperatures and can be used for density separations up to 6000 kg m^{-3}. After a separation has been carried out the molten salt is allowed to cool and solidify. The column of salt is then broken near its centre and the upper and lower portions are immersed in separate beakers of hot water. The salt dissolves readily and the separated minerals can then be recovered by filtering away the liquid.

Further details of these unusual fluids can be found in earlier mineralogical texts (see the Bibliography) but, before contemplating their use, the present-day mineralogist should seriously consider whether their use is unavoidable. The question has to be asked: "is the separation absolutely necessary?" If the answer is "yes", then the simplest and safest method of carrying out the separation must be sought and it will often be found that there is a better, simpler method of achieving the separation than by using heavy liquids.

3.3.6. Recovery of heavy liquids

All heavy liquids used for mineral fractionation are expensive and potentially dangerous, and every effort must be made to recover them after they have been used. Ethanol-based wash-solutions containing small portions of heavy liquid are generally treated by shaking the solution in a separating funnel with a large excess of distilled water. The bulk of the ethanol preferentially dissolves in the water whilst the heavy liquid forms an almost pure, insoluble layer beneath the aqueous solution. The heavy liquid is tapped off (Fig. 3.9) and it is re-cleaned by repeating the process a few times to extract any residual small traces of ethanol. Before being re-used the density of the recovered liquid should always be checked by using a standard density float – a small block of plastic of known density which is chosen so that it teeters when the liquid is of the correct density. If these standard blocks are not available then the liquid density must be determined by using a pycnometer bottle (see Chapter 4, section 5.2.1: the recovered heavy liquid is used instead of the mineral grains mentioned in that section).

Propanone (acetone) washings can either be treated with distilled water, in the same manner as ethanol washings, or allowed to stand in a well-ventilated fume cupboard whilst the solvent slowly evaporates. A small plastic block of the required density can be immersed in the liquid and will teeter when all the solvent has evaporated.

Wash-water containing Clerici solution is treated by gently evaporating the excess water over a water-bath in a good fume cupboard: care must be taken not to overheat the solution, otherwise a dark-brown residue will form.

Many of the heavy liquids darken with repeated use. This darkening does not affect their densities but makes it difficult to follow the course of a separation and it may not be possible to see when a separation is complete. The dark colour can be removed by shaking the liquid with a little fuller's earth or by warming with a little charcoal. The contaminated fuller's earth or charcoal can be removed by filtration and should be carefully and safely discarded.

3.3.7. Fractionation in a density-gradient column

The mineralogical compositions of composite two-phase mineral particles can be determined by fractionation in a density-gradient column. This is a liquid column ranging from a few centimetres to a few tens of centimetres high, in which the density of the liquid increases from top to bottom of the column.

Density-gradient columns are prepared by the controlled mixing of two miscible liquids. The densities of the unmixed liquids determine the maximum and minimum densities that can be produced. These columns are stable for short periods, but thermal and other disturbances tend to destroy them within a few hours.

A graded-density column can be used to carry out a liberation analysis of composite two-phase particles (Fig. 3.10), i.e. it can be used to determine the distribution of particle compositions. This procedure is, strictly speaking, applicable only to particles that consist of, at most, two minerals; both these minerals should have densities which are lower than that of the densest liquid in the column. There is then a simple relationship between the position of a mineral particle in the column (and, hence, its density) and the mineralogical composition of that parti-

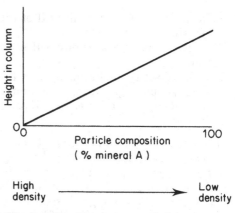

Fig. 3.10 Relationship between height in a graded density column and compositions of a two-phase mineral particles that settle at these positions.

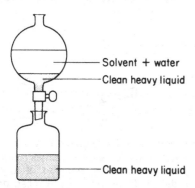

Fig. 3.9 Equipment for recovering heavy liquids from dilute "washings".

cle. This method of determining particle composition is especially useful for analysing coal samples (although, it is more usual to carry out a series of separate fractionations in a series of liquids of known densities). The coal samples consist, essentially, of two components: coaly material (with a density of about 1100 to 1200 kg m^{-3}) and shale (having a density in excess of 1800 kg m^{-3}). In this way the variations of shale (or "ash") contents in coal specimens can be determined and the "separating densities" needed in an industrial dense-medium plant for treating such a coal can be calculated.

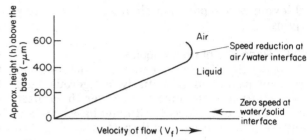

Fig. 3.11   Variations in velocity in a flowing film of liquid.

3.4. Hydraulic methods of mineral fractionation

These methods (often collectively called "gravity" methods) are useful for separating mineral particles that have large differences in densities. The hydraulic methods are especially valuable where the densities of some of the minerals exceed the values of the commonly used heavy liquids (see section 3.3). In the hydraulic methods the separations are generally (but, not exclusively) carried out in water, and if necessary many kilograms of material can be treated at a time.

3.4.1. Concentration criterion
A good indication of the ease with which a separation can be achieved by using a "gravity" method is given by a coefficient called the *gravity concentration criterion*. This is the ratio between the "effective" densities of the various mineral particles whilst they are immersed in the separating medium:

$$\text{Concentration criterion} = \frac{D_h - D_f}{D_l - D_f}$$

where D_h is the density of the mineral of higher density,

D_l is the density of the mineral of lower density,

D_f is the density of the immersion fluid – this is usually water but may, on occasions, be a "heavy" liquid, or just air.

If the concentration criterion between two groups of particles is greater than about 2.5 then separation is likely to be simple and can be achieved over a wide range of particle sizes; if the value is less than about 1.25 then good separation by hydraulic methods is likely to be difficult at any particle size.

The equation clearly shows that the denser the

separating fluid (up to the density of the less-dense mineral), then the greater the concentration criterion and the easier it will be to effect a separation. Of course, if the fluid density exceeds that of the lower-density mineral then the separation becomes a "heavy-liquid separation". If quartz (density 2650 kg m^{-3}) and diamond (density 3500 kg m^{-3}) have to be separated then the concentration criterion for these two minerals in water (density 1000 kg m^{-3}) is about 1.5: the concentration criterion for these same minerals in trichloroethene (density 1460 kg m^{-3}) is 1.7 and in air (density essentially zero) it is only 1.3.

3.4.2. Flowing-film separation
Differences in mineral densities form the primary basis of a number of separation procedures but the effects of density are often modified by the other factors, such as particle shape and size.

A film of liquid flowing over a smooth inclined surface has zero velocity at that surface and achieves its maximum velocity *near* to (but not *at*) the air–water interface (Fig. 3.11). When a mixture of mineral particles (comparable in size to the depth of the water) is introduced into such a flowing film of liquid the smaller grains will be under the influence of the slow, lower layers whilst the larger particles will protrude into the faster, upper layers of the liquid (Fig. 3.12). Since

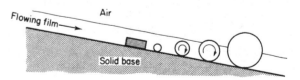

Fig. 3.12   Effect of the velocity gradient on small mineral particles. Small flat particles and very small rounded particles are relatively unaffected; particles of roughly the same size as the depth of the film are rolled down the incline by the velocity gradient. Very large particles are unaffected by the flowing film.

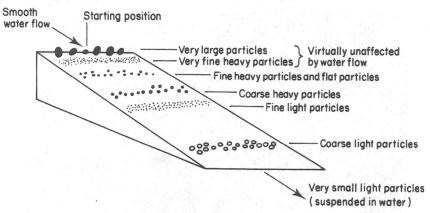

Fig. 3.13 Schematic diagram showing relative positions reached by various mineral particles after time (*t*) when separated under the action of a flowing film of water.

the liquid is moving down the slope, the larger grains will travel faster down that slope than smaller grains of the same shape and density. Furthermore, rounded, large grains will tend to roll and will travel faster than flat, large grains. In this way it is possible to use a flowing film of water to effect a separation of particles according to some combination of their mass, size and shape (Fig. 3.13).

It is obvious that all the hydraulic methods of mineral fractionation mentioned above work best when there are marked differences in the shapes and/or in the masses of the mineral particles. In general, it is an advantage if the feed materials are closely sized; in addition, the particles should be liberated, and the effective density difference $(D_h - D_f)/(D_l - D_f)$ should be at least 1000 kg m^{-3} to obtain good performance.

If the size differential is removed beforehand by screening, the particles can be separated on the basis of differences in their mass and shape and, since different minerals often show distinctive shapes, the flowing-film separator can be used to separate groups of minerals into a number of different fractions according to particle mass. The quality and speed of such a separation is seldom high but these features can sometimes be improved by subjecting the inclined surface to various asymmetrical bumping actions – as in the laboratory-scale "shaking table" (Fig. 3.14), and in the superpanner (Fig. 3.15).

Laboratory-scale shaking tables work best with mineral particles in the 25–250 μm size range, and they can treat many kilograms of material per hour. The superpanner is very suitable for separating even smaller mineral particles (in the size range 10–100 μm) but it can treat only a few tens of grams per hour.

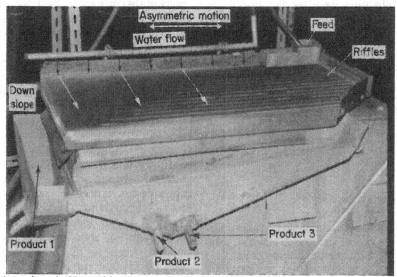

Fig. 3.14 Laboratory-size shaking table – used to carry out mineral separations on kilogram-scale specimens.

Fig. 3.15 Superpanner used in the mineralogical laboratory to separate small amounts of fine-grained minerals.

3.5. Flotation methods

Even if the densities of two minerals are the same it may still be possible to effect a "density" separation of these minerals if the effective density of *one* mineral can be reduced by preferentially attaching it to an air bubble. This procedure is called *flotation* and it is very widely used as an industrial process for separating minerals.

Most minerals in their natural state are *hydrophilic*, that is, they are readily wetted by water. A very small number of minerals, including graphite and molybdenite, are naturally *hydrophobic* (i.e. they are water-repellent). When air is bubbled through a mixture of hydrophobic and hydrophilic particles suspended in water, the hydrophobic particles will tend to adhere to the air bubbles (Fig. 3.16) and, provided the combined density of a particle plus its attached bubble is less than that of the water, the particle will float to the surface of the water where arrangements

can be made to remove it from the system. The hydrophilic particles will be wetted by the water; they will not attach to the air bubbles and, if their densities are greater than that of water (1000 kg m^{-3}), they will remain at the bottom of the container (or flotation cell).

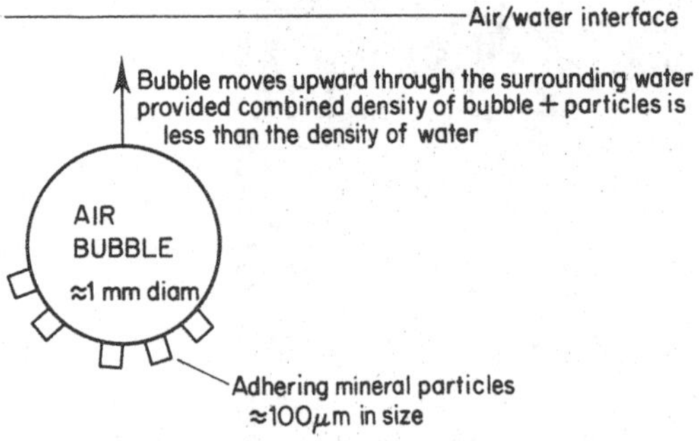

If, on the other hand, a mixture contains only minerals which are naturally hydrophilic then it may still be possible, by careful control of the chemical environment, to produce a "water-proof" coating on some selected mineral (or minerals) within that mixture. The artificially water-proofed mineral can then be separated from the other minerals by attaching it to air bubbles blown through the mixture.

The flotation method (as applied to industrial processes) is described in great detail in many textbooks on mineral processing (see Bibliography): the flotation methods which are used to fractionate mineral particles in the laboratory employ the same principles and the same chemical reagents. The laboratory methods work best on particles in the size range $1000–10 \, \mu\text{m}$: the upper size limit is set by the need to use large, unstable bubbles to lift the large particles; the lower size limit is set by the difficulty of obtaining *selective* water-proofing of very small particles.

Many of the laboratory-scale flotation cells can handle only a few grams of specimen at a time (Fig. 3.17), whereas others (Fig. 3.18) can handle a kilogram or more of material. As with all other mineral fractionating methods, the flotation methods work most efficiently when treating fully liberated particles. As mentioned earlier, it is not always easy to produce a completely liberated feed; even so, the flotation process can still be used to prepare upgraded products which can then be further purified by other, more suitable methods. For example, sulphide minerals, which

---
Air/water interface

Bubble moves upward through the surrounding water provided combined density of bubble + particles is less than the density of water

AIR BUBBLE
≈1 mm diam

Adhering mineral particles ≈100 μm in size

Fig. 3.16 Diagram showing hydrophobic mineral particles adhering to an air bubble suspended in water. If the combined density of the bubble plus particles is less than that of the water then the bubble will rise and carry the particles to the water surface, where they are collected.

FRACTIONATION OF MINERAL PARTICLES   **39**

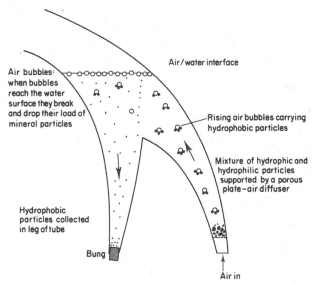

Fig. 3.17 Laboratory flotation cell (Hallimond cell) for treating small amounts (a few grams) of specimen. The mixture of mineral particles is placed on the air diffuser situated in one arm of the cell and air is blown in from a hand pump. The hydrophobic particles adhere to the air bubbles and rise to the water surface; the bubbles are unstable and break; the hydrophobic particles drop back into the other arm of the cell.

are naturally hydrophilic, can easily be made hydrophobic and they can be separated readily from naturally hydrophilic silicates and carbonates. It is also possible by choosing the appropriate chemical environment, to separate two sulphide minerals from each other or to separate

quartz from calcite. The various products can, if necessary, be further up-graded by small-scale density separation methods, by hand sorting and so on.

3.6. Magnetic methods

Minerals vary widely in their magnetic properties. There are several different magnetic properties but the one most commonly used in laboratory-scale separators is magnetic *permeability*.

In general terms, a mineral is said to be *magnetic* when it is visibly affected by a powerful magnet; on the other hand a *non-magnetic* mineral is not noticeably affected by a magnetic field (this definition is not strictly valid but, in the absence of a more rigorous and much lengthier explanation of the causes and nature of magnetism, it will suffice for the purposes of this book). By controlling the strength of the magnet (or, more correctly, by controlling the magnetic flux density) it is possible to fractionate minerals according to their magnetic permeabilities (Fig. 3.19).

3.6.1. The horse-shoe magnet

A simple separation of a highly magnetic mineral, such as magnetite (see Table 3.4) from a non-magnetic mineral, such as quartz, can be carried out with a small horse-shoe magnet. During such a separation the magnet should be enclosed in a thin plastic bag to prevent the magnetite particles from sticking firmly to the

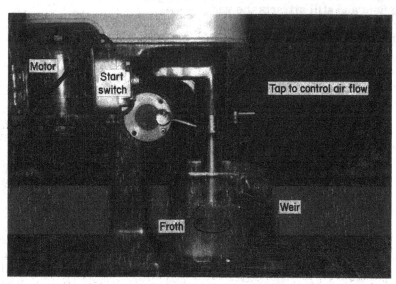

Fig. 3.18 Large-scale Denver laboratory flotation cell. The mixture of mineral particles is placed in the glass bowl; air is drawn down the impeller and "atomised" at the diffuser and hydrophobic particles are lifted by the air bubbles to the water surface. The addition of small amounts of a frothing reagent will produce a stabilised froth which is removed by hand-held paddles over the weir.

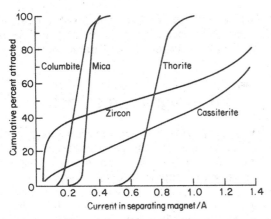

Fig. 3.19 Magnetic permeabilities of the "heavy" minerals in a Nigerian tin ore. A separation using a current of ≃0.4 Amps would separate columbite and mica from thorite. Zircon and cassiterite would occur in both the magnetic and non-magnetic fractions.

magnet: the quartz, of course, is not attracted to the magnet (see Practical no 2).

3.6.2. The "Eclipse" hand magnet

The "Eclipse" is a variable-strength permanent magnet (Fig. 3.20) which can be used to separate a population of particles into a number of magnetic categories. The strength of the magnetic field within the jaws of the magnet can be varied by aligning (or by misaligning) two separate magnets encased in the body of the instrument. In this way, the magnet can be used to produce at least four magnetic fractions, viz.:

(1) When the magnet is "off" it retains a small, residual magnetic field and still attracts the very highly magnetic minerals such as magnetite. (As with the horse-shoe magnet, it is advisable to

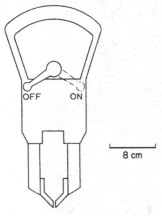

Fig. 3.20 Eclipse hand magnet. (Strength of magnetic field can be varied from "Fully off" to "Fully on" giving the maximum field strength when used to separate fine mineral particles.)

TABLE 3.4

Relative magnetic attractability of some common minerals. (These values are approximate only and and are subject to considerable variations.) (After Davis.)

Substance	Relative attractability
Strongly magnetic:	
Native iron	100
Magnetite	40
Franklinite	35
Ilmenite	25
Weakly magnetic:	
Pyrrhotite	7
Siderite	2
Hematite	1
Zircon	1
Limonite	1
Corundum	1
Pyrolusite	1
Manganite	1
Calamine	1
Non-magnetic:	
Garnet	less than 1
Quartz; rutile	(and decreasing
Cerussite	downwards)
Cerargyrite	
Argentite	
Orpiment	
Pyrite; sphalerite; molybdenite	
Dolomite; bornite	
Apatite; willemite; tetrahedrite	
Talc; magnesite; arsenopyrite	
Chalcopyrite	
Gypsum	
Fluorite	
Zincite; celestite; cinnabar	
Chalcocite	
Cuprite	
Smithsonite	
Orthoclase; stibnite	
Cryolite; enargite; senarmontite	
Galena; niccolite	
Calcite	
Witherite	

cover the magnet pole-pieces with a thin plastic bag to prevent the magnetite sticking to the jaws.)

(2) When the magnet is "half-on" it attracts moderately magnetic minerals such as ilmenite.

(3) When the magnet is "on" it attracts weakly magnetic minerals such as garnet and iron-rich mica.

(4) Minerals which are essentially non-magnetic, such as quartz, are not attracted to the magnet even when it is in its fully "on" position.

3.6.3. The isodynamic magnetic separator

Improved magnetic categorisation of minerals is possible with more sophisticated equipment. One

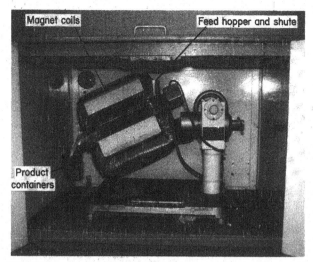

Fig. 3.21 Frantz Isodynamic magnetic separator showing feed hopper, feed control, separation zone, etc.

such piece of equipment is the isodynamic magnetic separator – a laboratory instrument (Fig. 3.21) for which there is no equivalent plant-scale device.

In the isodynamic separator a mixture of mineral particles is fed slowly into a vibrating chute which is held between the pole-pieces of a powerful electromagnet. The pole-pieces are so shaped that the magnetic field gradient is directed outwards – away from the magnet coils. The slope of the chute can be varied, both in the longitudinal and in the transverse directions: the longitudinal slope (usually 20–30°) determines the time taken by a particle to pass through the magnet and this controls the time available for the particle to be moved *up* the transverse slope (commonly 8–10°) by the magnetic field gradient. The strength of the magnetic field (and, hence, the steepness of the field gradient) is controlled by adjusting the current flowing through the coils of the electromagnet.

The feed rate is usually only a few tens of grams per hour and this is controlled by varying the amplitude of vibration in the feed chute. The feed to the instrument must be perfectly dry or the particles will tend to stick together: the feed must also be closely sized within the range 500–50 μm for optimum effect. Because of the magnetic remanence of the electromagnet the field gradient is never zero, and it is therefore important to remove all ferromagnetic minerals, such as magnetite, with a hand magnet before using the isodynamic separator: otherwise, these particles stick to the magnet pole-pieces, quickly clog the feed chute and bring the separation to a halt.

The isodynamic separator can be used not only

to fractionate mineral grains but also to determine, semi-quantitatively, the magnetic susceptibilities of the various mineral fractions. Thus, the magnetic susceptibilities of the minerals attracted to the magnet at different field strengths (i.e. at different coil amperages) can be calculated from the formula:

$$K_m \text{ (mass magnetic susceptibility)} = \frac{C \sin \alpha}{I^2}$$

where α is the transverse slope and I is the current flowing through the magnet coils. Magnetic susceptibility thus varies with the square of the current passing through the magnet and also with the angle of the transverse slope of the feed chute. Figure 3.22 is a nomogram that relates side slope, magnet current and mass susceptibility values.

3.6.4 The roll-type magnetic separator

This is a much less sensitive fractionating device than the isodynamic separator (section 3.6.3 above), but it has the advantage of being able to treat many kilograms per hour of material in the size range 50–500 μm. The separator consists of a rotating cylinder located around the curved pole-pieces of an electromagnet (Fig. 3.23). Material is fed from a hopper through a central gate on to the rotating cylinder. Non-magnetic particles are not attracted to the magnet and are thrown off the cylinder by centrifugal force into a collecting chute; the magnetic particles tend to adhere to the cylinder, but after rotation out of the magnetic field they fall off into a second collecting chute. As with the isodynamic separator, the mineral particles must be dry (so that they do not stick together) and they must be closely sized so that the larger particles do not "shield" the smaller particles (Fig. 3.24). Furthermore, the ferromagnetic particles must again be removed with a hand magnet (otherwise, "trains" of magnetic particles will trap non-magnetic particles and, in addition, the ferromagnetic particles will permanently adhere to the cylinder).

The operating variables of this type of separator include (1) the rate of feed – the slower the feed, the cleaner the products, (2) the cylinder rotation speed or, more correctly, its peripheral speed which affects the centrifugal force acting on a particle, (3) the size of the air gap between the rotating cylinder and its pole-pieces, and (4) the magnetic flux of the electromagnet which is controlled by the current flowing through the coils.

3.6.5. Magnetic separation – general

Simple magnetic separations in which a highly magnetic mineral is separated from a non-

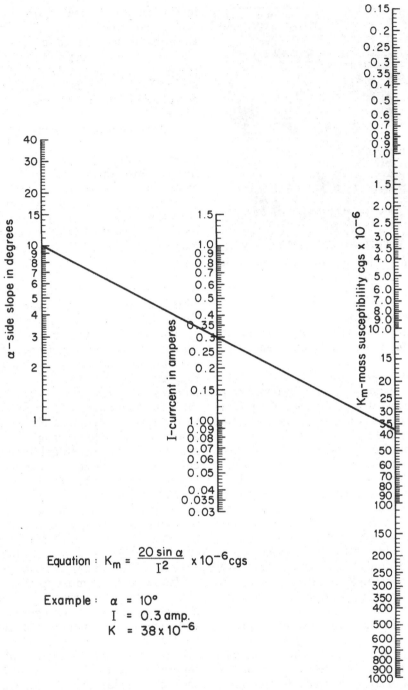

Fig. 3.22 Nomogram relating the magnetic susceptibilities of the minerals reporting to the "concentrate" chute on the Isodynamic separator to the side slope and the current flowing through the magnet (Frantz Separators Inc.).

magnetic mineral can be carried out with a bar magnet or a horse-shoe magnet. Minerals with closely similar magnetic properties can be separated in more sophisticated equipment such as isodynamic separators. These can also be used to *measure* the magnetic permeabilities of individual mineral particles.

If two minerals have the same (or very similar) magnetic permeability value it may still be possible to separate them by magnetic methods provided the magnetic property of *one* mineral can be changed whilst that of the other mineral remains unchanged. Hematite and quartz are both ordinarily non-magnetic but they can be readily sep-

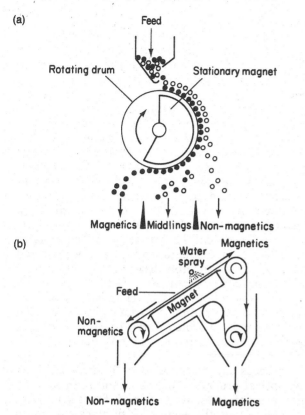

(a)

Feed

Rotating drum

Stationary magnet

Magnetics   Middlings   Non-magnetics

(b)

Water spray

Magnetics

Feed

Magnet

Non-magnetics

Non-magnetics   Magnetics

Fig. 3.23 (a) A roll-type magnetic separator. (b) A typical wet magnetic separator.

arated by magnetic methods after a reducing roast (i.e. after being heated at high temperature in a reducing atmosphere). The hematite changes to the highly magnetic mineral magnetite:

$$3Fe_2O_3 + C \rightarrow 2Fe_3O_4 + CO$$
$$3Fe_2O_3 + CO \rightarrow 2Fe_3O_4 + CO_2$$

Detailed information on the magnetic properties of minerals is difficult to obtain from the literature because these properties are greatly influenced by the presence of small amounts of magnetic impurities. Consequently, the magnetic properties of the minerals within any particular ore body must be *measured* and the results of these measurements need only apply to that single ore. The true magnetic properties of a min-

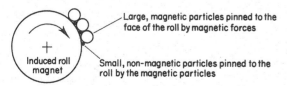

Large, magnetic particles pinned to the face of the roll by magnetic forces

Induced roll magnet

Small, non-magnetic particles pinned to the roll by the magnetic particles

Fig. 3.24 Shielding effect of large amounts of non-magnetic particles on an over-loaded magnetic separator.

eral such as cassiterite may be masked by the presence of thin skins of impurities such as iron oxide; these coatings can often be removed by leaching the grains in a suitable solvent or by using a simple scrubbing action in which the particles are tumbled together in a small rotating mill to abrade the coatings.

3.7. Electrical methods

Electrical methods of separation are based on the well-known attraction of unlike electric charges for one another and the repulsion that occurs between similar charges. The electrical properties of different mineral species can vary greatly and, in principle, these properties can be used in a variety of fractionating procedures. Unfortunately, large differences in electrical properties also frequently occur within a single mineral species: these variations are usually caused by the presence of small, unseen or undetermined proportions of impurities, thin films of alteration products or moisture adsorbed on to the mineral surface.

3.7.1. Surface conductivity
The most widely used electrical property for mineral fractionation is surface conductivity and the most effective means of carrying out such an electrically based separation is by means of the *electrodynamic* (or *high-tension*) *method*. This technique is widely used, both in the laboratory and in mineral processing plants, to separate the mineral components of beach sands and of alluvial deposits.

In a high-tension (or electrodynamic) separator dry (and, where possible, liberated) mineral particles are fed as a layer, one particle deep, on to a rotating earthed roll (Fig. 3.25). *All* the particles are electrically charged (at the same polarity) by an electrical spray discharge from an adjacent high-tension electrode: this electrode has a potential difference of up to 30 kV with respect to the earthed roll. Particles of high surface conductivity rapidly lose their induced electric charges to the earthed roll and then fall off freely along a path which is determined by the peripheral speed of the roll and by the mass and shape of a particle. Poorly conducting particles retain their induced charges and, since these charges are of opposite polarity to the roll, the particles tend to stick to the roll surface until the charges slowly leak away. The trajectories of the good conductors will be different from the paths taken by the poor conductors (see Fig. 3.23(a)) and a separation is achieved by setting a splitter at the appropriate

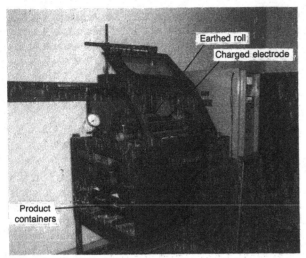

Fig. 3.25 Rapid high-tension (electro-dynamic) separator.

position. Many minerals, including some silicates, have such poor surface conductivities that their induced electric charges leak away so slowly that they have to be mechanically removed from the earthed drum by a stiff brush.

3.8. Hand-sorting methods

This is the oldest and, in some instances, still the best method of separating *small* amounts of fine-grained minerals. Unfortunately, it is a very slow and very tedious method and if possible, it should *only* be used to check on (or, if absolutely necessary, to improve on) the qualities of nearly pure products obtained from other, non-manual separation procedures. Hand sorting is generally applied to liberated particles and in many instances these are so small that they can only be seen clearly through a microscope. This sorting method can even be used to separate particles down to about 10 μm in size but only very small masses of material can be treated at these minute sizes.

When carrying out a hand sorting operation the mineralogist relies on a number of mineral properties – colour, shape, refractive index, cleavage, isotropism and so forth – to discriminate the various minerals. In fact, any optical feature that serves to distinguish one mineral from another may be used (see Chapter 4 for a description of these features). Once a mineral has been identified it is then carefully picked out with tweezers or separated from the other minerals by using a needle or a small brush. The qualities of the resulting products depend on the experience of the operator in identifying a required mineral,

and on his or her manipulative skill and patience in separating that mineral from the others. The over-riding requirement is usually the production of a high-grade concentrate of the selected mineral: the proportion of that mineral that is recovered is often of little importance (once again, "grade is more important than recovery").

The microscope should always be used at the *lowest* convenient magnification when hand sorting fine-grained material – this will minimise eye fatigue. The separation procedure can be somewhat simplified by using closely sized mineral fractions so that large particles do not conceal smaller particles and the particles must be washed free of any dust coatings because these will tend to conceal the true colour of the mineral particles. It is convenient to form the unsorted particles into a narrow band (Fig. 3.26) and a selected mineral is then displaced from this band by using a needle or a brush. These simple devices are often better than the various gimmicks (suction bottles, turn-tables, conveyor belts) that have been advocated to reduce the tedium of a hand-sorting operation.

On occasion, the natural differences between the optical properties of the minerals in a specimen are not adequate to permit hand sorting to take place (e.g. the separation of black columbite from very dark brown cassiterite). It may then be possible to alter the colour of a selected mineral by some chemical staining method. Thus, when the cassiterite in the above example is "tinned" (see Appendix 2, section 3.10(c)) its colour changes from dark brown to silvery-grey and it can then easily be differentiated during a hand-sorting operation from the unaltered, black columbite.

3.9. Miscellaneous fractionating methods

Any fractionating method is permissible – providing only that it works and that it can be carried out at an acceptable cost in time and patience.

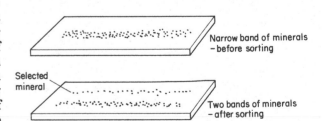

Fig. 3.26 Hand sorting very fine particles.

3.9.1. Dielectric separation

Mineral particles differ in their dielectric properties and this difference can be used as the basis of an unusual separating procedure. If two minerals of differing dielectric properties are immersed in a fluid of intermediate dielectric properties then the mineral with dielectric constant higher than the liquid will be attracted to a hand-held (insulated) electrode immersed in the liquid, and that of lower dielectric constant will be repelled from it.

This separation method is occasionally used, but it is potentially dangerous since most of the suitable liquids (kerosene, ethanol, methanol and nitrobenzene, among them) are highly inflammable and/or explosive. Its use is therefore not recommended – either for students or for mature mineralogists. There is, most probably, a better way of fractionating your particular mineral mixture!

3.9.2. Selective solution

Groups of minerals can sometimes be separated from each other by using selective chemical solution techniques. Columbite and ilmenite, for example, are often extremely difficult to separate by any of the techniques described above, but the ilmenite is much more reactive than the columbite towards hot concentrated hydrochloric acid. Consequently, if a mixture of the two minerals is leached for about twenty-four hours in the acid, the ilmenite will dissolve and the columbite will be left almost completely unaffected.

Selective solutions of this kind can be very useful. However, one component is always lost in the solution. The solvents and the reaction conditions must be carefully chosen to suit the requirements of the particular set of minerals concerned. Examples are given in Chapter 10 for the recovery of cassiterite by the selective solution of the enclosing silicates and sulphide minerals in hydrofluoric and nitric acids, respectively.

4. Conclusion

Small-scale mineral separations are often tedious and always very time-consuming. Therefore, before carrying out such a procedure, the mineralogist and the mineral engineer should always establish whether such a separation is really necessary. If it is necessary, then (and only then) he or she should proceed by using the simplest method that will provide the required product – screening is simpler than classification, which in turn is simpler to carry out than a heavy-liquid separation, and so on.

Modern methods of analysis do not always need pure fractionated mineral grains. For example, the X-ray diffraction spectrum of a mineral can be established without the need to prepare a pure mineral specimen. Similarly, the chemical composition of a single unknown mineral grain can be established in an electron probe X-ray microanalyser without either removing the grain from its matrix or preparing a pure fraction of that mineral species (see Chapter 9, section 9).

Minerals are fractionated (both in the laboratory and in mineral processing plants) by processes which make use of differences in their properties. If the natural differences are insufficient to produce good results then it is sometimes possible, and quite permissible, to change the properties of one or more minerals so as to improve the separation.

Chapter 4

Mineral identification

1. Introduction

There are approximately 3500 known minerals. This number is slowly increasing as new minerals are discovered by modern methods of study and as new mineralogical environments, such as the Moon, are explored. However, some substances which had previously been accepted as being true minerals are being discredited by careful studies that show that a material that had once seemed to be a single homogeneous phase is actually a complex, fine-grained mixture of more than one mineral.

The mineral engineer is fortunate in that, for most of the time, he or she need only be concerned with a small fraction of the known mineral species. Many minerals are exceedingly rare, being seldom seen outside museums, and these are not considered further in this book. Others do not figure among the minerals of commerce, whilst many more of the named minerals are merely subspecies of some major mineral species – for example, the feldspars are subdivided into eight or ten different subspecies. These subdivisions are vitally important to the petrographer and the geologist, but generally speaking are of little concern to the mineral engineer. Only some 250 to 300 minerals form common articles of trade or occur in significant proportions in ores and in mineral products. However, a single mineral deposit may contain a dozen or more of these minerals (see Table 1.6, Chapter 1) and it is extremely important that the mineral engineer can identify the minerals he is treating.

2. Diagnostic properties

Minerals are not easy to identify. In fact, the unambiguous identification of a mineral can, on occasion, demand a great deal of mineralogical skill and the use of elaborate analytical equipment (as described in later chapters). However, it is not always feasible (and, in a teaching environment, it may not even be desirable) to use the most sophisticated equipment in order to collect the diagnostic mineralogical data that are needed. The mineral identification scheme described in this book (Appendix 2) has been deliberately restricted to the use of readily available and easy-to-use equipment – i.e. facilities that can reasonably be expected to be available in a mineral laboratory and that can be used by trained engineers as well as by skilled mineralogists. The scheme makes no attempt to subdivide a single, solid-solution series (such as the plagioclase feldspars) into a number of separate subspecies, as is commonly done in many other determinative schemes.

2.1. Crystal structure

The single most diagnostically important feature of almost all minerals is their crystal structure (although a few minerals are very poorly crystalline and have no readily discernible crystal structures). The detailed structure of a mineral can be directly determined by X-ray diffraction techniques (see Chapter 9, section 6.1) but useful, *indirect* information on crystal structure can be obtained by much simpler analytical methods which determine a number of structure-dependent properties. For example, the examination of a mineral in polarised light can establish various aspects of its crystal symmetry (see Chapter 5), whilst measurements of physical properties such as grain shape, hardness, magnetic permeability and so on, can provide additional clues regarding the crystal structures of minerals and are discussed later in this chapter.

2.2. Chemical composition

This is a mineral property that often has great diagnostic value. Mineral compositions can be quantitatively determined either by bulk analyses of carefully prepared pure specimens or by microanalyses of single mineral grains. Some of these microanalyses can be carried out, if necessary, on single mineral grains whilst these grains are still embedded in their rocky matrices (Chapter 9, section 9.1). Qualitative chemical information about the presence or the absence of various elements and radicals in a mineral is also very useful and this information can often be obtained by simple, specially devised spot tests (see Appendix 2). However, as explained in Chapter 1, section 2.2.6, the permissible range of chemical composition within a single mineral species can be large and for this reason it may be difficult (and, at times, impossible) to make unambiguous identifications on the basis of chemical data alone. In addition, polymorphic minerals (see Chapter 1, section 2.2.5) are *different* minerals which have the *same* chemical compositions and these minerals can only be differentiated from one another by using data related to their crystal structures.

2.3. Preliminary identification

When a skilled mineralogist is confronted with an unknown mineral he (or, of course, she) can often quickly identify that mineral by using his considerable mineralogical knowledge and experience. He may not need to use any ancillary equipment at all. Even when he cannot identify some mineral immediately, he can quickly reduce the possibilities to a small group. He will then select a few confirmatory tests so that he can firmly establish the identity of the mineral.

The experienced mineralogist's identification is sometimes made after what may seem to be only a brief, superficial examination. Such an examination is, in fact, very thorough: all the salient features of the unknown mineral will have been duly noted: features such as colour, variations of colour, shape, size, heft (i.e. density) and so on. These features will be compared rapidly with the extensive "data-bank" which has been accumulated and stored in the mineralogist's memory over many years. It is the same kind of process as that which allows one to recognise a friend in a large crowd of people – a procedure that is difficult to explain but easy to operate! Attempts are being made to emulate the mineralogist's complex "information retrieval" system by using computer-based "expert systems" for mineral identification purposes – but, as yet, with only very limited success.

The student engineer and the practising mineral engineer, on the other hand, may have had very little mineralogical experience or training. However, equipped with little more than great confidence in their own limited ability they may, nevertheless, try to use the same techniques as the expert. But without the appropriate observational skills and lacking the mineralogist's extensive "data-bank", this procedure often degenerates into uncontrolled guesswork which may *never* provide the correct answer.

The best method (in fact, the only method) for a beginner is to use a simple, well-designed, systematic identification scheme. Several such schemes are available; all rely on the *sequential* measurements of specified mineral properties. The results obtained are compared with those obtained by other observers from standard specimens and these values are given in suitable "determinative tables". These tables are the student's mineral identification data-bank (see, for example, Appendix 2).

From the student's point of view, *any* systematic determinative scheme should be preferable to no scheme at all. The scheme in Appendix 2 has been specifically designed for identifying the kinds of small mineral grains which concern the mineral engineer. Even so, it is well to remember that no single determinative scheme is necessarily the best in all circumstances and any scheme may have to be modified to suit the equipment and the skills that are available, and the mineral assemblages that are being studied.

During the early stages of a student's training any new mineral should always be consciously *identified*: the mineral should not be unconsciously *recognised* (or guessed). In fact, the less experienced the student the easier it often is for him to "recognise" a mineral incorrectly. Thus a student who is acquainted with galena but little else in the way of lead-bearing minerals may feel certain that any dense, lead-rich mineral which is blue-grey in colour *must* be galena (the sulphide of lead). A mineralogist who is asked to examine such a mineral would quickly establish that the mineral is isotropic, has a hardness of 2.5–3, a marked cubic cleavage, a high optical reflectivity and a dark streak – further evidence that the mineral may be galena. But the experienced mineralogist might feel that all was not well (perhaps the density was a little high for galena) and, therefore making a final decision, would check the mineral for the presence of sulphur. It might then be discovered that the mineral was not a sulphide but a selenide of lead, and was in fact clausthalite (PbSe) rather than galena (PbS).

Whichever determinative scheme a student uses, it is essential to remember that significant variations in mineral properties can, and do, justifiably exist because of the common occurrence of solid-solution series in minerals (see Chapter 1, section 2.2.6). These compositional ranges, and the property variations that accompany them, are well known, however. The effects of the compositional variations are predictable, and are taken into account in any good determinative scheme.

3. Selecting a specimen for identification purposes

No simple determinative scheme will work successfully if the unknown mineral contains unknown proportions of foreign matter because this foreign material will give rise to quite unpredictable variations in the measured properties – variations that *cannot* be taken into account in any determinative scheme. It is vital, therefore, that the specimens used in the scheme consist of pure, liberated mineral grains. Such a specimen is not always easy to obtain and especial care must be taken to avoid carrying out determinative tests on certain types of material, since these may not always be recognised as being made up of mineral mixtures. In the rest of this section we shall look at these materials.

3.1. Fine-grained exsolution mixtures of minerals

As mentioned in Chapter 1, section 2.2.6, many materials exist as a single solid phase at elevated temperatures, but at lower temperatures this single phase exsolves (or precipitates) into two or more distinct mineral phases. One of the precipitated minerals is often finely disseminated in the

other (see Fig. 1.3) and may not be easy to see, even with a good microscope. It may be very difficult to separate the two phases from each other in order that determinative tests can be carried out on a *single* phase. For example, hematite can form thin lamellae within a host of ilmenite – the location, proportion and size of the hematite grains will depend on the composition and the history of the original "ilmenite" system. Variations in the hematite–ilmenite ratios can give rise to quite unpredictable variations in the properties of the combined ilmenite–hematite particles, and a property such as magnetic permeability may therefore have little value for identification purposes.

3.2. Solid–liquid mixtures

These are mineral gels which contain widely varying amounts of moisture and of adsorbed heavy elements. For example, the mineral chrysocolla is thought to be a silica gel containing unknown (but variable) amounts of copper and other metallic ions; the moisture content varies even with changes in the relative humidity of the surrounding atmosphere. The presence of adsorbed metal ions affects the colour of the mineral, and the varying amounts of adsorbed water affect the results obtained by quantitative chemical analysis.

3.3. Solid–gas mixtures

These are the porous minerals. The pores are usually filled with air but they can also trap moisture; more particularly, they can also trap any halogenated heavy liquids or Clerici solution which might have been used to prepare pure mineral fractions (see Chapter 3, section 3.3). The presence of these halogen elements (chlorine, bromine and iodine) can often be detected during electron probe microanalysis (Chapter 9, section 9) and may mistakenly be considered to be trace components of the host mineral. Furthermore, the presence of large numbers of air-filled pores greatly affects mineral properties such as density, hardness and chemical reactivity.

3.4. Solid–liquid–gas mixtures

Some minerals (especially quartz) contain large numbers of very small, sealed, and unconnected fluid inclusions. These inclusions contain a variety of components which were physically trapped in the host mineral during the time of its forma-

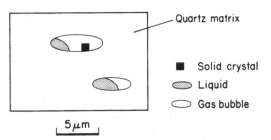

Fig. 4.1 Typical fluid inclusions in quartz.

tion. The inclusions can consist of:

(1) a liquid such as water or liquid carbon dioxide,

(2) a solid crystal, such as sodium chloride, with a gaseous component such as air or carbon dioxide, or

(3) a solid crystal in a mixture of gas and liquid (Fig. 4.1).

An individual inclusion is seldom more than about $10 \mu m$ in diameter, but these inclusions sometimes occur in such profusion that they can significantly affect mineral properties such as electrical and thermal conductivities. Furthermore, some of the gases contained in these inclusions are held under considerable pressures and these set up internal, bursting pressures that affect the strength and cohesiveness of the host mineral.

3.5. Coatings on otherwise homogeneous minerals

These coatings need bear no genetic relationships to the underlying minerals. In fact, several minerals occurring in the same environment may be coated with films of the same material. For example, all the minerals in an alluvial deposit may be coated with films of iron oxides deposited from the local stream, or the mineral grains in a beach sand may all be coated with a film of salt (NaCl). These films often affect the colours of the minerals (quartz coated with iron oxide ranges from white to yellow or even red in colour); they can also impede various chemical reactions (see the "tinning test" in Appendix 2, section 3.10(c)). Films of this kind also affect other mineral properties such as electrical conductivity and hydrophobicity. These extraneous films can often be removed in the laboratory, either by attrition (in a small tumbling mill) or by chemical solution, without affecting the nature of the host mineral.

3.6. Coatings of alteration products

These coatings are derived directly from the host mineral and are commonly formed by oxidation of that mineral. If the coatings are opaque they can conceal the presence of deleterious impurities within the host mineral; they also affect the colour, conductivity, reactivity and hydrophobicity of the host. These coatings can often be removed by attrition or by solution.

3.7. Leached surfaces

Some of the components of a mineral grain may be leached preferentially from its surface layers by natural processes. Consequently, that surface will show unusual colours and unusual concentrations of the remaining elements. For example, the surface of an alluvial grain of electrum (a naturally occurring gold–silver alloy) will often be leached of some or all its silver content: the grain will then have the typical "golden" colour of native gold and it will produce the streak associated with high-grade gold. Furthermore, an analysis of the surface layers of the mineral in an electron probe device (Chapter 9, section 9) would suggest that the grain consisted of pure gold.

3.8. Additional precautions

In addition to the above sources of variability in the properties of individual mineral grains, other, longer-term, property variations often occur within an ore body. These variations were originally caused by the geothermal and geochemical gradients that existed during the formation of the deposit. The gradients gave rise to progressive changes in mineral composition (and, therefore, changes in mineral properties) from one part of a deposit to another. For example, both the silver content of galena (PbS) and its hydrophobicity can change markedly from one level of a mine to another; consequently, the performance of that mineral in a laboratory flotation circuit and within a treatment plant can also vary.

In view of all the problems that can arise because of impurities in, or on, mineral grains it is no wonder that great care must be used to produce fresh, homogeneous grains for use in any mineral identification procedure. These grains are usually obtained, in the last resort, by careful manual selection of fractionated particles under a stereoscopic microscope (see Practical no 3).

4. The ideal identification scheme

Any mineral can be identified if its crystal structure and its chemical composition match those of a previously recorded mineral. Direct measurement of these features (for example, by the X-ray fluorescence and X-ray diffraction methods discussed in Chapter 9) may not always be feasible but mineral properties such as colour, density, hardness, cleavage, refractive index, and magnetism are easily measured and all are related to crystal structure. Properties such as taste, fusibility, reactivity and radioactivity, on the other hand, are closely related to chemical composition. Careful measurement of these physical and chemical properties will therefore, in most instances, allow a mineral to be identified.

The *ideal* identification scheme would, therefore, swiftly and cheaply measure these diagnostic properties and, from the results obtained, it would always provide the correct answer; furthermore, it would do so after only a small number of measurements. It is doubtful whether such a scheme exists – or is ever likely to be developed.

5. A practical identification procedure

In practice, a determinative scheme must always rely on the available skills and facilities and, whenever possible, it should be based on robust mineral properties. These are properties which are not significantly affected by the presence of small amounts of impurities or by small variations in mineral composition (for example, mineral density is a robust property whilst surface conductivity is not – it is very sensitive to the presence and to the location of trace impurities). It is also preferable to use scalar properties, such as density and specific heat, which are not dependent on mineral orientation, rather than rely on vectorial properties which may vary considerably with crystallographic direction. Scalar properties are also much better to use than vector properties when it is difficult to establish the crystallographic direction in which a measurement has been carried out. Hardness, refractive index and electrical resistance are vectorial properties.

It is advantageous if each test can provide quantitative, or at least semi-quantitative, information rather than a simple "yes/no" answer; in this way, the number of tests needed to arrive at an unambiguous identification can be kept to a minimum. Appendix 2 gives full details of a simple determinative scheme which was specially designed for identifying single, small, clean mineral fragments. This scheme was originally drawn up for the use of students of applied mineralogy at the Royal School of Mines, London, and has been extensively and successfully tested by them. More sophisticated methods of mineral identification, such as those based on careful examination of thin sections, are given in standard textbooks on mineralogy (see Bibliography), and are not described in detail in this volume.

5.1. Specimen selection

The suggested determinative scheme starts with the careful selection of a small number of grains of pure mineral: a good stereoscopic microscope (see Practical no 3) is useful for this vital task. The mineral is then subjected to a pre-determined sequence of simple, semi-quantitative physical and chemical tests which progressively eliminate more and more minerals until, in the usual case, only a single mineral remains and its identity is therefore established. Sometimes, although the standard tests eliminate a large number of minerals, the final identification has to be based on a small selection of specialised confirmatory tests (see Table 4.1).

5.2. Density (or specific gravity) of mineral grains

The *density* of a mineral is its mass per unit volume and this is usually given as kilograms per cubic metre ($kg\ m^{-3}$). *Specific gravity* is the ratio of the mass of a mineral to the mass of an equal volume of water at 277 K: this value is a dimensionless number. Quartz, for instance, has a density of 2650 $kg\ m^{-3}$ and a specific gravity of 2.65.

Density is a fundamental, scalar characteristic of all solids and the density of a mineral is a most important diagnostic property. Density depends on the kinds of atoms that make up the mineral and also on the manner in which these atoms are packed together, i.e. density depends on both the chemical composition of a mineral and on its crystal structure. Small variations in the proportions of the cations within an isomorphous series of minerals will not significantly affect the packing density of the atoms but major changes in composition may well affect the crystal structures and, as a consequence, the densities of the mineral phases. For example, as shown in Fig. 1.3, there is a significant increase in density between columbite $(Fe,Mn)Nb_2O_6$ which has a density of

TABLE 4.1

Flowsheet representation of a practical scheme for identifying the minerals commonly encountered in the mineral industry

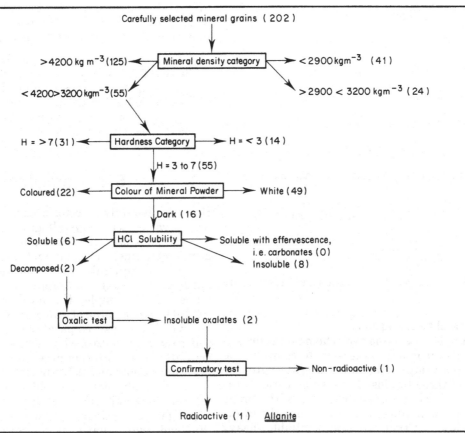

(Note: numbers in parentheses giving the approximate numbers of minerals do not always balance because many minerals occur in more than one category)

5300 kg m^{-3} and tantalite (Fe,Mn)Ta$_2$O$_6$, density 7300 kg m^{-3}.

The effects on mineral density due to variations in crystal structures, *without* change in chemical composition, are shown in Table 1.5. This lists groups of polymorphous minerals which have the same chemical compositions but different crystal structures. Thus, diamond (density 3500 kg m^{-3}) and graphite (density 2230 kg m^{-3}) have identical compositions, but in the diamond structure the carbon atoms are tightly packed together whilst in graphite the atomic packing gives rise to loosely held plates of carbon atoms.

Density can be measured in several ways. The method used on any specific occasion will depend on (1) the amount of material available, (2) the nature of this material and (3) the accuracy required. It often happens that only very small amounts ranging from a few milligrams to a few grams of clean, pure mineral are available.

Furthermore, these small amounts of clean, liberated material are likely to be very fine-grained and, the particle size is frequently less than 100 μm. The most suitable density-measuring method for material of this kind is the pycnometer method.

5.2.1. The pycnometer method

A pycnometer is a small bottle (Fig. 4.2) fitted with a long, ground-glass stopper. The stopper is pierced along its length by a small-diameter hole. The dry, empty bottle is first weighed (P); the mineral grains are put into the bottle and the bottle plus the mineral are then weighed (M). The difference ($M - P$) is the mass of the mineral. The bottle, still containing the mineral grains, is filled with water and again weighed (S). Finally, the bottle is emptied and then filled with water only before being weighed once more (W). The volume of the mineral specimen is

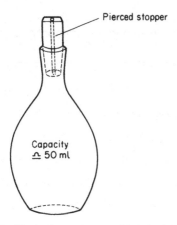

Fig. 4.2 Typical pycnometer (density bottle).

$W + (M - P) - S$ and its specific gravity is

$$\frac{\text{mass}}{\text{volume}} = \frac{M - P}{W + (M - P) - S}$$

For further details of this method see Practical no 4.

5.2.2. Use of heavy liquids

In the simple determinative scheme described in Appendix 2, it is only necessary to compare the density of an *individual* mineral grain with one or more of three liquids of known densities. This method can be used successfully with single grains less than 100 μm in size. It has the drawback, however, that all the available heavy liquids have density values below 4200 kg m^{-3} and the method therefore cannot provide much useful information about those metallic minerals which are of even higher densities and which are often of great interest to the mineral engineer.

In this semi-quantitative method of density determination a clean, small fragment of mineral is classified into one of four density classes. This is done by comparing the density of the mineral with the densities of *single* drops of liquids of known densities held on a waxed or greased glass

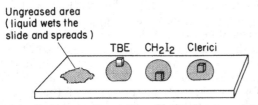

Fig. 4.3 Standard glass slide. Standard glass microscope slide showing how the heavy liquids form globules on the greased surface. The globules can then be used to establish the density category of a mineral grain; e.g. a light particle floats on the tetrabromoethane.

slide (Fig. 4.3). See Practical no 4 for further details.

This simple group of density tests categorises every mineral into one of the following classes which form the first sub-division of the determinative scheme given in Appendix 2:

Class 1, density less than 2900 kg m^{-3},
Class 2, density equal to or greater than 2900 kg m^{-3} but less than 3200 kg m^{-3},
Class 3, density equal to or greater than 3200 kg m^{-3} but less than 4200 kg m^{-3},
Class 4, density equal to or greater than 4200 kg m^{-3}.

5.3. Hardness

Hardness is really a vector property, the value of which varies with crystallographic orientation. The methods used to measure hardness are usually very crude, however, and only a comparatively few minerals show markedly different hardnesses along different crystallographic planes. For example, the hardness of kyanite (Al_2SiO_5) has a value of 4 to 5 on Mohs' scale (see below) when measured parallel to the long axis of the crystal, but 6 to 7 when measured across the crystal. It is useful, therefore, to establish which crystallographic face is being tested, but unfortunately this information is seldom available when a small, broken grain is being tested.

In general, minerals, such as quartz and diamond, which have strong covalent bonding, are difficult to deform; minerals, such as talc and graphite, are held together only by the weak van der Waals bonds and are, therefore, easily disrupted.

5.3.1. Mohs' hardness

The scale of hardness traditionally used by mineralogists is Mohs' scale (Table 4.2). This scale categorises any mineral into one of ten classes from 1 (very soft) to 10 (very hard). Mohs' scale is approximately geometrical in character and each succeeding hardness category is roughly twice as "hard" as the preceding one; for instance, a mineral of hardness 6 on this scale is about twice as hard as a mineral of hardness 5 (Fig. 4.4).

The Mohs' hardness of a mineral grain measures its resistance to being scratched (other "hardness" values – see below – measure resistance to being deformed or indented). If the specimen is large and can be held between the fingers then the hardness determination procedure is simple. The mineral is pressed on to and drawn across a smooth surface of a standard, low-hardness mineral; if this standard is scratched

TABLE 4.2
Mohs' scale of mineral hardness

Mineral standard	Hardness value
Diamond	10
Corundum	9
Topaz	8
Quartz	7
Orthoclase feldspar	6
Apatite	5
Fluorite	4
Calcite	3
Gypsum	2
Talc	1

(See also Fig. 4.4 showing relationship between Mohs' hardness values and the more accurately quantitative Vickers indentation value of hardness.)

then the specimen is drawn across the next hardest mineral in Mohs' scale and so on, until a standard specimen is *not* scratched. The hardness of the mineral specimen then lies between that of the last mineral which was scratched and that of the unscratched standard.

The hardness of a small mineral grain that cannot be held between the fingers is best determined by first placing the grain on the smooth surface of a standard mineral: it is then pressed firmly into that surface with a flat wooden

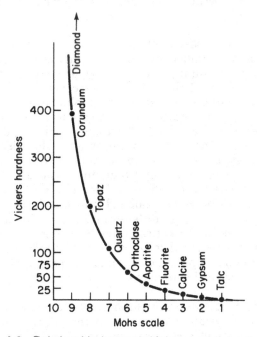

Fig. 4.4 Relationship between Mohs' scratch hardness values and hardness values from the Vickers indentation test.

implement, such as the handle of a mounted needle, and steadily drawn across it. As before, if the standard is scratched then the mineral has a hardness value greater than that of the standard.

5.3.2. Vickers hardness
Other values of "hardness", more sophisticated and more quantitative than Mohs', are measured in some determinative schemes. The most commonly used method measures the VHN (Vickers Hardness Number) and it is especially useful for measuring the hardness of small mineral grains which are still embedded in a rocky matrix. The VHN value measures the resistance of a mineral to indentation by a standard diamond "point" applied under a known load for a fixed time. For example, a diamond, shaped like an inverted pyramid, is pressed onto the polished mineral surface under a steady load of between 10 and 100 grams for several seconds. The depth of the indentation produced is an indication of the hardness of the mineral. In practice, it is easier to measure the surface area of the indentation rather than its depth. The VHN is then given by 1854.4 F/d, where F is the gram force applied to the surface and d is the area of the indentation (in μm^2).

The surface of a mineral is often "work-hardened" by being polished and the surface layers of a polished specimen may be harder than the bulk material. Consequently, the load on the diamond "point" must be large enough to allow it to penetrate beneath the altered surface layers into the unaltered bulk material.

Further details of this method of measuring mineral hardness are given in Zussman (see Bibliography).

5.3.3. Hardness – a general note
In order that a hardness determination shall provide the maximum amount of useful information it is important that the measurement is carried out on the mineral of interest, and not on any inclusion within that mineral. Nor must the measurement be made on some surface coating that does not truly represent the mineral. Furthermore, it is important not to confuse the true hardness of a single mineral grain with the brittleness, or lack of granular cohesion, of a fine-grained mineral aggregate. Thus, the hardness of quartz is, by definition, 7 on Mohs' scale, but the quartz in a fine-grained, friable sandstone may nevertheless seem to be much softer; the aggregate may appear to be scratched when, in reality, it is only the loosely held grains that are being torn apart. It is important, therefore, to ensure that a definite scratch is formed on the

standard mineral by the unknown mineral (or vice versa): the presence of the scratch, which must be an elongated indentation and not just a linear accumulation of powder, must be confirmed by examining the specimen with a hand lens or with a stereoscopic microscope.

5.4. Colour

The body (or bulk) colour of a mineral is, perhaps, its most obvious property. Unfortunately, minerals can vary markedly in colour and this property is therefore of only limited value when trying to identify an unknown mineral.

5.4.1. Absorption of white light
The colour of a material is caused by the absorption of specific wavelengths from the illuminating beam of light and this absorption can vary markedly with the presence of trace amounts of certain elements. Thus, a small amount of chromium in beryl (a beryllium-aluminium silicate) will impart a deep green colour; the mineral quartz can be colourless, pink, blue or red depending on the presence of small amounts of such elements as iron, vanadium, chromium or manganese. The mineral tourmaline can vary in colour from dark green to red; it is even possible for an unbroken crystal of tourmaline to show one green end and one red end. Such a distinctive colour variation within a single grain has great diagnostic value, but once such a grain is broken, it is no longer obvious that the red fragment and the green fragment are both examples of the same mineral.

The colours of the metallic ore minerals are reasonably constant; pyrite is always yellow, chalcopyrite is brass-yellow, bornite is brownish-bronze. However, some metallic minerals tarnish quickly in moist air; for example, bornite rapidly oxidises to form an iridescent violet-blue surface film that conceals the true colour of the mineral. This "peacock" colour can itself be a useful pointer to the identity of the underlying mineral.

It does, however, sometimes happen that the colour of a mineral is reasonably constant; malachite is *always* green, azurite is *always* some shade of blue. It also frequently happens that the colour of a particular mineral is constant within a single geological deposit; cassiterite, for example, can show a wide variety of colours but may nevertheless only show a single colour within a particular deposit. However, when minerals from a variety of primary geological sources accumulate in an alluvium or in a beach sand then a single species can show a wide range of colours,

e.g. alluvial cassiterite ranges from black to brown to honey-yellow to near-white.

5.4.2. Fine-grained impurities
Various colours can also arise in mineral grains because of the presence of very fine-grained impurities; for example, small amounts of fine-grained iron oxide inclusions in an otherwise colourless quartz will impart a yellow or reddish-brown colour. Colour differences within a single mineral species can also arise because of physical discontinuities and structural imperfections: for instance, the mineral opal ($SiO_2.nH_2O$) is made up of myriads of small silica spheres and the discontinuities between these spheres give rise to the multi-coloured "opalescence" of the mineral.

5.4.3. Colour changes by thermal effects
The colours of some minerals that contain structural imperfections can be changed by heat treatment or by bombarding them with electromagnetic radiation. For example, heat treatment is sometimes used by unscrupulous merchants to change the colour, and so increase the value, of semi-precious minerals like topaz or zircon. A change of colour by controlled heat treatment can sometimes be of considerable diagnostic value but, more often, the colour change merely causes confusion. These colour changes are especially confusing in an industrial environment where minerals may be passed through a variety of heating operations as part of the separation procedure.

5.4.4. Fluorescence effects
A small number of minerals fluoresce under ultraviolet light or when they are bombarded with X-rays. This optical fluorescence can have great diagnostic value: for example, both calcite ($CaCO_3$) and scheelite ($CaWO_4$) are usually white in ordinary light and are not easy to distinguish. However, the minerals can readily be distinguished under ultraviolet light because the scheelite shows a bright blue fluorescence whilst the calcite is unaffected (Fig. 4.5, see colour section). Similarly, diamond and quartz cannot always be easily differentiated but diamond fluoresces in the optical wave-band under X-ray bombardment whilst quartz is unaffected (this is the basis of a plant-scale mineral separation process used in the diamond industry).

5.4.5. Fluorescence in electron beams
Some minerals, particularly if they contain certain types of structural imperfections, fluoresce under electron beam bombardment – this phenomenon is called *cathodoluminescence* (see

Chapter 9, section 9.1). Different minerals show different cathodoluminescent colours and these can be viewed in modern electron probe X-ray microanalysers. The phenomenon is occasionally a useful diagnostic feature but is not, as yet, widely used.

5.4.6. Streak (colour of mineral powder)

Although the body colour of a mineral is on the whole of dubious diagnostic value, the colour of a mineral powder (usually called the *streak*) is much more constant than the body colour and is therefore of greater value. Traditionally, the streak of a mineral is the colour produced by rubbing the mineral on a *streak-plate* – a small white tile of unglazed porcelain which has a Mohs hardness value of about 6. Minerals that are harder than the porcelain will not produce a streak but will scratch the plate instead. The powders of such hard minerals are produced by crushing the minerals. Details of the colours of mineral powders are given in Appendix 2.

It is difficult to describe the colour of a mineral powder accurately and the determinative scheme in Appendix 2 therefore only requires that the colour of a mineral powder be classified into one of three groups: (1) white (or colourless), (2) "dark" (brown to black) and (3) "coloured" i.e. red, yellow, green or blue.

5.5. Mineral composition

Accurate determinations of the chemical compositions of small mineral grains require sophisticated, expensive equipment (such as, for example, an electron probe microanalyser, discussed in Chapter 9, section 9). However, it is comparatively easy to determine the presence (or, where appropriate, the absence) of various elements and radicals within a mineral. The results of these determinations, when used in conjunction with information regarding mineral properties such as density and hardness, are very useful for mineral identification purposes.

5.5.1. Reactions of minerals in hot flames

Although they are now considered to be old-fashioned, the "blow-pipe" tests (given in older textbooks on mineralogy) can, on occasion, be highly informative. In a blow-pipe test a very hot flame is produced by blowing a steady stream of air into an ordinary methane or alcohol flame. This hot flame is then used to fuse or to decompose a small fragment of mineral. For example, the mineral stibnite (Sb_2S_3) fuses readily in the flame

and this can be a useful confirmatory test of the identity of the mineral. Other minerals can be decomposed by the flame and the products of the decomposition will produce coloured condensates in the cooler zones around the flame. The *colour* of the flame which is produced during a blow-pipe test may also provide indications of the major elements within the mineral. Thus, an intense yellow flame shows the presence of sodium, a crimson flame shows the presence of strontium or lithium, and so on.

5.6. Crystal habit

The term *habit* is used to describe the shape which is most commonly assumed by the grains of a specified mineral within a rocky matrix. This shape is controlled by the crystallographic structure of the mineral and by the space available during growth; for example, a mineral such as mica, which has a well-marked layered arrangement of atoms, always tends to show a platy habit whilst minerals such as galena and pyrite, which have three-dimensionally regular crystallographic arrangements, tend to show cuboid and octahedral habits respectively.

Mineral grains from eluvial deposits (deposits which are left behind by the removal of soluble, or readily decomposed, rocky matrices) will retain their natural shapes. Minerals which are intentionally broken in a mineral processing mill or in a laboratory will, however, only retain remnants of their crystal habits and, therefore, habit cannot often be used by the mineral processing engineer for mineral identification purposes. In addition, care must be taken in interpreting shape and form because some minerals grow as replacements for earlier minerals: that is, they form *pseudomorphs* ("same shapes") of other minerals by growing in the space vacated by a mineral that has decomposed. The outward shape of such a pseudomorphic grain clearly need bear no relation to its internal crystallographic structure and is of no help for mineral identification purposes.

5.7. Cleavage

Some minerals break readily along weak crystallographic planes: this property is called *cleavage* and its effects are often seen in the broken mineral grains which are produced during mineral treatment products. Minerals such as galena and halite cleave in three orthogonal directions i.e. in three directions at right angles to each other (Fig. 4.6). These minerals tend to produce cubic or

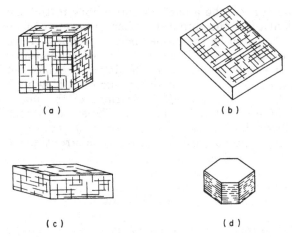

Fig. 4.6 Cleavage traces in minerals: (a) three cubic cleavages at right angles to one another (e.g. galena); (b) three rhombohedral cleavages, not at right angles to one another (e.g. calcite); (c) two cleavages – the angle between them can vary (e.g. 124° for amphibole, 93° for pyroxene); (d) single perfect cleavage (e.g. mica).

cuboid fragments that show numerous right-angled corners. Other minerals cleave in three directions which are not orthogonal but which are fixed relative to one another, e.g. broken calcite grains tend to form rhombohedra. Other minerals, such as the amphiboles and the pyroxenes, cleave in two directions: the angle between these two cleavage directions can sometimes be measured (Practical no 6) and can be used for diagnostic purposes. Some minerals, such as graphite, molybdenite and mica, cleave very readily in one direction only and, therefore, they break into very thin sheets.

Traces of cleavages tend to be retained in broken mineral fragments, even when the crystal habit is almost totally destroyed. Cleavage can, therefore, be a useful property for the engineer in identifying broken mineral particles.

5.8. Magnetic permeability

All minerals are affected by magnetic fields. Diamagnetic minerals such as calcite, halite and native gold have small negative magnetic permeabilities and are very slightly repelled by a strong magnet. Paramagnetic minerals such as beryl and rutile show small positive magnetic permeabilities and are weakly attracted to a strong magnet. Other paramagnetic minerals such as monazite are much more strongly attracted by a strong magnet whilst ferro-

magnetic minerals, such as magnetite or γ-hematite, are so strongly magnetic that they are attracted even by very weak magnets (see Chapter 3, section 3.6).

The magnetic permeability of a single mineral grain can be roughly assessed by using a strong, hand-held magnet and this value can be of considerable diagnostic importance. However, the magnetic permeability of a diamagnetic mineral grain (or of a weakly paramagnetic mineral) can be greatly affected by the presence of small amounts of ferromagnetic inclusions. Thus, the essentially "non-magnetic" mineral zircon becomes brown in colour and significantly magnetic when it is contaminated by about 1%, by weight, of micrometre-size inclusions of magnetite. These inclusions can only be seen with great difficulty even with a good optical microscope and, unless their presence is recognised, the mineral might easily be wrongly identified. Consequently, as mentioned repeatedly in this chapter, great care is needed to ensure that only pure mineral grains are used when measuring magnetic properties for mineral identification purposes.

5.9. Taste

It is possible to identify a few minerals by their distinctive taste. These minerals, which include alum ($KAl(SO_4)_2.12H_2O$) and halite (NaCl), are readily soluble in water. Therefore, they dissolve on the tongue and give characteristic "sharp" or "salty" tastes. However, because they are so readily soluble, they are often lost during specimen preparation procedures that involve the use of water.

6. Conclusion

Mineral identification can be a difficult and time-consuming task. This work is made easier (especially for the beginner) by following a well-defined, systematic procedure.

Most minerals can be positively identified provided their crystal structures have been accurately determined. Unfortunately, all the methods available for directly measuring these structures are expensive and demand considerable interpretive skills. Consequently, attempts are made to *deduce* the crystal structures of minerals by measuring their structure-dependent properties.

Many minerals can also be unambiguously identified provided their chemical compositions are accurately known. Again, it is often difficult to provide suitably comprehensive analyses and, therefore, the student must try to determine the presence of major chemical components by suitable tests or by measuring composition-dependent properties.

All determinative schemes demand that measurements be carried out on a *single* mineral phase and every attempt must be made to achieve this. A full determinative scheme is given in Appendix 2.

Chapter 5

The polarising microscope in applied mineralogy

1. The optical properties of minerals

Visible light is a wave-like electromagnetic phenomenon with wavelengths between about 400 nm (the blue end of the visible spectrum) and 700 nm (the red end). *Monochromatic light* consists of only a single wavelength within this range, and it is, on occasion, of special value to the mineralogist. Most mineralogical observations, however, are carried out under white light, which is a mixture of all the visible wavelengths.

According to the wave theory of light, the electromagnetic fields which are associated with an ordinary light beam are directed at right angles to the direction of travel of that beam: in other words, the electromagnetic field associated with a light beam vibrates, at random, about its direction of travel (Fig. 5.1). If "ordinary" light is passed through a special *polarising* filter (a sheet of Polaroid, for example), then the light becomes *plane-polarised*, i.e. the electromagnetic field is constrained to vibrate in one direction only, and this single direction is still at right angles to the direction of travel of the beam (Fig. 5.2). Other materials, including many minerals have the ability to constrain the complex vibrations of ordinary light into *two* planes which are at right angles to each other and also are at right angles to the direction of travel of the beam (Fig. 5.3).

The *refractive index* of a mineral is the ratio between the velocity of light in air and its velocity in the mineral, i.e.

$$n = V_a/V_m$$

where n is the refractive index, V_m is the velocity of light in the mineral and V_a is the velocity of light in air. In isotropic substances light is transmitted with the same velocity in all directions and, therefore, the refractive index is the same in all directions. In anisotropic materials, light is transmitted at different speeds along different

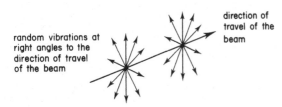

Fig. 5.1 An 'ordinary' light beam showing the random vibrations that exist at right angles to the direction of travel.

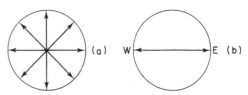

Fig. 5.2 Cross sections (a) of an ordinary light beam; (b) of the same beam after passing through a polarising filter (direction of polarisation is, in this case, E–W).

58

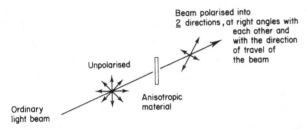

Fig. 5.3 Effect of passing an ordinary (unpolarised) beam of light through an anisotropic material.

crystallographic directions and there are corresponding variations in the refractive indices of these materials.

2. The polarising microscope

This is one of the most useful of the instruments found in the mineralogical laboratory. It is very commonly used by mineralogists and mineral engineers to determine the optical properties of a wide range of transparent minerals and of opaque minerals. It is also used to examine and measure the geometrical features of rocks, ores and mineral products – features such as mineral proportions, grain sizes and so on.

Figure 5.4 is a diagram of a typical petrological or polarising microscope. Instructions for setting up and using such a microscope are given in Practical no 6. When necessary, this type of instrument can be employed as a non-stereoscopic microscope using ordinary non-polarised light to examine rocks, minerals, ceramics, slags, cements and other mineral-like materials. However, the primary function of the petrological microscope is in the examination of thin sections and polished sections of rocks and mineral products under polarised light in order to establish various crystallographic features of the individual minerals.

Thin slices of many minerals transmit light and the *transmitted* rays can be analysed to obtain useful mineralogical information. Other minerals are opaque and will not transmit light even through very thin slices; these can be studied only by analysing the light which is *reflected* from their surfaces.

2.1. Use of transmitted light

The research petrologist may spend a great deal of time studying a single specimen in great detail with the transmission, polarising microscope. The mineral engineer, on the other hand, seldom needs such a wealth of optically based information about a specimen, nor can he usually afford to wait long for the type of information that he requires. Consequently, much of the optically based detail that is normally collected by the classical petrographer has been omitted from this discussion. Instead, the aim in this book is to show how the polarising microscope can be used to provide the engineer with rapidly produced, quantitative information regarding mineral identities, mineral proportions, rock textures, grain sizes and so forth.

The main difference between a polarising microscope and the "ordinary" variety is the presence in the former of *two* polarising devices. One of these devices (called the *polariser* and made from a sheet of Polaroid) is positioned in the light path between the light source and the specimen (Fig. 5.5). It is orientated so that the light which passes through it is constrained to vibrate along the so-called East–West direction only (i.e. in a direction from right to left as viewed in the eyepiece of the microscope). The second polarising device (called the *analyser*) is positioned in the light path in such a way that it can be used to analyse the light beam *after* it has passed through the specimen. The analyser is orientated so that it will only transmit light which is vibrating in the North–South direction (i.e. from top to bottom when viewed through the eyepiece).

2.1.1. Ordinary transmitted light
This is a very useful way of using the petrological microscope, though it is often neglected. The instrument, like the ordinary stereoscopic microscope, can be used to distinguish the minerals in thin sections and also those occurring in loose, unmounted, fine-grained transparent particles.

2.1.2. Linearly-polarised light
When the polariser is inserted into the light path, the light entering the specimen is said to be *linearly-polarised*, and can be used to determine certain mineral characteristics which can be useful for mineral identification purposes. For example, some minerals change colour (or change the intensity of their colour) when they are rotated in plane-polarised light. This effect, called *pleochroism*, is comparatively rare but, for this very reason, it can often be a useful diagnostic feature when identifying minerals such as biotite (which has a white-to-brown pleochroism) and tourmaline (brown-to-black or -green pleochroism) – see also Table 5.1. Pleochroism is caused by the absorption of different wavelengths of light along different crystallographic directions within a crystal.

60

(a)

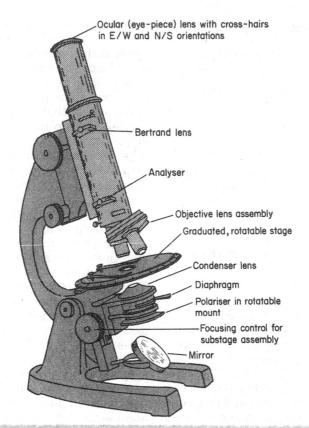

Ocular (eye-piece) lens with cross-hairs
in E/W and N/S orientations

Bertrand lens

Analyser

Objective lens assembly

Graduated, rotatable stage

Condenser lens

Diaphragm

Polariser in rotatable
mount

Focusing control for
substage assembly

Mirror

(b)

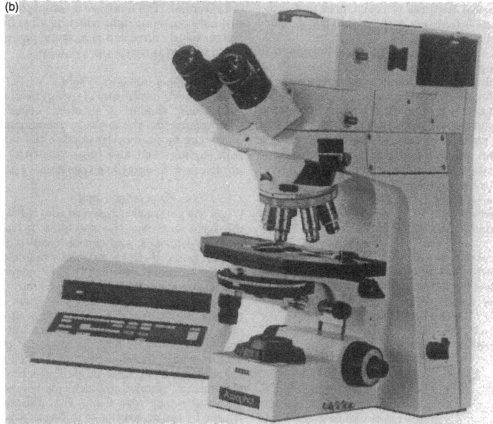

Fig. 5.4 (a) Principal features of a polarising microscope; (b) illustration of modern example where many of the features are boxed in (courtesy of Zeiss Instruments, Oberkochen, BDR).

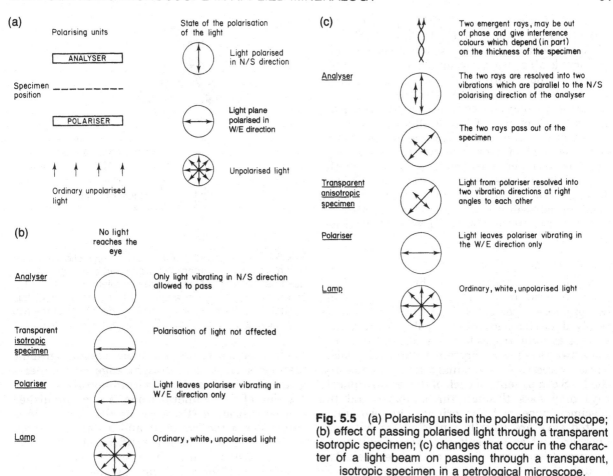

Fig. 5.5 (a) Polarising units in the polarising microscope; (b) effect of passing polarised light through a transparent isotropic specimen; (c) changes that occur in the character of a light beam on passing through a transparent, isotropic specimen in a petrological microscope.

Other minerals, notably carbonate minerals such as calcite and dolomite, are colourless and non-pleochroic but show unusually high *birefringence*, i.e. there is a large difference between the high and the low values of the refractive indices of a single crystal. If the high value of refractive index is higher than the refractive index of the mounting medium whilst the low value is lower than that of the mounting medium then, when the mineral is rotated, its relief (the prominence of its outlines) will vary from "high"

TABLE 5.1
Some examples of pleochroism in minerals

Mineral	Pleochroic colours
Hornblende	dark green to light blue-green
Biotite	reddish-brown to dark brown
Hypersthene	pink to green
Andalusite	pink to pale green or white
Cordierite	yellow to pale blue or white
Glaucophane	lavender to blue or yellow-green

to "low" and back again to "high". The mineral will thus show a twinkling effect when rotated in linearly-polarised light. This twinkling phenomenon can be made more easily visible (especially for the inexperienced student) by rotating the polariser whilst keeping the specimen stationary.

2.1.3. Use of crossed polarisers

(i) *Isotropic minerals*
When both the polariser and the analyser are inserted into the optical path of the microscope it is said to be operating with *crossed polars*. In this condition the directions of polarisation of the two polarisers are set at 90° to one another (see Fig. 5.5). Consequently, if there is no specimen on the stage to affect the polarisation of the light beam then no light can reach the ocular lens: this is because only light that vibrates in the E–W direction passes through the polariser and, since this light has no N–S component, it is *all* rejected by the analyser. Similarly, if an isotropic transparent specimen (i.e. one that does not affect the

direction of polarisation of light) is placed on the
stage then, once more, no light can reach the eye
when the polarisers are crossed. Isotropic materials include all gases, most liquids, glass, isometric
minerals (those with crystal structures of cubic
symmetry), and *some* of the sections cut through
anisotropic minerals (see below). Rotation of such
specimens through 360° has no effect on the polarisation of the transmitted light beam and, therefore, the field of view through the microscope
remains dark for a complete revolution of the
stage. Parts of a specimen that contain no minerals (e.g. pores in the rock, or the epoxy mounting
medium between mineral particles) are also isotropic and must not be confused with isotropic
minerals.

(ii) *Anisotropic minerals*

When polarised light passes through an anisotropic mineral (in any direction other than along
its optic axes – see below) then the light beam is
resolved (or changed) into two polarised rays that
vibrate at right angles to one another. Neither of
these two rays need vibrate parallel to the polarisation directions of the analyser. Consequently,
the N–S components of each of these two separate
rays may pass through the analyser and the
specimen appears to be bright. The two transmitted rays are usually out of phase with one
another, and as a result the viewer will see a
coloured interference image of the specimen. The
actual colours vary with the refractive indices of
the minerals, their orientations and their thickness. Although the *section* thickness is standardised at 30 μm (see Practical no 7) the *mineral*
thickness within the section can vary from 30 μm
to zero and the interference colours can vary
accordingly (Fig. 5.6). When an anisotropic
specimen is rotated in plane-polarised light the
vibration directions of the light emerging from it
will momentarily coincide with those of the
polarisers. When this happens, the light from the
polariser is not modified by the specimen and the
light is, therefore, completely eliminated by the
analyser: the specimen is temporarily dark and is
said to be at its *extinction* position. Extinction
occurs every 90° and, at extinction, the specimen
remains dark for a few degrees of rotation. The

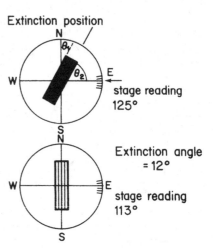

Fig. 5.7 Measurement of an oblique extinction angle in
an anisotropic specimen: the prominent cleavage trace is
used as the datum. Note that the smaller value of θ (θ_1) is
taken as the extinction angle since, by convention, this
cannot exceed 45°. (The polarisation directions of the two
polarisers are N–S and E–W in the microscope.)

angle between some prominent crystallographic
feature, such as the straight edge of a mineral
grain or a cleavage trace within a grain, and the
position of that mineral when it is extinguished
can be measured (Practical no 6). This angle is
called the *extinction angle* and it has some diagnostic value in mineral identification (Fig. 5.7).

On occasion, a number of grains of the same
mineral will form regular, symmetrical intergrowths which are called *twin grains*. In such a
grain the crystal orientation of one part is different from, but directly related to, the crystal orientation of another part. When this happens, the
different "twin" sections extinguish at different
times and in different positions. This *twinning*

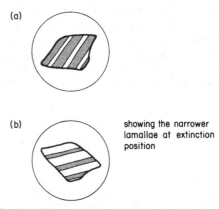

Fig. 5.8 (a) Polysynthetic (repeated) twinning in the
plagioclase feldspar – the wider, dark lamellae are at
extinction position; (b) same specimen showing the
narrower lamellae at extinction.

Fig. 5.6 Possible variation of mineral thickness within a
standard 30 μm thin section.

effect can be a very useful diagnostic test to distinguish between, say, quartz (which does not normally form twins) and plagioclase feldspar (which almost invariably forms multiple twins) (see Fig. 5.8).

(iii) *Uniaxial minerals*

Minerals of hexagonal or tetragonal crystal symmetry have a *single* direction, called the *optic axis*, along which the polarisation of light is not affected. A section of such a uniaxial mineral cut at right angles to the optic axis will not affect polarised light, therefore; furthermore, if both the polariser and the analyser are inserted in the light path, no light will reach the ocular lens. Thus, minerals from the hexagonal and tetragonal classes will, in effect, be "at extinction" under these conditions. Rotation of the specimen produces no change, since the crystallographic orientation remains the same. Consequently, the specimen remains dark throughout a full 360° rotation of the stage, and the same response is obtained as from an isotropic mineral. When the light beam passes along any direction other than the optic axis then the mineral will extinguish four times during a complete revolution and the extinction angle is always either symmetrical (i.e. it occurs at the 45° position) or "parallel" (i.e. the extinction angle is zero).

Since there is only a small probability that a mineral will be cut exactly at right angles to its optic axis, an experienced mineralogist will sometimes use a special microscope fitted with a complex specimen stage called a universal stage. This allows any mineral section to be re-orientated relative to the light beam; for example, any section can be orientated so that its optic axis is, in effect, along the axis of the microscope. This procedure is complex and is not recommended for beginners.

(iv) *Biaxial minerals*

Minerals that show orthorhombic, monoclinic, or triclinic crystal symmetry are optically *biaxial*. These minerals have *two* distinct axes along which the polarisation of light beams is not affected. Sections cut through these minerals will appear to be bright or coloured in transmitted polarised light except when these sections are normal to *either* of the optic axes. The extinction angles for these minerals are usually "oblique" – they are rarely parallel or symmetrical (cf. uniaxial minerals). The *maximum* extinction angle obtained from a number of differently orientated sections of the same mineral has some diagnostic value in the identification of minerals such as the plagioclase feldspars and other rock-forming minerals. (*Note*: by convention the extinction angle is always measured to the *nearest* cross-wire in the microscope and it can, therefore, never exceed 45°.) It is also possible to determine the angle between the optic axes (the *optic axial angle*). This angle has some value in the identification of certain minerals, but its determination is difficult and not recommended for beginners.

(v) *Summary of the use of crossed polars*

All sections through minerals that belong to the isometric symmetry class are isotropic and, as a consequence, remain dark throughout a complete revolution of the stage. Minerals of the hexagonal and tetragonal crystal systems are anisotropic and most sections show either straight or symmetrical extinctions as the stage is rotated. However, the very small number of sections which are cut normal to the single optic axis of these minerals will behave like sections through isotropic minerals, i.e. they remain dark throughout a 360° revolution of the stage. Minerals that belong to the remaining crystal systems (orthorhombic, monoclinic and triclinic) show oblique extinction positions except on those rare occasions when the sections are cut normal to either of their *two* optic axes (see Table 5.2).

2.1.4. Use of convergent light

The petrological microscope has facilities for producing highly convergent beams of light that can be focused into the specimen (see Practical no 6). With this light arrangement it is possible (but often quite difficult) to produce optical "interference figures". These figures can be used to differentiate certain categories of uniaxial and biaxial minerals (these minerals can be either "positive" or "negative", but for further details see Bibliography).

TABLE 5.2

Summary of effects during examination of minerals in crossed polars

Crystal symmetry	Class	Comments
Isometric (cubic)	Isotropic	All sections isotropic
Tetragonal Hexagonal	Uniaxial Uniaxial	Optic-axial section isotropic: others anisotropic: extinction straight or symmetrical
Orthorhombic Monoclinic Triclinic	Biaxial Biaxial Biaxial	Optic-axial sections isotropic: others anisotropic; extinction oblique
Amorphous	Isotropic	All sections isotropic

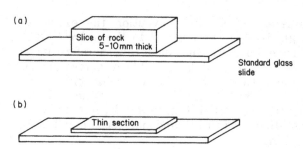

Fig. 5.9 Preparation of a thin section: (a) a slice 5–10 mm thick is cut from the rock specimen and fixed onto a standard glass slide; (b) rock slice is reduced to 30 μm by careful grinding.

2.1.5. Preparation of thin sections

Rock specimens that contain significant proportions of transparent minerals are prepared in the form of thin sections in order that they can be examined with transmitted light. Details of the methods used to prepare these sections are given in Practical no 7.

The preparation of good thin sections requires great skill, exceptional cleanliness and considerable patience. It is very easy to introduce deleterious artefacts into the specimen and, by careless handling, it is even possible to lose the specimen entirely. Briefly, a rock specimen is cut with a diamond-armoured saw (Fig. 5.9) into a parallel-sided slice a few millimetres thick: this slice is stuck on to a standard glass slide (75 mm × 25 mm). It is carefully ground down to a standard thickness of 30 μm and is then covered by a thin glass sheet, or cover glass.

Thin sections of particulate specimens are made in a similar manner. Closely sized particles are stuck on to a *thin* glass sheet and are ground down to produce a flat surface. An ordinary (standard) glass slide is stuck on to this flat surface and the original thin sheet of glass is ground away. A layer of the particles is then removed until each particle has two parallel faces and is 30 μm thick.

2.1.6. Summary of information that can be obtained using the transmitted-light polarising microscope

(1) *Ordinary (unpolarised) light is used to determine:*
    (*a*) mineral habit,
    (*b*) particle and grain shapes,
    (*c*) mineral cleavages,
    (*d*) mineral colour – whether uniform or variable,
    (*e*) the relative refractive indices of one mineral and another, and of a mineral and its mounting medium (the shadow test),
    (*f*) alteration of minerals at their surfaces and along cracks,

(*g*) intergrowths and inclusions: their shapes and arrangements,
(*h*) size ranges (and, if possible, size distributions) of mineral grains and/or mineral particles,
(*i*) mineral proportions,
(*j*) mineral associations,
(*k*) mineralogical compositions of mineral particles.

(2) *The following* additional *features can be examined with plane-polarised light:*
    (*a*) pleochroism,
    (*b*) twinkling (mainly in carbonate minerals),
    (*c*) relative refractive indices (the Becke line).

(3) *Crossed polars are used to establish:*
    (*a*) isotropism or anisotropism,
    (*b*) extinction angles of anisotropic minerals,
    (*c*) interference colours.

(4) *Convergent light can also be used to determine:*
    (*a*) uniaxial or biaxial symmetry,
    (*b*) positive or negative sign, but the methods for doing so are not given in this book.

2.2. The reflected-light polarising microscope

Some minerals are opaque – even in very thin slices – and therefore they cannot be identified by their behaviour in transmitted light. These minerals can, however, be studied in detail by using light which is reflected vertically from their polished surfaces. This study is carried out in a *reflected-light microscope* which is usually a modification of (or an attachment to) a transmitted-light microscope.

In the reflected-light microscope the light from a high-intensity source enters the instrument through a side tube (a high-intensity light is essential because of the numerous, inevitable light losses that occur in the optical system). The light is reflected by a system of mirrors *perpendicularly* on to a polished specimen held on the microscope stage. Some (usually only a small part) of this light is reflected from the specimen surface and passes vertically back through the microscope to the ocular lens (see Fig. 5.10).

2.2.1. Reflectance

In ordinary, unpolarised white light the colour and the brightness of a mineral image will vary

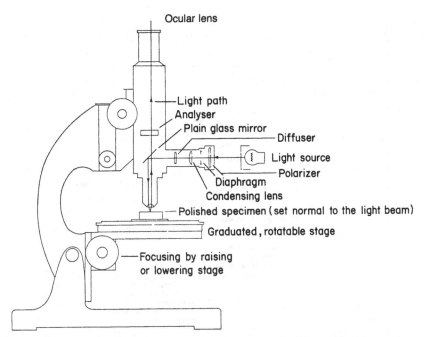

Fig. 5.10 Principal features of a reflected light polarising microscope (for illustration of a modern example, see Fig. 5.4(b)).

with the spectral reflectance characteristics of the specimen. The brightness (or reflectance) of a well-polished mineral is the ratio of the intensity of the reflected beam to that of the incident light beam (this value is expressed as a percentage). The reflectance value is high for the native metals (see the data in Table A.2.4 in Appendix 2); it is also quite high for the sulphide minerals; it is low, however, for the silicate minerals (since the silicates are almost all transparent minerals, the reflectance of such a mineral is necessarily low because most of the incident light is transmitted). With a little experience it is possible to make a good qualitative assessment of the value of reflectivity, but this reflectance value can also be quantitatively measured if the appropriate equipment is available (see Practical no 8).

Most of these reflectance measurements are carried out under white light. For some minerals the reflectance varies markedly with the wavelength of the incident light and the reflected image then shows a distinctive colour in white light. However, most minerals appear as shades of grey when using white light and some experience is necessary to distinguish one subtle shade of grey from another.

2.2.2. Isotropism
Reflected polarised light can be used to determine some of the crystallographic features of an opaque mineral. For example, polarised light can be used to determine whether a mineral is isotropic or anisotropic. With crossed polarisers an isotropic mineral remains dark throughout a complete 360° rotation of the specimen (cf. transmitted polarised light). As with transmitted polarised light, however, the rare isotropic sections that have been cut through *anisotropic* minerals at right angles to their optic axes will also remain dark throughout a complete rotation. Most sections cut through anisotropic minerals show bright (and, very occasionally, coloured) images. Extinction then occurs every 90° when the section is rotated in the polarised light.

Many other optical effects can be observed and measured by the skilled mineralogist with the polarising reflected-light microscope but this kind of work is best left to the specialist petrographer and is outside the scope of this book.

2.2.3. Mineral identification
Opaque minerals can often be identified by *qualitative* assessments of their colours (and change of colours on rotation), reflectivities (or reflectances), isotropism, internal reflections and polarising effects in reflected light. These qualitative assessments can sometimes be supplemented by *quantitative* measurements of reflectivities and mineral hardness values. Figure 5.11 shows the relationships between the reflectivities and the hardnesses of the more commonly occurring opaque minerals. Because of the effects of anisotropic crystal structures and because of the wide

Fig. 5.11 Vickers micro-indentation hardness data for a number of opaque minerals (after Bowie and Simpson in Zussman, 1977).

range of allowable variations in the chemical compositions of many minerals, most minerals do not show up as "point" values on Fig. 5.11. Instead, each mineral occupies a small rectangle and these rectangles frequently overlap so that confusion can arise when using reflectivity and hardness values to identify similar minerals.

The value of the reflectivity in a mineral identification scheme is enhanced if the reflectivity values are measured in a series of monochromatic light beams. For example, Fig. 5.12 shows how the reflectivities of a number of minerals differ when measured at a number of

wavelengths. Table 5.3 shows this information for a small group of minerals.

The reflectivities of small regions (a few square micrometers in area) on a polished mineral surface are measured using a selenium cell, or other suitable measuring device. The reflectivity results are then compared with those obtained from carefully prepared, standard specimens under the same measuring conditions. The hardness values of the *same* regions are measured by a micro-hardness device in which the resistance offered by a mineral to a small, weighted diamond "point" is established.

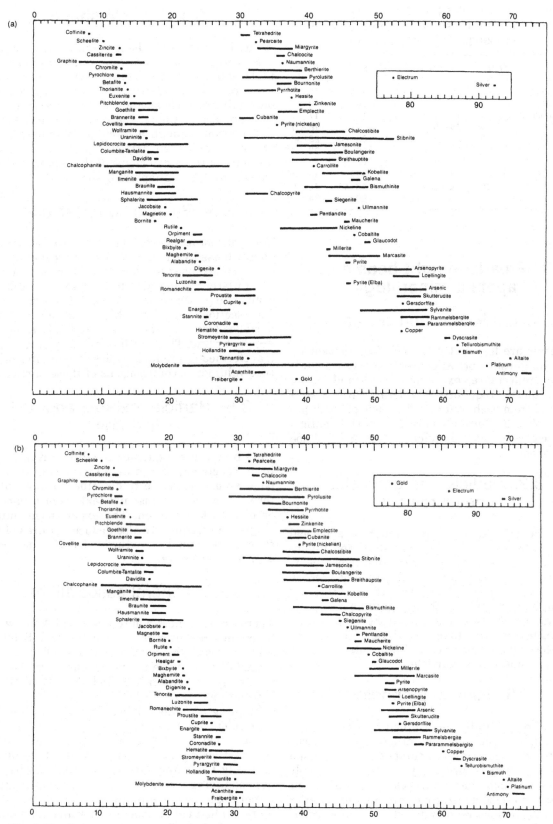

Fig. 5.12 Reflectance data at (a) 470 nm and (b) 546 nm (after Bowie and Simpson in Zussman, 1977).

TABLE 5.3
Variation in reflectivities of minerals at different wavelengths of light (%, in air)

Mineral	470 nm	546 nm	589 nm	650 nm	White
Scheelite	10	10	10	10	10
Chromite	13	12	12	12	12
Wolframite	16	16	14	15	17
Ilmenite	15–20	16–20	16–20	17–20	18–21
Hematite	27–33	26–31	25–30	23–26	25–30
Chalcocite	36	33	31	29–30	32
Pyrite	36	39	42	44	46
Galena	47	43	42	42	43
Chalcopyrite	31–34	42–45	44–47	46–48	42–46
Gold	39	78	86	90	74

3. Polarised light beams in applied mineralogy

3.1. Mineral identification

Minerals are frequently *identified* by measuring their diagnostic optical properties: these include colour, reflectance, extinction angle and so on. Even more frequently, minerals are *distinguished* from each other on the basis of combinations of the more obvious qualitative optical characteristics such as colour, brightness, shape and associations.

3.2. Particle and grain size measurement

The size of a section cut through a spherical particle (or grain) can be accurately and unambiguously measured in a polarising microscope by comparing its diameter with the divisions on the calibrated eyepiece. The "size" of a section cut through an irregularly shaped particle is usually defined by its two-dimensional Feret diameter (see Chapter 7 on image analysis). This is the distance between two vertical tangents drawn on opposite sides of the particle.

3.3. Mineral proportions

These are generally determined by *distinguishing* and then counting the minerals which occur at a pre-determined set of points on the specimen surface. It may be necessary to carry out simple checks on such characteristics as isotropism or refractive index for each measured point in order to distinguish between "wanted" and "unwanted" minerals.

3.4. Particle shape

As with other methods of examination, such as the stereoscopic microscope and the electron probe microanalyser (see Practical no 3 and Chapter 9, section 9, respectively), it is possible to determine particle shape with the polarising microscope. For example, the random chord length distribution through a randomly mounted population of mono-size particles provides a unique measure of the mean shape of those particles (see Image Analysis, Chapters 6 and 7).

3.5. Particle compositions

The polarising microscope can also be used to obtain biased estimates of particle compositions (i.e. it can be used to determine the proportions of the different minerals that appear to occur in composite particles) from the examination of random sections cut through populations of those particles (see Chapter 6, section 4.2.3), either by area or by line. These estimates can only rarely be transformed to produce unbiased estimates of the true volumetric compositions of those particles.

3.6. Particle textures and rock textures

Once he has distinguished the different minerals in a specimen, the mineralogist can use the polarising microscope to describe, in qualitative terms, the two-dimensional texture of a particle or of a rock specimen. The experienced mineralogist can frequently also deduce features of the *three*-dimensional texture from its two-dimensional image.

3.7. Mineral associations

There are instances where a mineral A is always found in close proximity with mineral B but is seldom associated with mineral C. The mineralogist can often use the polarising microscope to obtain good qualitative indications of these associations (Fig. 5.13). For example, after examining a number of sections of a tin ore, the mineralogist might claim that it contains very fine-grained cassiterite and that virtually all of the cassiterite is associated with silicate minerals and only trace amounts are associated with the accompanying sulphides. More quantitative mineralogical determinations demand the use of more sophisticated image analysing techniques as described in Chapter 6.

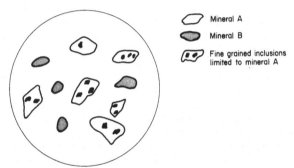

Mineral A

Mineral B

Fine grained inclusions
limited to mineral A

Fig. 5.13 Mineral associations in polished mineral
particles.

4. Conclusion

The polarising microscope has been a most
important tool in petrology and in mineralogy for
well over a hundred years. It enabled mineralogists to gain good indications of the crystallographic structures of minerals long before it was
possible to measure these structures by direct
methods. It has also been widely used as a principal tool for the identification of minerals.

Even with the advent of modern methods of
mineralogical analysis the polarising microscope
is still of very great value and it can be used, even
by the novice, to distinguish one mineral from
another and to measure industrially important
mineralogical characteristics such as mineral
proportions, grain sizes, grain locations, particle
compositions and rock textures.

Thin sections of transparent minerals are
examined in transmitted light whilst polished
sections of opaque minerals can be examined (on
the same instrument) using reflected light.

Chapter 6

Image analysis – theory

1. Introduction

1.1. What is image analysis?

Image analysis is concerned with the measurement of geometrical features that are exposed on two-dimensional images. In geological and in mineralogical studies these images are frequently derived from plane sections cut through rock specimens or from sections cut through mounted mineral particles. Whenever possible, these images are measured by methods that provide fully quantitative results.

1.2. Plane sections

In the context of image analysis the term *plane section* is applied, strictly speaking, only to polished sections. The term is not usually applied to the ordinary geological thin section because many of the mineralogical features that exist within these sections are smaller than the thickness of a section and, consequently, they are difficult to measure accurately. Thus, features such as mineral grains or pores that are less than $30\ \mu m$ in size may overlap within the confines of the specimen and, when viewed in transmitted light, two or more of these features can appear to be one, much larger, feature (see Fig. 6.1). However, it is permissible (in the context of image

analysis) to treat a thin section as a plane section if the features within it are much larger than its thickness. But even here, the effects of sloping boundaries between adjacent minerals can greatly confuse the results obtained during the analysis.

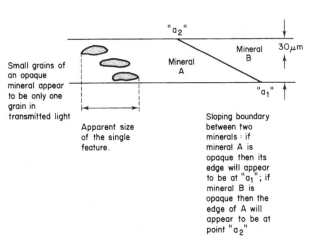

Fig. 6.1 Diagrammatic illustration of a normal "thin" section (30 μm thick) from a rock specimen. This shows how very small mineral grains can overlap and appear to be a single, much larger, grain when viewed with transmitted light; it also shows how a sloping boundary between adjacent comparatively coarse grained mineral grains can confuse the results of a grain size analysis.

2. Producing suitable images

Images of plane sections cut through geological materials are usually produced by viewing the sections (1) through a microscope, (2) with a television camera, (3) with an electron microscope or (4) with an electron probe X-ray microanalyser. These images are often so complex that, in the past, it was not possible to analyse them in detail because it took much too long to carry out the necessary measurements by the manual methods which were then available. Instead, the mineralogist tended to "describe" these complex images, as best he could, in qualitative terms. Nowadays these descriptions are not suitable for the mineral engineer because he requires quantitative numerical data for process design and process control purposes. Many attempts have been made to simplify the methods used to collect quantitative information from rock sections, and gradually over the years it has become possible to quantify even complex images by using computer-controlled automatic measuring systems. These new systems swiftly produce large amounts of accurate data from suitable specimens and, when correctly interpreted, the data are proving to be of enormous value to the geologist, the mineralogist and the mineral engineer (see Chapter 7).

3. Analysis of images

Figure 6.2 (see colour section) is an image of an ore that contains many suphides and it illustrates the fact that images obtained from mineralogical specimens can be very complex. Such images can show a number of different mineral phases, the mineral grains in a single specimen can cover a wide size range, the spatial relationships between the mineral grains are often complicated (i.e. the textures are complex), and the optical contrast between the various minerals is often poor. Consequently, many mineralogical images are difficult to analyse but it is nevertheless often absolutely essential for the mineral engineer to obtain as much quantitative information as possible from them.

In addition to these practical problems, the data collected from a two-dimensional section only occasionally provide direct, unambiguous information about the nature of the features that exist in the three-dimensional bulk of the original specimen. Many measurements from two-dimensional images are biased but they can sometimes be used to derive unbiased estimates of the three-dimensional values that they represent. These transformations of low-dimensional measured data into their appropriate higher-dimensional values often involve the science of stereology (see Chapter 8, section 2).

4. Historical

From very early times geologists and mineralogists have been greatly interested in the geometrical properties of rocks and minerals – properties such as the sizes, shapes and proportions of the minerals in a rock, or the shapes, sizes, compositions and textures of mineral particles (see below). This kind of information has been used to (1) improve our understanding of the origin and the nature of rocks, (2) assist in the design of mineral treatment operations, (3) assess the performances of these operations, (4) determine the products of the reactions between minerals and various reagents, (5) design and produce mineral-like materials (such as ceramics) that have pre-arranged physical properties, and so on.

4.1. What needs to be measured?

The detailed mineralogical information required by the mineral engineer about a mineral specimen includes the following features:

(1) the identities of all the minerals of potential interest,

(2) the crystal structures of the important minerals,

(3) the chemical compositions (and, where necessary, details of the variations in compositions) of selected minerals,

(4) the volumetric and/or the weight proportions of all the minerals of interest,

(5) the physical and chemical properties of all the minerals which are likely to affect the processes used during the treatment of an ore,

(6) the distribution (or partitioning) of selected elements within the various minerals,

(7) the textures of the rock (or the unbroken ore),

(8) the natural (i.e. the unbroken) grain-size distributions of selected minerals,

(9) the size distribution of the mineral particles in broken rocks or in particulate materials such as beach sands or alluvials/deposits.

(10) the shapes of mineral particles produced or used in mineral processing operations, e.g. the ferrosilicon particles used in dense–medium separations,

(11) the mineralogical (rather than the chemical) compositions of mineral particles,

(12) the textural relationships between the minerals in the composite particles, and

(13) other features such as the number, location and sizes of pores, the specific surface areas of mineral grains and/or mineral particles, and so on.

4.2. Traditional methods of measurement

At first, geologists and mineralogists only collected *qualitative* or semi-quantitative information about the features listed in section 4.1, but, more recently, they have made great efforts to devise measuring methods that *quantify* this information. For example, it is possible to use various mineral separation techniques (see Chapter 3) to determine mineral proportions; selective solution of some of the components in a specimen allows the grain sizes to the stable (unreacted) minerals to be determined (Chapter 10, section 3); and traditional manual image analysing methods have been used to determine other mineralogical characteristics such as particle compositions (see below). Of these measuring methods, only the image analysis techniques can be automated in any significant degree.

4.2.1. Stereoscopic images

When opaque features such as mineral particles exist in a transparent matrix, such as air, then it is easy to establish the three-dimensional (spatial) relationships between them and and the surrounding matrix. This is done by *stereoscopically* viewing the three-dimensional specimen – as we do, for instance, when we walk into a room and see a chair in the middle of the floor, or a lamp hanging from the ceiling. Unfortunately, most rocks are essentially opaque to radiation of visible wavelengths (except in very thin sections) and it is seldom possible to measure any three-dimensional rock features by these direct stereoscopic methods. Various attempts have been made to use electromagnetic radiations other than visible light for carrying out the stereoscopic examination of bulk specimens but none of these methods has been successful: X-rays are totally absorbed by a few millimetres of rock, and the wavelengths of the radiations of the more penetrative radio frequencies are too long for them to be used to measure the fine mineralogical details that most rocks contain.

4.2.2. Serial sectioning

Another possible method for determining the three-dimensional features of minerals in an opaque rock is by *serial sectioning*. In this method a large number of closely-spaced parallel sections are prepared from the rock specimen (Fig. 6.3) and these sections are examined sequentially. The results from each individual section are then used to "re-create" the original three-dimensional structure. This procedure is extremely tedious, especially when the features of interest are small, and large numbers of sections must be cut very close to one another. In fact, this method is so tedious that its use can only be recommended for very special research purposes or for studying a material that contains only a *single* feature of great interest – a mineral vein cutting through a rock, for example. (In medical research, on the other hand, the procedure is commonly used to study the details of a single organ such as a diseased kidney.)

4.2.3. Random sectioning

Happily, the mineralogist can deduce much of the information he requires concerning three-dimensional rock features from indirect, image-analysing methods. These methods involve the examination of only a very small number of sections cut in random orientations through a specimen. Even so, the early methods of analysing these images were so tedious and so boring that mineralogists were reluctant to adopt them: it took too long to carry out an analysis and it required too great a mental effort to produce accurate results.

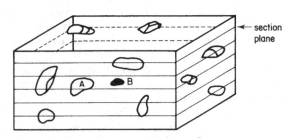

Feature A would be disclosed, while feature B would not, by the planar spacing used in the diagram

Fig. 6.3 Serial-sectioning procedure for determining the three-dimensional features of a rock specimen. The sections must all be parallel and the distance between adjacent sections must be constant. If the specimen contains very small features then the sections must be cut very close together and large numbers of such sections are needed to provide all the required information concerning that specimen.

5. Measurement techniques

5.1. Area measurements – De Lesse

The first record of the use of a quantitative image-analysing technique dates back to 1848. De Lesse (a French mineralogist) established the *volumetric* proportions of a selected mineral within a rock by measuring the *area* occupied by that mineral on a random section cut through that rock. His method of measuring the area proportion was very, very tedious: he traced the outlines of all the grains of the selected mineral on to a piece of oiled, translucent cloth which was of the same area as the specimen surface. He then pasted the cloth on to a sheet of tinfoil of the same size and the pieces that represented the selected mineral were cut out with scissors and weighed on an "analyst's balance" (Fig. 6.4).

The ratio of the weight of those tinfoil pieces that represented the selected mineral to the total weight of all the tinfoil gave a good estimate of the *area* proportion of that mineral on the measured surface. De Lesse then *assumed* (and it has since been shown that his assumption was correct) that this area proportion is equivalent to the volumetric proportion of the mineral in the bulk specimen, provided that:

(1) the measured section is a random section through the rock (i.e. the direction of the section does not match any texture, banding or patterning within the rock),
(2) the measured area on the section is much larger than the individual grains of the selected mineral, and
(3) the random section is large enough to include a statistically adequate number of those measured grains.

The principle behind the work of De Lesse was sound, but there was no way in his day of simplifying or reducing the vast amount of work involved in making the measurements. Consequently, it is not surprising that his original method of area measurement did not gain favour with geologists and mineralogists. However, recent developments in computing and in electronic sensing devices now enable De Lesse's measuring principle to be used in modern, automated image-analysing devices which do not need much (if any) human involvement (see Chapter 7).

5.2. Linear measurements – Rosival

In 1898 Rosival developed a much simpler method than De Lesse's for measuring the volumetric proportions of the minerals in a rock. This system measures lines instead of areas. The sectioned specimen is covered by a diffuse (widely spaced) grating of parallel measuring lines (Fig. 6.5); the proportion of the total length of these lines which falls on a specified mineral gives the *linear ratio* of that mineral:

$$\frac{L_A}{L_T} = \text{linear proportion of mineral A}$$

where L_A = total length on mineral A and L_T is the total traverse length across the specimen.

The linear ratio of a mineral can be shown to be numerically equivalent to its volumetric ratio in the bulk specimen. As with area measurements,

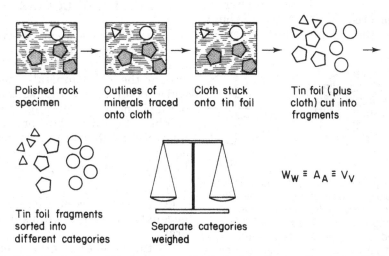

$W_W \equiv A_A \equiv V_V$

Polished rock specimen

Outlines of minerals traced onto cloth

Cloth stuck onto tin foil

Tin foil (plus cloth) cut into fragments

Tin foil fragments sorted into different categories

Separate categories weighed

Fig. 6.4 Illustration of De Lesse's method of mineral proportion determination.

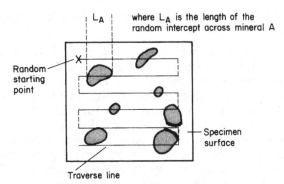

where L_A is the length of the random intercept across mineral A

Fig. 6.5 A grid of widely spaced parallel lines covering the specimen surface. Total length of sampling line = ΣL_T; length of sampling line on mineral $A = \Sigma L_A$; linear ratio of mineral $A = \Sigma L_A/\Sigma L_T$. (If the specimen surface is a random surface then this linear ratio is equivalent to the area proportion of mineral A on the measured surface, and also the volumetric proportion of mineral A in the specimen.)

however, this relationship only holds under certain conditions, i.e. when

(1) the measured section is a random section through the rock,

(2) the section is sufficiently large to include an adequate number of grains of the selected mineral,

(3) the measuring line is long compared with the "linear" sizes of those grains,

(4) no grain is crossed by more than one measuring line, and

(5) the orientation of the traverse line does not coincide with any lineation or texture in the specimen.

Rosival's method is mathematically sound and it is also comparatively simple to carry out. Coarse-grained material, such as porphyritic granite, can be measured using a metre rule but a microscope and the appropriate eyepiece

  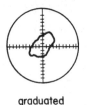

coarse grid   fine grid   graduated

Fig. 6.6 Typical eye piece graticules that can be fitted into a microscope. Only those minerals which are found under the intersections of the grids should be measured during a point-counting exercise. These graticules can, however, also be used to measure particle or feature sizes (once they have been calibrated against a standard, stage micrometre scale).

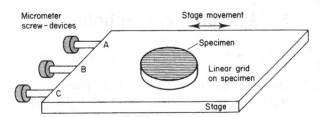

Fig. 6.7 Representation of a multi-micrometre linear measuring system. Each screw-device is allocated to a separate mineral: ΣL_A is the total traverse length on mineral A, and $\Sigma L_A/\Sigma L$ gives the linear proportion of mineral A (and, under appropriate conditions, the volumetric proportion of A).

graticules are necessary to measure fine-grained specimens (Fig. 6.6). Various mechanical aids have been developed to reduce the manual effort needed to make these measurements. These aids include microscope stages which are controlled by up to six separate micrometer screw threads: each of these is used to measure the total length across a specified mineral during a traverse over a specimen (Fig. 6.7).

It has recently become possible to use computers to provide an almost fully automated version of this line-measuring procedure (see Chapter 7). The modern version can be used not only to determine mineral proportions, but also to obtain estimates of particle- and/or grain-size distributions (Chapter 8, section 2.2). The method can also be used to compare the grain-size distributions of minerals in similar materials and in the various products obtained from a mineral treatment operation. Linear data can also be used to determine the specific surface areas of minerals and particles (Chapter 8, section 2.3).

5.3. Point counts – Glagolev

A point-counting system for determining mineral proportions was introduced about 1930 by Glagolev and, separately, by Thompson. In this system the only minerals measured are those which occur at the intersections of two sets of parallel lines set (generally) at right angles to one another (Fig. 6.8). The proportion of the grid intersections that falls on a specified mineral is equivalent to the volumetric ratio of that mineral in the bulk specimen, provided that:

(1) the measured section is a random section,

(2) the section is sufficiently large to include an adequate number of grains of the selected mineral,

(3) the number of grid intersections is large, and

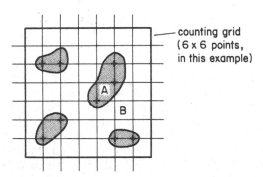

counting grid
(6 x 6 points,
in this example)

A
B

minerals at each
intersection are
measured.

+ measured points on
"selected" mineral

Fig. 6.8 Point-counting procedure: 9 points on shaded mineral A, 27 points on unshaded mineral B. Point proportion of mineral A = 9/36 = 0.25 (but total number of points is too small for high statistical accuracy).

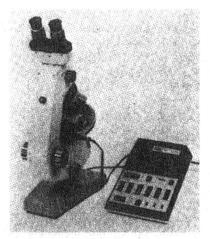

Fig. 6.9 Automatic point counter (courtesy of Hacker Instruments, Inc.).

(4) the orientations of the grid of points is random relative to any pattern or lineation in the specimen.

Thus,

$$\frac{P_A}{P_T} = V_A$$

where P_A is the proportion of measured points that fall on mineral A; P_T is the total number of measured points, and V_A is the volume proportion of mineral A.

Ordinarily, the minerals at the grid intersections are identified, and their presence is recorded, by the mineralogist. Some reduction of manual effort is possible by using an electro-mechanical counting device (Fig. 6.9) to record the results of the observations. When the appropriate button is pressed the number of points occupied by the selected mineral is incremented by one and the next intersection is automatically brought into view. The mineral at that intersection is identified and the appropriate button pressed, and the procedure is repeated until a sufficient number of points has been counted (see section 7.2 of this chapter on the statistics of point counting).

When a solid-rock specimen is being measured then virtually all the intersections will fall on mineral grains (if, occasionally, a "point" falls on a pore then this pore should be counted as a separate phase). When a particulate specimen is measured then many of the counting points will

fall on to the mounting medium: these points should be ignored except when the object of the analysis is to discover the "particle density" of the specimen.

With the development of computers, point measurements can now be carried out by fully automatic equipment (see Chapter 7). These systems first identify the mineral at some arbitrary (random) starting point, record the result and then move the specimen to the next point, repeating the operation until a sufficient number of points has been analysed.

5.4. Particle (or grain) counting

Particle-counting methods can also be used to determine the proportions of the various minerals in some mineral specimens – especially, those minerals which occur as liberated particles in alluvial deposits and in beach sands; for example, in the tin-mining areas of south-east Asia and in the titanium-rich beach sands of India and Australia.

It is often much more convenient to count unmounted particles than to expend time and effort in preparing properly mounted and polished sections for line analysis or point analysis. In the particle-counting method the particles are presented to the viewer in the form of a layer, one particle deep, contained in a shallow flat-bottomed dish. Certain precautions are necessary when interpreting the results of a particle count analysis; these are discussed below.

5.4.1. Screened particles

If all the counted particles are of equal volume then the *particle* proportion of a specified mineral is equivalent to its *volumetric* proportion – provided that sufficient particles are counted so as to achieve statistical validity (see section 7.2 below and Practical no 10). That is, under suitable conditions,

$$G_{g(A)} = V_{v(A)}$$

where $G_{g(A)}$ is the particle proportion of mineral A and $V_{v(A)}$ is the volume proportion of A in the specimen. Unfortunately, even closely screened mineral particles are seldom of equal volume because particles differ markedly in shape due to differences in their crystallographic structures (Fig. 6.10). Thus, a cubic mineral such as garnet tends to form squat, roughly spheroidal particles whilst tetragonal minerals like zircon tend to form elongated, ovoid particles with length-to-breadth ratios of 2:1 or even 3:1. As a result, an individual zircon particle within a closely screened size fraction will be significantly larger in volume than an accompanying garnet particle and, although the densities of the two

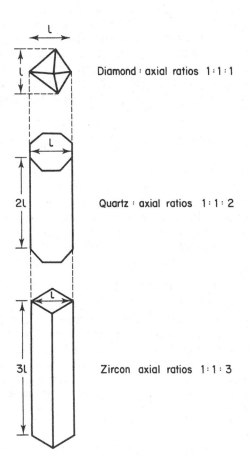

Fig. 6.10 Variation in volumes of particles of identical screen size.

particles are roughly the same, the zircon particle will weigh a great deal more than the garnet. If we were to assume (improperly) that the zircon particles and garnet particles have the same volume, then their mass ratio would be 4700 kg m^{-3} (the density of zircon) divided by 4000 kg m^{-3} (the density of garnet), i.e. the mass ratio would be 1:1.75. If on the other hand, the garnet particle has axial lengths in the ratio 1:1:1 whilst the zircon has axial lengths of ratio 2:1:1 then the zircon particles will weigh 2.35 as much as the garnet.

5.4.2. Classified particles

As shown above, it is obvious that great care must be used when trying to relate particle proportions to volumetric ratios. The problem is further compounded if the mineral particles are fractionated by a classification procedure prior to measurement. As shown in Chapter 3, section 3.2, there can be wide variations in the volumetric (and the mass) ratios between mineral particles that have the same settling velocities, but are of different densities. For example, a quartz particle can weigh nearly three times as much as a galena particle which reports in the same classified fraction. This (often unexpected) difference in particle mass must be taken into account when calculating the mass proportions of the various particles which occur in classified specimens.

5.4.3. Counting procedures

In practice, particulate specimens – preferably in the form of closely screened fractions – are analysed by one of the following procedures:

(1) *all* the particles in the specimen are identified and counted – this procedure is only practicable with small specimens, but when it can be used it eliminates any sampling error at the counting stage (but see section 6 of this chapter for other possible errors);

(2) *all* the particles in selected fields of view are counted: these fields are taken in a regular pattern across the whole specimen. In practice, a grid pattern is often drawn on the outside of the base of the specimen dish, using a wax pencil: each intersection of this grid is then taken as the centre of a measured field of view;

(3) count only those particles which coincide with the intersections of the fine grid formed by a graticule fitted to a suitable microscope; many fields of view must, of course, be measured.

Particle counting is comparatively easy but the results are often difficult to interpret. Ideally, all the particles should be liberated (i.e. should consist of only one mineral) and they should all be of the same volume. If these ideal conditions are not met then the results of a particle-counting analysis must be treated with great care.

5.4.3. Gross counting

It is sometimes permissible to use a comparatively fast variant of the ordinary particle counting technique. This procedure, known as *gross counting*, is especially useful when the mineral of interest occurs in only very small proportions and it is therefore necessary to count very large numbers of particles for accurate results (see section 7 below for a discussion of statistical errors). In the gross-counting method the *average* number of particles in a field of view is determined by counting *all* the particles (irrespective of mineralogical composition) in a number of randomly selected fields. Once this average value has been determined, then a succession of such fields can be examined swiftly (and without being counted) in the search for the rare mineral particles of interest. For example, suppose that twelve randomly positioned fields of view contain a total of 1200 particles, then the average number of particles per field is 100. If only five particles of the mineral of interest are found in a total of 1000 fields of view then it is reasonable to assume that the proportion of that mineral in the specimen is approximately $(1000 \times 100) \div 5 = 0.005\%$ (by particle number).

6. Operational limitations and errors in image analysis

There are many operational difficulties that must be taken into account whenever an image-measuring method is used. In addition, all image-analysing methods suffer from problems of spatial resolution, misidentification of minerals and so on. These problems must be understood and appropriate action must be taken to minimise their adverse effects on the accuracy of an analysis. The accuracy achieved during analysis is difficult to quantify except by keeping close checks on the precision and reproducibility of a large set of similar analyses. However, the following sources of error must always be taken into account whenever any kind of image analysing method is used.

6.1. Spatial resolution

All mineral grains tend to look alike at sizes that approach the wavelength of light. Thus, in an optically based measuring system all features that are smaller than 1–2 μm are very difficult to *identify* even though their *presence* can easily be established. At these sizes a dark ring appears within the boundary of each grain and restricts the area of mineral that can be seen and which can be used for identification purposes. Again, in these circumstances, if two grains are separated by a distance that approaches the wavelength of light then it becomes difficult to observe the separation between them and the two grains tend to appear as one, larger grain.

Spatial resolution problems of this kind can be minimised by (1) using the best available optical equipment, or (2) using non-optical equipment, such as electron microscopes, which have better resolving power than any optical device.

6.2. Misidentification

Misidentification of a mineral can occur because of the limited resolution of the measuring equipment (see above). Misidentification can also be due to incorrect or hasty observation by the operator. Minerals can also be misidentified because of the ambiguous nature of the data obtained from the measuring system (see Chapter 7 on automatic image analysers).

The risk of misidentifying a mineral can be minimised by making careful measurements with the best possible equipment.

6.3. Misrecording results

Correct measurements may be incorrectly recorded, especially by a tired and/or bored observer. During a manual image analysis the experimenter may have to make an observation and record a result every second or two over an extended period. In these circumstances it is very easy to make errors – errors of identification and/or errors in recording the results of those observations. These errors can be minimised by increasing the amount of time devoted to an analysis and by taking as much care as possible during that analysis. Misrecording errors can, of course, be virtually eliminated by replacing the human observer by an automated measuring and recording system (see Chapter 7).

6.4. Miscalculation of results

Pocket calculators and microcomputers are almost invariably available, nowadays, for calculating the results of an analysis. Consequently, there is little reason for calculation errors: however, these errors still occur.

Miscalculation can be minimised by greater care and attention to detail on the part of the experimenter; it can be virtually eliminated by using automatic data-handling systems (see Chapter 7).

6.5. Effects of composite particles

When particulate materials are analysed by a particle-counting method (described in Chapter 6, section 5.4, above) the presence of composite particles (which cannot properly be allocated to any single mineral phase) will distort the results.

This possible source of error can be reduced by restricting the use of particle-counting methods to the measurement of *liberated* particles. It can be eliminated by using the point, line or area methods of image analysis to determine mineral proportions.

6.6. The use of incorrect mineral densities

Mineral engineers desire some types of mineralogical information in the form of *mass* proportions (rather than number proportions or volume proportions). Calculation of mass proportion from measured particle, point, line or area measurements is subject to error if the mineral densities are not accurately known. The engineer usually uses textbook values of mineral density for these calculations, but whenever possible it is better to *determine* the required density values by carrying out the necessary measurements (e.g. the pycnometer method described in Chapter 3).

7. Calculation of statistical errors

In addition to the operational errors described in section 6 above, all measurements which involve a sampling procedure are subject to statistical errors. These errors (unlike operational errors) can easily be assessed and can be kept within desired limits by making a sufficient number of independent observations.

7.1. Binomial distribution

Mineralogists usually use the binomial distribution to estimate the statistical errors that are involved during particle-counting or point-counting experiments. The presence of a mineral of interest at the measured point or in the counted particle is taken to be a "success" and its absence is a "failure". As long as the measured particles or points are independent of one another then the binomial distribution can be used to calculate the number of observations needed to achieve a result of specified accuracy.

The standard deviation of the binomial distribution is:

$$\sigma = \sqrt{\frac{pq}{N}} \qquad\qquad 6.1$$

where σ = standard deviation, p is the proportion of the selected mineral, $q = (1 - p)$, and N is the number of measured particles (or points).

The acceptable error in most mineral treatment operations is the 95% confidence limit (i.e. $\pm 2\sigma$) and the formula for calculating the acceptable *absolute* error (e) is:

$$e = \pm 2 \sqrt{\frac{pq}{N}} \qquad\qquad 6.2$$

Thus, the number of particles (or points) that should be measured so as to achieve a specified absolute error value of e at the 95% confidence level is:

$$N = \frac{4pq}{e^2} \qquad\qquad 6.3$$

The *relative* error (E) on p, the proportion of a selected mineral, is often a more useful measure of precision. It is defined as:

$$E = e/p \qquad\qquad 6.4$$

The number of observations needed to achieve a specified relative error at the 95% confidence level is obtained by substitution in equations 6.3 and 6.4, giving

$$N = \frac{4q}{pE^2} \qquad\qquad 6.5$$

where E is the specified relative error on p.

In practice, neither N nor p is known at the beginning of an experiment, but they can be found by an iterative procedure. Suppose, for example, that it has been decided that the measured value of p should have a relative error (at the 95% confidence level) of 0.1 (i.e. 10% relative error); what is the minimum number of observations that should be made to achieve this accuracy?

(1) *Assume* some value of p; use previous

knowledge, current observation or just guess-work. Let this assumed value be 0.5.

(2) Calculate N on the basis of the assumed value of p: thus,

$$N = \frac{4q}{pE^2} = \frac{4 \times 0.5}{0.5 \times 0.01} = \frac{4}{0.01}$$

$$= 400 \text{ observations.}$$

(3) Carry out 400 particle (or point) analyses. Suppose that the result showed 100 "successes" on the selected mineral; substitute this value in the equation:

$$N = \frac{4q}{pE^2} = \frac{4 \times 0.75}{0.25 \times 0.01} = \frac{12}{0.01}$$

$$= 1200 \text{ observations}$$

i.e. this new (and better) estimate of p demands that 1200 particles (or points) be determined.

(4) Carry out a further 800 measurements. Suppose the overall result from the 1200 measurements was 380 "successes"; then

$$N = \frac{4q}{pE^2} = \frac{4 \times 0.68}{0.32 \times 0.01} = \frac{2.72}{0.003}$$

$$= 907 \text{ observations.}$$

This shows that the number of observations that has now been made is adequate to achieve the required statistical accuracy. The *actual* relative error on p, at the 95% confidence level, that has been achieved after 1200 observations is:

$$E = \sqrt{\frac{4q}{pN}} = \sqrt{\frac{4 \times 0.68}{0.32 \times 1200}} = \sqrt{0.0071} \simeq 0.08$$

viz. the relative error on p is ±8%. This is well within the required value of 10% and no further measurements are needed.

If in stage 4, above, only 280 successes had been recorded after 1200 measurements then the new value of N would be $4 \times 0.767/0.233 \times 0.01 = 1317$, and at least 117 more measurements would have been needed to achieve the required accuracy (see below).

$$p = \frac{280}{1200} \quad \text{and} \quad q = 1 - \frac{280}{1200}$$

$$N = \frac{4q}{pE^2} = \frac{4 \times 0.767}{0.233 \times 0.01} = 1317$$

It can be seen that the total number (N) of particles, or points, that must be counted is inversely proportional to p, the proportion of the selected mineral, and it follows, therefore, that if p becomes very small, then N becomes very large. For example, if p is 1% (i.e. the proportion of p is 0.01) and the required relative error E on p is 10%, then

$$N = \frac{4q}{pE^2} = \frac{4 \times 0.99}{0.01 \times 0.01} = \frac{4 \times 99}{0.01} = 39\,600$$

When N is very large, it is often advisable to concentrate the mineral of interest (see Chapter 3) before carrying out the counting procedure. In this way, p is significantly increased and N is significantly reduced. Unfortunately, it is not always possible to achieve a suitable concentration of the selected mineral, and if that is the case the gross-counting method may have to be used instead of the individual particle- or point-counting method.

7.2. Statistics of gross counting

In the "gross counting" method (Chapter 6, section 5.4.3) the comparatively small number of particles (or points) of interest (y) that occur in N_F fields of view are counted. If the average number of particles, of all kinds, in each field of view is $\bar{x}$ then the proportion of the mineral of interest (p) is:

$$p = \frac{y}{N_F \bar{x}} \qquad \qquad \textbf{6.6}$$

i.e. the mineral proportion is the number of "successes" of that mineral, divided by the *total* number of observations. As before, with the conventional particle counting procedure, the relative error, E_1, on p is given by

$$E_1^2 = \frac{4q}{pN}$$

$$= \frac{4(1-p)}{pN_F \bar{x}} = \frac{4(1-p)}{y} \qquad \textbf{6.7}$$

However, if the value $y/N_F\bar{x}$ is to be used as an estimate of p then the error on $\bar{x}$ must also be considered. The value $\bar{x}$ is estimated by counting the *total* number of particles in each of N_c fields of view, so that

$$\bar{x} = \frac{1}{N_c}\sum_{i=1}^{N_c} x_1 \quad \text{and} \quad \sigma^2 = \frac{1}{N_c - 1}\sum_{i=1}^{N_c}(x_i - \bar{x})^2$$

The relative error on p is given by E_2 where:

$$E_2 = \frac{2\sigma}{\bar{x}\sqrt{N_c}} \quad \text{(with 95\% confidence)}$$

(where σ is the standard deviation). E_2 is independent of E_1, i.e. the number of particles in a field of view is not influenced by p, the mineral proportion. Therefore,

$$E_{\text{total}} = E = \sqrt{E_1^2 + E_2^2}$$

Therefore

$$E^2 = \frac{4(1 - p)}{y} + \frac{4\sigma^2}{\bar{x}^2 N_c}$$

and

$$y\left(E^2 - \frac{4\sigma^2}{\bar{x}^2 N_c}\right) = 4(1 - p)$$

Hence

$$y = \frac{4(1 - p)}{E^2 - \dfrac{4\sigma^2}{\bar{x} N_c}}$$

Strictly speaking, the statistical errors described in sections 7.1 and 7.2 apply only to the *number* proportions that are counted: these errors apply only indirectly to the volume proportions or the mass proportions which are calculated from those number proportions.

8. Conclusion

Practical mineralogists, especially those working in the mineral industries, require vast amounts of quantitative numerical mineralogical data. These data are usually needed in great haste since they are used for some or all of the following purposes: (1) assessing the value of mineral "prospects", (2) designing mineral treatment processes, and (3) controlling the operation of mineral treatment plants.

The traditional manual methods of providing the required data include image analysis of suitably prepared specimens which can establish proportions of the various minerals by counting particles or points and by measuring lines or areas.

The manual methods of measurement are too slow for many industrial uses and the mineralogist is, increasingly, being forced to rely on automated (and even fully automatic) methods of image analysis for the required information (see Chapter 7).

Chapter 7

Modern image analysers

1. General

The chemical, physical and optical methods traditionally used to measure some of the mineralogical characteristics of ores and plant products have been described in earlier chapters. These methods, when used together, can provide good, quantitative measurements from all kinds of mineralogical specimens. However, most of them are too slow to provide the large amounts of routine mineralogical data needed by the modern mineral industry. Furthermore, the traditional methods are often unable to provide the quality of information now being demanded.

As shown in Chapter 6, manual image-analysing methods can be used to produce some of the required data but these methods demand constant care and dedicated involvement from a trained mineralogist. They are, therefore, slow and expensive. However, many of the basic principles used in the manual image-analysing methods can be adopted for use in fast, modern, *automatic* systems. In particular, these new devices can readily (and rapidly) measure values such as mineral proportions, grain-size distributions and particle compositions, which are routinely needed by the industrial mineralogists.

2. Modern methods of image analysis

All the modern image-analysing systems analyse two-dimensional images which have been derived from sections cut through three-dimensional specimens. These images can be formed:

(1) by using the human eye;
(2) with the aid of an optical microscope;
(3) in an electron microscope;
(4) in an electron-probe X-ray microanalyser;
(5) with a television camera.

Images of suitably prepared sections are then automatically measured in ways that are directly analogous to those which are used during the manual analysis of similar images. First, the mineral of interest is distinguished from the accompanying phases by the most appropriate means (in some instruments the mineral of interest is distinguished by its optical brightness, in others it is distinguished by its chemical composition). A pre-determined set of measurements is then carried out; these may be point analyses, linear measurements, area measurements and so on. The measurements are "filtered" and "adjusted" so as to eliminate redundant and ambiguous data before they are electronically recorded by computers, usually on to disc or magnetic tape. Finally, the filtered data are used to calculate the analytical results and to deduce statistical parameters such as, for example, the mean sizes of mineral grains or specific surface area values.

2.1. Selecting an automated image-analysing system

There is no doubt that with many complex materials adequate discrimination between minerals can best be provided by an experienced, conscientious mineralogist but, unfortunately, such a person is often very difficult to find and, in any case, is usually too slow to provide the large amounts of routine mineralogical data required by the mineral engineer for process control purposes in an operating mineral processing plant.

Any automatic image-analysing system must, therefore, be able to discriminate a selected mineral from its associated minerals and it must do this swiftly and without ambiguity. Furthermore, when a particulate specimen is being studied, the analysing system must also be able to distinguish each mineral from the mounting medium. This ability to provide unambiguous phase discrimination is a primary, vital requirement and without it no automatic system of measurement (nor, for that matter, any manual system) can function properly or produce accurate results.

2.2.1. The ideal image analyser
The ideal image-analysing system would be able to produce area analyses, line analyses, point analyses and particle counts, as desired. It would also be able to produce accurate information about each of any number of selected minerals within a specimen and it would be capable of doing this almost instantaneously.

This ideal system would also have adequate spatial resolution for the work in hand and it would be able to measure mineralogical features of a wide size range within a single image – features from, say, a centimetre to a micrometre in size. The procedures needed to prepare specimens for this analyser would be simple and quick to carry out; the prepared specimen would be viewed in air (rather than, say, in a vacuum) so that specimens could be changed quickly and easily. The ideal, automatic measuring instrument would also be robust, reliable and versatile; the capital cost of buying the instrument and the

running costs would be small; and it should be possible for unskilled personnel to maintain and operate the instrument (i.e. the presence of the mineralogist would not be essential to get the instrument to work or to keep it working).

Unfortunately, there is no such system – nor is there ever likely to be one. However, many of the recently developed automatic systems have at least some of the attributes mentioned above, and a few of the very latest systems show quite a number of the qualities demanded of the ideal system.

2.2. Comparison between manual and automatic systems

How does one select the measuring system needed for a specific task? How does one decide whether to use the point, linear or area method of measurement? Table 7.1 compares the merits (and also the disadvantages) of the two basic kinds of image-analysing systems – the manual and the automatic. The advantages and the drawbacks of using the point-, line-, area- and grain-measuring systems are given in section 3.

TABLE 7.1
Comparison of manual and automatic systems of image analysis

Manual	Automatic
Commonly restricted to grain-counting and/or point-counting procedures ($G_g + P_p$ only)	Can be used for grain counting and for point counting: in addition, can be used for line and area measuring (G_g, P_p L_l and A_a)
Tends to produce subjective measurements	Once set, an automatic analyser produces objective measurements
Tedious and slow: accuracy likely to vary from start to finish of a set of measurements because of operator fatigue	Fast and capable of making large numbers of observations – each having the same accuracy
Good for high-quality research work on complex materials where time is of secondary importance and the total number of specimens is small: not suitable for the routine measurement of numerous specimens	Excellent for routine work where accurate information is required speedily from large numbers of similar specimens: not suitable for measuring small numbers of complex specimens because of the long time that may be needed to determine the optimum measuring conditions
High running costs; low capital costs	High capital losts; low running costs

The quality of the information provided by a manual method of image analysis depends entirely on the skill, attention and dedication of an experienced mineralogist. On the other hand, in an automatic system of measurement the function of the mineralogist is partly, or even sometimes, almost totally, replaced by a computer-controlled measuring device. But, even the automatically produced data still depends, in large measure, on the way in which the measuring system is set up by the mineralogist, and on the nature of the software used in the instrument. However, the increasing availability of automated systems of measurement offer the industrial mineralogist several advantages, and can eliminate the tedium of making large numbers of repetitive (and, basically, very simple) measurements. The time and the energy saved by using these systems is being employed by the mineralogist for other, more highly skilled work.

3. Comparison between different types of automatic system

Table 7.2 lists the criteria that should be used by a mineralogist when deciding on the suitability of a particular automated image analyser for a specified measuring task. In brief, the automated

TABLE 7.2
Criteria for selecting an image analysing system

| Selection criterion | Image-analysing system: | | | |
	Automatic area measuring	Automatic line measuring	Semi-automatic	Manual
Signals used to discriminate minerals	Optical or electron-optical	Optical, electron-optical, or X-rays	Optical	Optical
Typical measuring device	TV system	Photodensitometer or microanalyser	Planimeter; light pen	Microscope
Specimen preparation requirements	Very demanding: specimen must be polished and flat	Very demanding for optical and electron-optical devices, less demanding for X-ray systems	Less demanding than for automatic measuring systems	Less demanding than for automatic or semi-automatic systems
Specimen-viewing conditions	In air for optical devices, in vacuum for electron-optical systems	In air for optical devices, in vacuum for electron-optical and X-ray systems	In air	In air
Quality of mineral discrimination	Fair	Fair to good	Excellent	Excellent
Speed at which information is collected	Very fast; about 1 second per raster	Fast; from 1 to 10 ms per observation	Slow; about 1 second per observation	Slower even than semi-automatic systems
Kind of information obtained	Zero-, one- and two-dimensional	Zero- or one-dimensional	Zero-, one- and two-dimensional	Zero-dimensional (usually)
Availability of stereological transforms	Not often available for two- to three-dimensional transformations	One- to three-dimensional sometimes available	One- to three-dimensional sometimes available	
Nature of material that should be measured	Simple materials, where optical discrimination is easy	More difficult materials, where optical discrimination is difficult	Complex materials, where mineral discrimination is very difficult	Single, very difficult specimens
Types of specimen that should be analysed	Large numbers of routine, simple specimens from operating plants	Large numbers of more difficult routine specimens from operating plants	Small numbers of difficult specimens	Individual complex materials (for research purposes)
Analysing costs per specimen	Small	Large	Larger still	Very large

systems should be used if the task involves the measurement of large numbers of comparatively simple, routine specimens; manual systems should only be used for examining small numbers of complex, non-routine materials.

The automated systems fall into two main, but overlapping, categories:

(1) systems that are designed primarily to measure areas, and

(2) systems designed, in the first instance, to measure lines.

All the area-measuring systems can also be used to measure lines, points and particles, whilst many of the line-measuring systems can be used to measure points and particles as well (but only a few of these systems can measure areas).

3.1. Area-measuring systems

All area-measuring systems are based on television scanning technology. Most of the commercially available area-measuring systems distinguish one mineral from another by the strengths of the *optical* signals obtained from carefully prepared polished – or thin – sections.

In these optical systems the specimen (or a preliminary optical image, such as a photograph or a photomicrograph of that specimen) is viewed through a television camera. The light reflected (or when necessary, the light transmitted) from the specimen produces a pattern of electric charges on the vidicon (or plumbicon) detection unit of the camera. These charges are proportional to the brightness at points in the image – i.e. a bright area on the specimen produces a large electric charge on the equivalent area of the vidicon (Fig. 7.1). The effective part of the vidicon is about 1 cm by 1 cm in size and the amount of charge on each of a large number of small areas on that vidicon is measured, in sequence, by a narrow electron beam which repeatedly scans its

back surface. The electron beam traces a series of horizontal lines across the vidicon and each succeeding line is moved "down" by a distance equal to the size of the electron beam (usually about 15 μm). In this way, all the charges on the whole of the vidicon are measured during each frame or *raster*. There can be up to 1000 lines per frame and up to 1000 contiguous small areas (called *picture-points* or *pixels*) along each line: a single frame thus consists of a dense array of up to a million contiguous pixels. The charge intensity at each pixel can be measured in about a microsecond and a complete frame is commonly measured in less than a second.

The charge values from each pixel are passed to a dedicated computer which carries out the following basic measurements:

(1) The brightness at each pixel: each mineral has a small range of brightness and these brightness values are used to discriminate the various minerals.

(2) The size of each mineral grain is calculated. A grain consists of a number of contiguous small areas that show the same small range of optical brightness. The *size* of such a feature can be defined in several ways (Fig. 7.2) but it is often taken to be the length of the longest chord across the feature in the direction of the television scan. The *area* of any grain can be established by counting the number of pixels within its boundary.

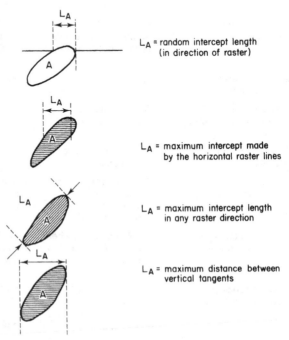

L_A = random intercept length (in direction of raster)

L_A = maximum intercept made by the horizontal raster lines

L_A = maximum intercept length in any raster direction

L_A = maximum distance between vertical tangents

Fig. 7.2 A variety of ways of defining the "size" of a feature measured by a television type analyser.

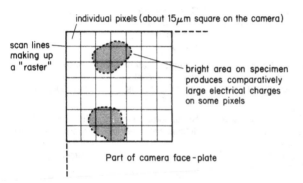

individual pixels (about 15μm square on the camera)

scan lines making up a "raster"

bright area on specimen produces comparatively large electrical charges on some pixels

Part of camera face-plate

Fig. 7.1 A section of television camera face-plate.

(3) The proportion of the total frame area that is occupied by each mineral is calculated; this is equivalent to the number of pixels occupied by all the grains of that mineral divided by the total number of pixels in the frame (cf. the point-counting method described in Chapter 6, section 5.3).

(4) The number of separate grains of a specified mineral that fall within a frame is calculated.

All television-type image analysers incorporate television monitors, on which the mineralogist can conveniently check the measurements that are being carried out by the instrument.

The spatial resolution of a television-type image analyser depends on the type and quality of the television camera and on the quality of the image which is presented to the camera. When the original image is produced by a good quality optical microscope the resolution is about 4 or 5 μm. It cannot be better because of the inevitable degradation of the image in even the most advanced television cameras.

The smaller the mineralogical features that need to be resolved, the larger the overall magnification that must be used in the measuring system. It follows that the smaller the mineral features, the smaller will be the effective frame area and the larger the number of frames which have to be measured in order to cover a given area of specimen. In addition, the higher the magnification, the smaller the depth of focus of the microscope and the greater the care needed to prepare a really flat, well-polished specimen.

Specimens measured in optically based television-type image analysers can be illuminated with transmitted and/or with reflected light. It is normal practice to use bright-field, polychromatic illumination but it is also permissible to use monochromatic, polarised or dark-field illumination. In fact, any combination of light sources can be used as long as it provides sufficient optical contrast for the features of an image to be unambiguously discriminated from one another.

However, even good specimen preparation procedures and careful selection of the optimum type of illumination will not always provide adequate optical discrimination of the minerals in a specimen. This poor discrimination may be due to the fact (see Chapter 4) that the optical characteristics of many minerals vary with their orientations and with their chemical compositions. Furthermore, the optical characteristics of many minerals overlap those of others (see section 3.2.1 below). The *automatic* discrimination of one mineral from another can also be affected by instru-

mental deficiencies, such as variations in sensitivity across the television camera, but these difficulties have now been reduced to very low levels by the instrument manufacturers.

Some of the mineral discrimination problems encountered with optically based automatic measuring systems can sometimes be overcome by using a limited amount of manual control (see below), but there are other occasions when optical discrimination is not possible (or at least is impractical). It may then be more appropriate to use automatic systems that rely on electron beams to "illuminate" the specimen (see section 3.2.2).

3.1.1. The Quantimet image analyser

This instrument is typical of a number of television-based, optical image analysers. It was developed in the early 1960s and was the first commercially available instrument of its kind. Figure 7.3 shows a general view of a modern (1986) model and Fig. 7.4 shows diagrammatically how the system works.

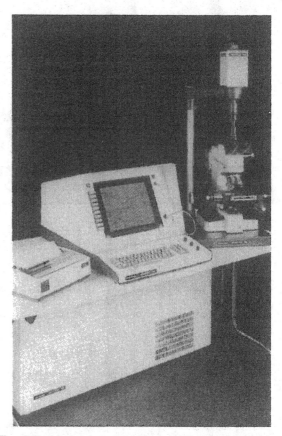

Fig. 7.3 Modern model of automatic television-screen image analyser. This shows the TV camera and microscope on the right, the monitor centre and printer on the left (courtesy of Cambridge Instruments).

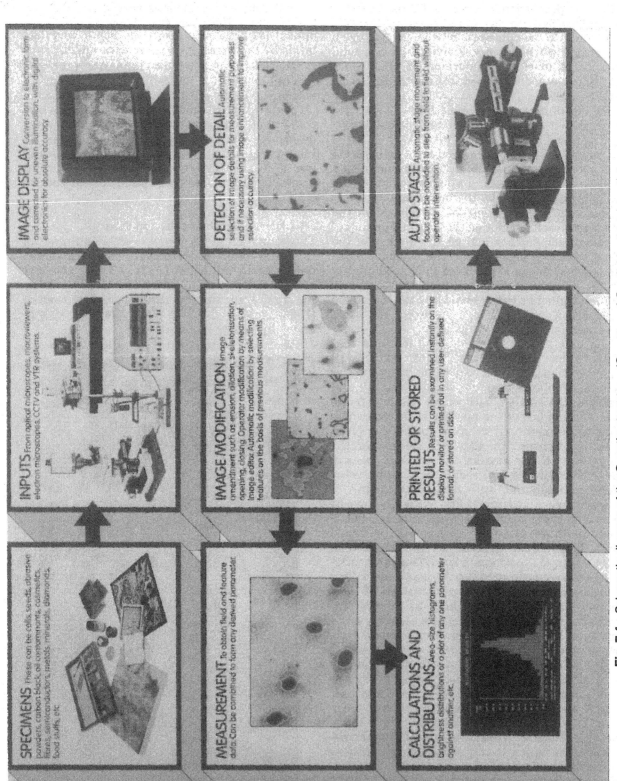

Fig. 7.4 Schematic diagram of the Quantimet system. (Courtesy of Cambridge Instruments.)

In the Quantimet, an image of a suitably prepared specimen is usually formed by the microscope and the television camera but if necessary an image can also be produced from a photomicrograph or from the back-scattered electron signals in a scanning electron microscope (see Chapter 9, section 10.2). The image is subdivided into a large number of pixels and each is assigned a value which depends on its brightness (called its *grey-level*). These grey-level values are temporarily stored in the control computer and they are used for making sets of measurements – either on a complete raster or on specified features within that raster. The results are usually presented in tabular form on a computer-operated printing device (see Table 7.3).

Whenever an area-measuring system like the Quantimet is used, there is always a chance that some interesting mineralogical feature will be cut by the boundary of the television raster. When this happens, its "size" may be incorrectly measured; this difficulty has been largely overcome by making the measured raster smaller than normal and by using a *guard frame* (Fig. 7.5) around that original, measured raster. Features which overlap on to the guard frame can then be measured in

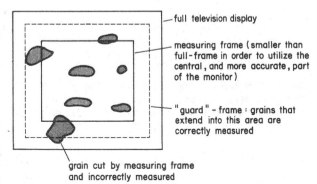

Fig. 7.5 Quantimet-type image analyser with guard frame.

TABLE 7.3

Presentation of measured data: a typical result of a linear analysis of a rock specimen consisting of a host mineral (A) enclosing very fine-grained minerals (B, C and D) (the measurements were carried out on a specially adapted Camebax electron probe X-ray microanalyser – see Chapter 9, section 9)

Intercept length (μm)	Number of intercepts on:			
	Mineral A	Mineral B	Mineral C	Mineral D
0–5	965	39	1247	30
5–10	262	8	332	3
10–15	59	1	39	—
15–20	48	—	15	—
20–25	24	—	3	1
25–30	28	—	2	—
30–35	20	—	—	—
35–40	21	—	1	—
40–45	17	—	—	—
45–50	10	—	—	—
50–55	9	—	—	—
55–60	14	—	—	—
60–65	9	—	—	—
65–70	9	—	—	—
70–75	5	—	—	—
75–80	10	—	—	—
>80	212	—	—	—
Total number of measured grains	1722	48	1639	34
Total traverse length on each mineral (μm)	193 212	172	6494	122
Mean intercept length on each mineral (μm)	112	4	4	4
Linear proportion (and also volumetric proportion) of each mineral (%)	96.6	0.1	3.2	0.1

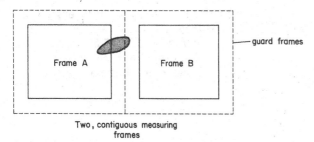

Two, contiguous measuring frames

The feature is measured in <u>both</u> guard frames and provision must be made to "remember" this fact

Fig. 7.6 Problem encountered when the size of the measured feature approaches that of the frame. The feature is measured in both guard frames and provision must be made to remember this fact.

full. However, when two or more contiguous frames are measured (Fig. 7.6) their guard frames will overlap, and provision must be made for the control computer to "remember" details of those features that occur in the overlap area. If necessary, the whole of a specimen surface can be analysed by measuring a complete set of contiguous frames. On the other hand, it is more usual for values such as the proportions of the various minerals in a specimen to be determined by measuring discrete, unrelated small areas (or frames).

The following sequence of operations must be carried out when an automatic image analysis is performed on the Quantimet:

(1) The specimen is polished to an "optical" finish: it should be flat so that no re-focusing need be carried out as the specimen is moved beneath the light source (some instruments now have automatic focusing arrangements which can accommodate slight deviations from flatness).

(2) The instrument and the light source are carefully set to obtain the best measuring conditions and the best degree of mineral discrimination.

(3) The appropriate computer program is selected so that the instrument automatically performs a set of measurements and produces the information in readily understood form.

(4) The results obtained must be analysed and assessed in terms of the appropriate mineralogical context. This can only be done by the mineralogist or the mineral engineer (no machine can yet do this). Only a skilled human being can relate the analytical results to the mineral treatment process, or to the rock sample, from which they were obtained.

3.1.2. Semi-automatic area-measuring systems

For many purposes it may be necessary or advantageous to use a measuring system which combines the best features of the highly discriminatory (but exceedingly tedious) manual methods of analysis, with the high-speed performance of an automatic method which may not provide an adequate degree of mineral discrimination. In these semi-automatic systems the minerals or other features of interest are *discriminated* by the mineralogist but they are *measured* automatically. In this way, the skill and the experience of the mineralogist can be successfully coupled with the speed and objectivity of the automatic measuring instrument.

In the semi-automatic system, minerals that differ only very slightly in optical characteristics can, nevertheless, be distinguished by the mineralogist on the basis of such characteristics as colour, refractive index, birefringence, polarisation effects; twinning, etc. (see Chapter 4, section 5). In fact, *any* property (or any combination of properties) may be used to differentiate one mineral from another: even the presence of polishing artefacts may be useful in this respect. Figure 7.7 shows how galena can almost invariably be distinguished by the triangular etch pits which are inevitably produced during the polishing procedure. Furthermore, the mineralogist can sometimes improve mineral discrimination by ignoring artefacts such as metal fragments derived from pumps or sieves, air bubbles trapped in epoxy mounts and so forth.

When using a semi-automatic measuring system the mineralogist first indicates the feature of interest by pointing at it, or by inscribing around it, with an electronic pen or computer "mouse" (Fig. 7.8). The instrument is then instructed to carry out a series of measurements on that feature and to record the results. Another feature is then pointed out by the mineralogist and the measuring and recording operations are repeated until a sufficient number of measurements have been carried out.

As with all area-measuring methods, the semi-automatic systems can, in theory, be used to measure the whole of a specimen surface. In this way, no sampling errors need to be introduced during the measurement of a specimen.

The semi-automatic systems are comparatively cheap to buy and to run. They are particularly useful in research establishments where they can be used to study complex materials in which it is difficult to differentiate the various features. The speed of such an analysis is of only secondary importance. However, all the semi-automatic systems demand the close and constant attention of a

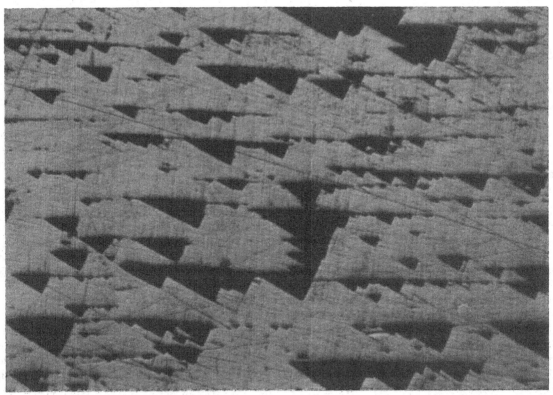

Fig. 7.7 Photomicrograph of a badly polished galena grain showing the distinctive triangular polishing pits.

Fig. 7.8 Electronic pen (or electronic "mouse") to indicate the outlines of a mineral grain.

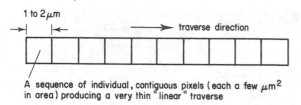

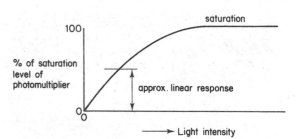

A sequence of individual, contiguous pixels (each a few μm^2 in area) producing a very thin "linear" traverse

Fig. 7.9 Basic principle of linear analysis as carried out on television type analyser.

Fig. 7.11 Typical response of a photomultiplier unit to light beams of various intensities.

skilled operator throughout a potentially protracted measuring procedure.

3.2. Automatic line-measuring systems

In an automatic line-measuring system the measurements are carried out along widely spaced, parallel traverse lines, and these lines are themselves made up of a large number of contiguous "points" (Fig. 7.9). In practice, the specimen is mechanically moved along an S-shaped path whilst it is illuminated by a small beam of light or whilst being irradiated by a focused beam of electrons. No attempt is made to measure *all* the specimen surface but every effort is made to ensure that the measuring grid adequately *samples* the whole surface.

The linear sampling grid should be arranged so that the spacing between adjacent lines is greater than the size of the largest feature that will be measured in the specimen. In this way, no feature will be measured more than once during a traverse and the randomness of the linear measurements will be preserved. Unfortunately, the size of the largest feature may not be known at the beginning of an experiment – in fact, the object of the study may be to establish this value.

In such circumstances, it is prudent to adopt a large line spacing at the start of an investigation: if the initial, coarse spacing fails to reveal any large features in the specimen then the original grid can be in-filled with a second (and, if necessary, a third and fourth) set of parallel lines (Fig. 7.10).

3.2.1. Optically based line-measuring systems

These systems use narrow beams of light to illuminate a flat, polished specimen. The amount of light reflected (and/or transmitted) from a "point" on this specimen is measured by a light cell or photomultiplier device. Variations in "brightness" generally indicate changes in mineral content at the contiguous, small measured areas, but may also be due to artefacts introduced during the specimen preparation procedures.

A good photomultiplier can distinguish a range of light intensity levels between zero (complete darkness) and the very bright saturation level of the device (Fig. 7.11). Despite this wide range of response an optical line-measuring system is rarely able to distinguish more than five or six minerals within a single specimen. This is mainly because the combinations of minerals generally found in mineral specimens do not cover the whole range of response available from a photomultiplier (Fig. 7.12). In addition, the natural

Initial widely-spaced traverse to determine approximate sizes of the mineralogical features

Subsequent, correctly-spaced traverse for features of maximum size of L_A

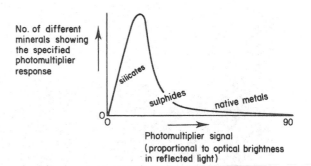

Fig. 7.10 Spacing of traverse lines on a specimen surface.

Fig. 7.12 Illustration of the "bunching" effect of minerals over a small range of optical brightnesses.

(and expected) spread in the optical properties of a single mineral may add to the difficulties of discrimination and these difficulties can be further compounded by variations in optical responses from a mineral due to variable specimen preparation effects.

The speed of a linear analyser (whether it is based on the use of optical signals or on X-rays – see below) is always limited by the need to make accurately controlled, mechanical movements between each measurement, and the normal working speed of such an instrument is about 100 measured "points" per second (cf. area-measuring systems where the speed can be up to 10^6 pixels per second).

The spatial resolution of an optically based line-measuring system is limited by the effective size of the light beam (usually, no better than 3 to 5 μm) or by the size of the "step" taken between each measured "point" (commonly 3 to 5 μm).

3.2.2. Non-optical line-measuring systems

Some line-measuring systems do not use light beams to illuminate the specimens. Instead, they

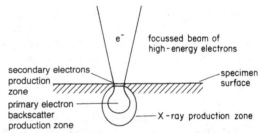

Fig. 7.13 Signals produced when a mineral specimen is bombarded by a beam of high-energy electrons. (All individual signals and combinations thereof are available for mineral discrimination purposes.)

differentiate one mineral from another by using combinations of the signals produced when a specimen is bombarded by a focused beam of high-energy electrons (Fig. 7.13). Different minerals produce different X-rays under these conditions (see Chapter 9, section 2). The X-rays can be measured in a number of separate crystal spectrometers, or they can all be measured virtually simultaneously by using solid-state, energy-

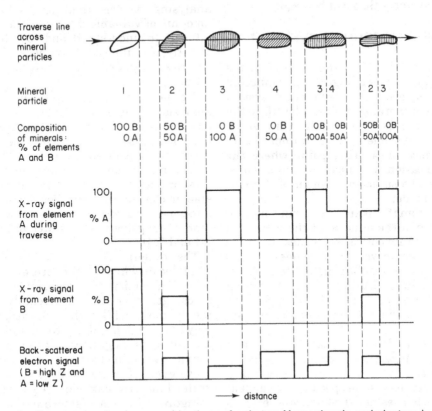

Fig. 7.14 Schematic diagram showing how combinations of only two X-ray signals and electron back-scatter signal can discriminate various minerals in a specimen during a linear traverse. The mineral particles are embedded in an epoxy matrix which produces no detectable X-ray signals and very weak electron signals.

dispersive X-ray detectors. The measured values are then used to discriminate one mineral from another, and good discrimination is often feasible, even in complex specimens. This good discrimination is possible because these instruments do not rely on a *single* signal (as in the optical devices) but use a *number* of largely independent, but simultaneous, X-ray signals (Fig. 7.14). However, the X-ray-based measuring devices are comparatively slow and counting times of 10 milliseconds or more are needed at each measured point to collect statistically useful numbers of X-ray quanta. The spatial resolution achieved by an electron-beam instrument of this kind is limited by the X-ray generation geometry (see Fig. 7.13) and is normally about 2 μm. Better resolution is possible by using the electron responses derived from the specimen (see Chapter 9, section 9).

Electron-beam instruments are generally much more expensive than optical instruments and, furthermore, they only operate when the specimen is held under vacuum. However, for many combinations of minerals (e.g. mixed sulphides or mixed silicates) these electron-beam systems are the only ones that can discriminate the various minerals, and they are therefore the only automatic measuring systems that can be used.

3.3. Point-measuring systems

In a point-measuring system the measurements are carried out at a series of regular, unconnected points. In one modern, partly automated system of analysis the mineralogist first identifies the mineral and records the result through a keyboard. The specimen then automatically moves to the next measuring point where the manual identification and recording procedures are repeated until a sufficient number of observations has been made.

In a fully automatic point-measuring system the identification of the mineral at the measured "point" is based on its optical brightness or on its X-ray response (see above). The specimen is then moved by computer-controlled motors to the next point, where the identification and the recording procedure is repeated. The cycle of operations continues until sufficient observations have been made.

Particle-count analyses are carried out in a similar manner. A beam of light (or a beam of electrons) locates a particle whilst traversing a pre-arranged path across the specimen. The identity of the particle is established and the result is recorded. The traverse is then continued until another particle is located, and so on until an adequate number of particles has been measured.

4. Preparation of specimens for automatic image analysis

It is often difficult for even the skilled mineralogist to distinguish between a true mineralogical feature (such as a pore) and a similar feature, such as a "polishing pit" which has been produced by the specimen-preparation procedures (cf. Fig. 7.7). It is not surprising, therefore, that differentiation of such features is even more difficult for an automatic image analysing system. Consequently, great care must be taken to ensure that polishing artefacts and other artificial features are kept to a minimum and this is only possible by the skilful use of good specimen-preparation methods (Practical no 9).

5. Presenting the data obtained by image analysis

The most exact way of presenting the results of any set of measurements is to provide the raw data in their measured form. In this way no information is lost. However, automatic image-analysing devices tend to produce such vast amounts of valuable data that they often become difficult to understand and to interpret. Consequently, it is almost always necessary to condense the raw, measured data into more manageable form.

A typical set of results from an automatic linear analysis is given in Table 7.3. The intercept lengths are tabulated in arithmetic categories but they can also, if necessary, be tabulated in geometric categories (Table 7.4). These tabulated data are sometimes further condensed so as to present the important results in a more acceptable or convenient form (for example, the calculated mean intercept length values may be presented alongside the tabulated data, as in Tables 7.3 and 7.4).

The mineralogist, or the mineral engineer, must decide which form of presentation should be used in a particular circumstance – the arithmetic categories are the more readily understood by the layman, whilst the logarithmic categories may be used to highlight the details that occur at the fine sizes. Logarithmic categories can also be useful in bringing to the notice of the engineer the prevalence of small numbers of very coarse-grained mineral particles (see Table 7.4).

Figure 7.15 is a histogram showing the data given in Table 7.3 and Fig. 7.16 is another plot showing the same data as Table 7.4. These two methods of presentation show, in an easily under-

TABLE 7.4

Same data as given in Table 7.3 but the intercept length categories follow a geometrical progression

Intercept length (μm)	Number of intercepts on:			
	Mineral A	Mineral B	Mineral C	Mineral D
2	665	26	883	22
2–4	300	13	364	8
4–6	133	4	194	1
6–10	129	4	138	2
10–16	86	1	46	—
16–25	45	—	11	1
25–40	69	—	3	—
40–63	55	—	—	—
63–100	42	—	—	—
100–160	39	—	—	—
160–250	25	—	—	—
250–400	28	—	—	—
400–630	28	—	—	—
630–1000	16	—	—	—
1000–1600	24	—	—	—
1600–2500	22	—	—	—
>2500	15	—	—	—
Total number of measured grains	1722	48	1639	34
Total traverse length (μm)	193 212	172	6494	122
Mean intercept length (μm)	112	4	4	4

Table 7.4 has the same number of size categories as Table 7.3 but Table 7.4 provides more details regarding both the largest and smallest intercepts.

stood, visual way, the "shapes" of the size distribution of the mineral grain in the measured specimen. Plots of this kind readily show whether a measured distribution is skewed, flat, double-peaked and so on. Histograms such as Fig. 7.15, where the size categories are of uniform range, can be used to show the *mode* of the distribution (i.e. the most commonly occurring value in the distribution): such a histogram can be very useful for the engineer and size distribution data may also have petrogenetic significance.

Sometimes histograms are, without justification, "smoothed out", i.e. a line is used to join the mid-points of the tops of all the histogram

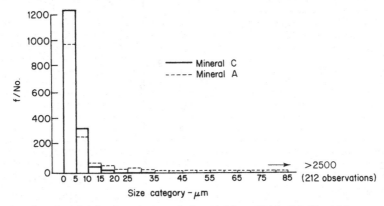

Fig. 7.15 Histogram showing size distribution of minerals A and C from Table 7.3. The histogram cannot readily present all the data without becoming unwieldy allowing space for each category. Furthermore, only two of the minerals can be shown conveniently.

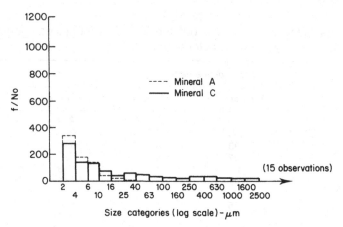

Fig. 7.16 Data from Table 7.4 plotted on a logarithmic scale (see Fig. 7.15 where same data are plotted in uniform arithmetic size categories). All sizes up to 2500 μm can be shown easily.

columns – such a line suggests to the mathematician that the measured values follow a *continuous* distribution function whereas, in practice, all measured mineralogical distributions are discontinuous.

Histograms are not always the most convenient ways of presenting large amounts of data (try plotting all the data from Table 7.3 on to a single pair of axes so that the size distributions of the different minerals can be compared). Some improvement in clarity can be achieved by using histograms of different colours for each mineral. However, it is often more advantageous to re-plot the data in *cumulative* form, where the cumulative fraction greater than (or smaller than) a stated size is plotted for a range of sizes (Fig. 7.17). In Figure 7.17 the cumulative size distribu-

tions of three of the minerals from Table 7.3 have been plotted on to a single pair of arithmetic axes. Figure 7.18 shows the same raw data plotted on to logarithmic axes. The first differential of a cumulative distribution curve is sometimes used to produce a "pseudo-continuous" distribution curve.

Although the "cumulative" curves are very useful, they tend to smooth out small variations in the measured data. This smoothing effect may not matter when the variations are due to small experimental errors but it can matter a great deal if the variations are real characteristics of the measured specimens. The use of logarithmic

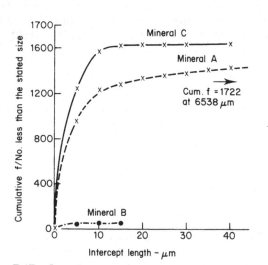

Fig. 7.17 Cumulative size distributions of three of the minerals from Table 7.4 (mineral D would closely follow the curve from mineral B).

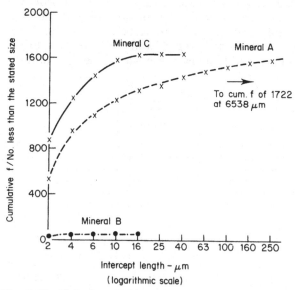

Fig. 7.18 Data from Table 7.4 plotted using cumulative frequency values and a logarithmic scale for the intercept lengths.

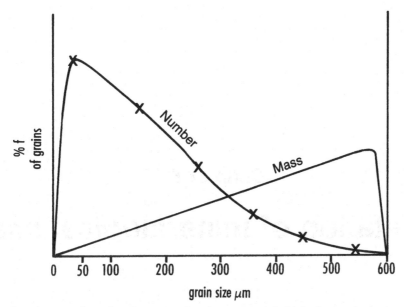

Fig. 7.19 The measured number frequency has been converted into the equivalent mass frequency.

scales tends to smooth out variable data even further. However, it is the mineralogist's responsibility to ensure that data manipulation (of any kind) does not conceal important features of the experimental results.

It is sometimes necessary to plot mass distribution values; at other times the engineer may want to plot numerical or volumetric data. The two curves in Fig. 7.19 show the same basic information: one curve shows the calculated mass frequency distribution whilst the other shows the original numerical frequency distribution. The number frequency curve might suggest that the analysed material was very fine-grained but the mass frequency curve shows that the material is, in fact, very coarse-grained, i.e. the bulk of the tin is in the form of coarse-grained cassiterite. The point to be remembered is that there is no purpose in drawing a graph unless it aids the interpretation of the data.

Even with the use of tables and graphs it is often difficult to make meaningful, quantitative comparisons between two (or more) sets of results. Many mineralogists, therefore, characterise sets of data by using distributional parameters such as the mean, the standard deviation, the mode and the median. The *mean* is the average value derived from all the measurements, the *standard deviation* is a measure of the dispersion (or spread) of the observed values, the *mode* is the measured arithmetic category with the highest frequency of values, and the *median* is the middle value when all the experimental values are set out in ascending (or descending) order.

6. Conclusion

Modern image analysers tend to be fully automatic in operation. Mineralogical images are especially difficult to analyse because of the problems involved in unambiguously differentiating one mineral from another and in excluding the artefacts introduced during specimen preparation.

Many automatic measuring systems are based on television technology and these are used to make area measurements of mineralogical features. Other systems, which make linear measurements, rely on accurate mechanical movement to move a specimen under a stationary optical or electron beam.

The choice of analyser for a particular task depends on the combinations of minerals that must be measured, the speed of analysis that the mineralogist requires, and the amount of money available for purchasing equipment. Maximum *speed* is provided by optically based, television-type image analysers; maximum automatic *discrimination* is provided by the comparatively more expensive, and slower, X-ray-based systems.

Chapter 8

Interpretation of mineralogical images

1. Introduction

Mineralogists and mineral engineers frequently derive much of their information from the examination and measurement of two-dimensional mineralogical images. For example, polished sections are often examined in order to establish the proportions of the mineralogical features which occur within blocks of ore and within populations of mineral particles. However, it often happens that the features seen on polished surfaces do not accurately represent the three-dimensional reality of the material that has been sectioned.

The study of the relationships between the features measured on two-dimensional sections and the real features that they represent in three-dimensional bulk materials is called *stereology*.

2. Stereology

Some of the mineralogical features in a rock, or in an ore, are three-dimensional in character; for example, a mineral grain has a volume, and is, therefore, a three-dimensional feature. Other features within a rock may have only zero, one or two dimensions. For example, four, or more, grains meet at a zero-dimensional *point* within a specimen; the linear junction where three mineral grains meet is a one-dimensional *line*, and the boundary between any two mineral grains is a two-dimensional *plane*.

Some of the mineralogical features known to exist in a three-dimensional specimen cannot be represented on any two-dimensional section. This is because a section cut through a feature will always produce an image which has one less dimension than that original feature. Thus, if a three-dimensional mineral grain is sectioned, the section can only show a two-dimensional representation of that grain; similarly, the boundary plane between two mineral grains is always represented on the section by a line. In the same way,

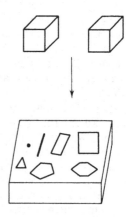

Fig. 8.1 Diagram illustrating the various two-dimensional shapes produced when randomly positioned and orientated cubes of uniform size are cut by a plane section. The section shows points, lines, triangles and 4-, 5- and 6-sided sections.

a linear feature within a specimen will be represented on the section by a point, and a zero-dimensional feature in the specimen will not show up at all on the section.

Because of this loss of one dimension during sectioning, a section always provides distorted information about certain important mineralogical features. Thus, if a rock contains uniformly sized, randomly orientated cubes of, say, galena then a section through this rock will show the galena as a variety of two-dimensional shapes and sizes (Fig. 8.1). In fact, the galena cubes can appear on the section as points, lines, squares, rectangles, pentagons and hexagons, and these two-dimensional shapes can appear in various sizes. Similarly, uniformly sized spherical bodies, such as the pores in metallurgical slags, will appear on a section as circles of different diameters (Fig. 8.2), while particles that are of uniform size, uniform shape and uniform mineralogical composition will appear on a random section in a variety of sizes, shapes and compositions (see section 2.3 of this chapter and Fig. 8.3, below).

As shown in Chapters 6 and 7, mineralogists use area-, line- and point-measuring systems for analysing rock and mineral sections. None of these methods can determine *all* of the structural detail that exists in the original three-dimensional specimen. In general, the lower the number of dimensions measured by the measuring system, the less information can be obtained by it. For example, area measurement can provide information on the shape, size and location of

a mineral grain, even though both the shape and the size information will be biased. A line-measuring system can also provide information about the size, the *mean* shape and location of a mineral grain – but again, the size information will be biased; a point analysis can provide information only about the location of the grain – it can tell us nothing about its shape or its size.

Stereological methods of data analysis can sometimes be used to transform the measured, low-dimensional data (obtained by area, linear or point analyses) in order to obtain statistical estimates of their equivalent higher-dimensional values (see below). All such stereological transformations are statistical in character and, require large amounts of accurate data of the kind that can only be produced by automatic methods of mineralogical analysis (Chapter 7).

It is obvious, therefore, that stereological transformations require that the measured features occur in large numbers within a specimen. It is fortunate that most mineralogical materials contain sufficiently large numbers of suitable features; others may not. A small hand-specimen of a fine-grained copper ore may contain tens of thousands, or even hundreds of thousands, of copper-bearing mineral grains: any standard-size polished section, say 2.5 cm in diameter, taken through that specimen is likely to expose many hundreds of these grains. On the other hand, a random, standard-size section through a typical gold ore seldom shows even a few gold grains; it is then necessary to analyse a number of sections to produce sufficient data for stereological analysis.

For example, if a pyritic copper ore contains 1% copper by weight, and if all this copper occurs as the mineral chalcopyrite, then the ore will contain roughly 3% chalcopyrite by volume. If this mineral occurs as uniformly distributed and uniformly sized cubic grains, 40 μm in size, then the ore will, on average, contain almost half a million chalcopyrite grains per cubic centimetre: thus:

Volume of ore specimen = $(10^4)^3 \, \mu\text{m}^3 = 10^{12} \, \mu\text{m}^3$
Proportion, by volume, of chalcopyrite = 3% = 3×10^{-2}
Volume of single chalcopyrite cube = $40 \times 40 \times 40 \, \mu\text{m}^3 = 6.4 \times 10^4 \, \mu\text{m}^3$
Total number of chalcopyrite cubes per cubic centimetre of ore = $10^{12} \times 3 \times 10^{-2}/6.4 \times 10^4 = 468\,750$

Any section cut through that cubic centimetre of rock is likely to expose a large number of those chalcopyrite grains.

On the other hand, a gold ore may contain only 1 part per million (by volume) of native gold (this

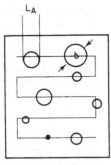

L_A

Section through uniformly sized, randomly distributed spheres.

(L_A = random linear intercept)

Fig. 8.2 Diagram illustrating the variously sized circles produced when randomly positioned spheres of uniform size are cut by a random plane section. The diameters of the circles will vary from *d* (the diameter) to zero. (Note that the diameters are also larger, on average, than the random intercept lengths.)

is roughly equivalent to 10 g t^{-1}). If all this gold is uniformly distributed as 40 μm cubes, how many cubes, on average, will there be in 1 cubic centimetre of ore?

Volume of ore = $10^{12} \mu m^3$
Proportion of gold = 10^{-6} (by volume)
Volume of single gold cube = $6.4 \times 10^4 \mu m^3$
Total number of gold cubes per cubic centimetre of ore = $10^{12} \times 10^{-6}/6.4 \times 10^4 \simeq 16$

A random section cut through a cubic centimetre of this gold ore is very unlikely to expose more than one or at most two grains of the gold.

Table 8.1 gives a list of the features within a mineralogical specimen that can be measured during an image analysis. Some of the stereological relationships between the measured values and three-dimensional reality are quite simple. More commonly, however, these relationships are complex (as shown below).

2.1. Modal analysis

Modal analysis is the name given to the determination of the volumetric proportions of minerals in a specimen. These volumetric proportions can be estimated readily from measurements made on randomly sectioned surfaces by area, line or point

TABLE 8.1
Real features that can be measured by various analytical procedures

Analytical procedure	Number of dimensions measured by the analytical procedure (D_m)	Number of dimensions of the real feature (D_f)	$D_m + D_f$
Point analysis	0	0 (points)	0
	0	1 (lines)	1
	0	2 (areas)	2
	0	3 (volumes)	3
Linear analysis	1	0	1
	1	1	2
	1	2	3
	1	3	4
Areal analysis	2	0	2
	2	1	3
	2	2	4
	2	3	5
Volumetric analysis (by serial sectioning)	3	0	3
	3	1	4
	3	2	5
	3	3	6

Only where the value of $D_m + D_f > 2$ can a feature be measured by image analysis.

analysis. Thus (as shown in Chapter 6, section 5),

$$V_v = A_a = L_l = P_p$$

where V_v is the volumetric proportion of the selected phase in the three-dimensional specimen,

A_a is the area proportion of that phase on the measured section,

L_l is the linear proportion of the required phase along an extended linear traverse across the section, and

P_p is the point proportion of the phase found at the intersections of a closely spaced grid on the section surface.

The measured values (A_a, L_l or P_p) provide *unbiased* estimates of V_v, irrespective of the spatial distribution of the measured phase in the specimen.

2.2. Grain-size distribution analysis

If mineral grains are randomly dispersed throughout a bulk specimen, then sections cut through that specimen will show random profiles through some of those grains. In addition, random lines drawn on the section will produce random linear intercept lengths across some of the grains. (Random points on the specimen surface can provide no useful information regarding grain-size distributions and will not be mentioned further in this context.)

The measured size distributions of the profiles (and linear intercepts) *always* provide *biased* estimates of the true, three-dimensional sizes of the grains, since the measurements invariably *overestimate* the proportion of small grains in the specimen. The magnitude of the bias is greater for random intercepts than for random profiles (see Fig. 8.3) and the amount of bias also depends on the shapes of the measured grains. These biases can be removed by stereological methods if the

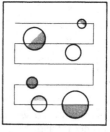

Fig. 8.3 Diagram illustrating the bias introduced by sectioning randomly orientated and positioned uniformly sized spheres, each of which contains 50% by volume of mineral A and 50% of mineral B.

TABLE 8.2

Transformation of random linear intercept measurements into three-dimensional values (for pores in metallurgical slag)

Random intercept length through circles seen on a two-dimensional section (μm)	Frequency	Calculated spherical diameters of the pores in the bulk specimen:	
		% by number	% by volume
2	5	60.00	0.0
2–4	6	28.00	0.1
4–6	3	5.75	0.1
6–10	5	1.64	0.1
10–16	9	2.45	0.4
16–25	10	0.44	0.3
25–40	21	0.84	2.4
40–63	26	0.47	5.2
63–100	31	0.30	13.3
100–160	22	0.08	14.3
160–250	16	0.29	17.9
250–400	11	0.01	21.9
400–630	4	0.00	12.1
630–1000	2	0.00	11.8
Totals	171	100.00	99.9
Mean size (m)	109		368

The stereological method that was used to make this transformation of data obtained from one-dimensional measurements into three-dimensional reality assumes that the very largest random intercept was derived from a sphere of the same diameter. In addition, a known proportion of the smaller intercepts were known to have been derived from the larger spheres. This method is often called the Saltykov method, after the man who first used it.

Note that the probability of the traverse cutting through a pore varies with the projected area of that pore.

grains are of simple regular shapes (cubes, spheres, parallelepipeds) (Table 8.2). Otherwise, stereological transformation is not possible and great care is needed to take into account the likely effects of the biases when interpreting or using grain-size data which have been derived from sectioned specimens.

2.3. Particle shape determination

If it were possible to measure a large number of random profiles (or random lines) through a single particle then the size distribution of the profiles (or lines) would uniquely define the shape of that particle. Similarly, if a population of particles of uniform size and shape were mounted, at random, in a matrix then the profiles (or lines) seen on a section surface would provide an estimate of the average shape of those particles.

Table 8.3 shows the length distribution of a set of random intercepts through a specimen containing closely screened mineral particles, all of which are of similar, but not identical, shape. The mean value of this intercept length distribution is, as expected, much smaller than the known

screen size of the particles. The presence of a number of very small intercepts indicates that the particles are angular (a zero frequency of zero size would show that the particles had no edges but were smooth and convex), and the length of the longest intercept shows the degree of elongation of the particles. In fact, the size distribution of intercept lengths provides a mathematical description of the particle shape.

2.4. Mineral associations

It is often important, from the mineral engineer's point of view, to establish the spatial relationships between two or more minerals. For example, when designing a tin-recovery operation it is necessary to determine whether the tin-rich mineral (usually cassiterite) is associated in the ore with, say, silicate minerals or sulphides: if the cassiterite is found only with the silicates then it becomes useful to establish whether it is associated with some *particular* silicate mineral.

These mineral associations can be quantified by using a parameter called the *proximity index* which can be calculated from the results of a

TABLE 8.3

Length distribution of linear probes through a population of randomly distributed monosize mineral particles

Intercept length (μm)	Frequency
2	346
2–4	86
4–6	137
6–10	169
10–16	174
16–25	150
25–40	177
40–63	183
63–100	192
100–160	47
160–250	1
Total	1662
Mean intercept length (μm)	26

The measurements were carried out on uniformly sized grains of cassiterite with a nominal screen size of 97.5 μm (i.e. all the particles passed through a 100 μm sieve and all were retained on a 95 μm sieve).

The large numbers of very small intercepts suggest that the grains are angular; the number of intercepts that are larger than 100 μm in size indicate that the particles are not equi-axial in shape but are significantly elongated. The parameters of the linear size distribution (for example, its moments) can be used to characterise the mean shape of the cassiterite particles. Note also that the mean intercept length is much shorter than the nominal screen size.

TABLE 8.4

Specific surface area values (from linear measurements) calculated per unit volume of the minerals in the left-hand column (μm^2 per μm^3)

Minerals	A	B	C	Total
A	—	0.0390	0.0084	0.0474
B	0.0873	—	0.0094	0.0967
C	0.1577	0.0789	—	0.2366
				0.3807

Mineral A represents the combined silicate minerals in a pyritic Cornish tin ore and mineral B the combined sulphide minerals, whilst mineral C is the tin-bearing cassiterite. The high specific surface area of the cassiterite shows that it is much finer-grained than the other two groups of minerals.

where $S_{V(A)}$ = specific surface area of mineral A,
$S_{(A)}$ = total surface of A,
$V_{(A)}$ = volume of A, and
$\overline{L}_{(A)}$ = mean intercept length on mineral A.

Table 8.4 gives some typical surface area values obtained from a linear analysis carried out with an automatic measuring system. Table 8.5 shows how the interfacial surface areas given in Table 8.4 can be converted into proximity indices which give the *proportions* of the surface of each mineral that are in contact with each of the accompanying minerals.

2.5. Particle compositions

Random profiles (or random line probes) through composite particles that contain more than one mineral will invariably provide the mineral engineer with biased information regarding the compositions of these particles. For example, the proportion of liberated probes (i.e. profiles or lines that are made up of a single phase) will always be greater than the proportion of liberated particles.

linear analysis. This index shows the proportion of the surface area of a selected mineral which occurs in contact with another specified mineral (or group of minerals). The proximity index of mineral A relative to mineral B is:

$$V_{(A;B)} = \frac{S_{(A,B)}}{S_{(A)}} = \frac{S_{v(A,B)}}{S_{v(A)}}$$

where $V_{(A;B)}$ is the proximity index,
$S_{(A,B)}$ is the surface area of A in contact with B,
$S_{(A)}$ is the total surface area of A,
$S_{v(A,B)}$ is the specific surface area of A in contact with B (i.e. the contact area per unit volume of A), and
$S_{v(A)}$ is the specific surface area of A.

The specific surface area of a mineral can easily be calculated from the results obtained during area or line analyses. Thus,

$$S_{V(A)} = \frac{S_{(A)}}{V_{(A)}} = \frac{4}{\overline{L}_{(A)}} \qquad 8.1$$

TABLE 8.5

Interfacial surface area per unit volume of the phases in the left-hand column (μm^2 per μm^3)

Minerals	A	B	C	Total
A	—	0.82	0.18	1.00
B	0.90	—	0.10	1.00
C	0.67	0.33	—	1.00

The specimen contained 66.7% of mineral A (mixed silicates), 29.8% of mineral B (mixed sulphides) and 3.5% of mineral C (cassiterite).

The results show that about two-thirds of the cassiterite in this specimen is associated with the silicates whilst the other third is associated with the sulphide minerals.

Furthermore, the grade distribution of the composite probes (i.e. those probes that have cut more than one mineral in a particle) is, generally, a poor estimate of the grade distribution of the composite particles within the specimen. The extent of these biases is illustrated in Fig. 8.3. Each particle is exactly the same as all the others and has a volumetric grade of 0.5 (i.e. it contains 50% of mineral A and 50% of mineral B – by volume). Figure 8.3 shows that:

(1) a significant proportion of both kinds of probe are liberated, even though the specimens are known to contain composite particles only;

(2) the probe grades cover the range from 0 to 1, even though all the particles are known to have the *same* volumetric grade of 0.5;

(3) the variations in probe grades not only reflect particle grade but also show the effects of particle structure – note that the probe grade distribution is very different for spheres with planar interfaces from that for concentric spheres;

(4) the area probes are less biased than the line probes. These biases in the measured data can, as yet, only rarely be removed and their removal involves the use of complex stereological procedures which are outside the scope of this book.

2.6. Particle textures

The mineral engineer is often greatly interested in particle textures. The various particles illustrated in Fig. 8.4, for example, would react very differently in different mineral treatment processes. However, random sections and linear probes through composite particles inevitably tend to present simplified versions of their textures (Fig. 8.5). At present, it is very difficult to reconstruct the true textures of particles from the results obtained from random area or linear probes, and all that the practising mineralogist can do is to remember that the observed particle textures are often poor representations of the real, three-dimensional textures that exist within the measured particles.

2.7. Other measurable parameters

It is possible to derive estimates of a number of other useful mineralogical parameters from the results of image analyses. For example, a linear analysis can provide information on:

(1) the average number of grains of a specified mineral per unit volume of rock,

(2) the mean distance, in three-dimensional space, between the grains of a specified mineral within a rock,

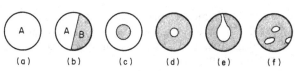

Fig. 8.4 How particle texture affects particle behaviour in different mineral treatment operations: (a) mineral A can be completely and swiftly leached by acid (or other) attack; (b) mineral A would be completely recovered by a leaching process; but either it would report to tailings in a density-based (or magnetic) separation treatment, or it would lower the grade of the concentrate; (c) mineral A is completely and swiftly recovered by a leaching process but such a particle will lower the grade of a flotation concentrate; (d) mineral A cannot be recovered by a leaching process; nor will it be recovered by flotation; (e) mineral A is slowly, but completely, recovered by a leaching process but it will not report into a flotation concentrate; (f) mineral A is only partly recovered by leaching, flotating or density-based methods of separation.

(3) the distribution of the nearest-neighbour distances between the grains of a specified mineral species,

(4) the way some of these parameters vary in different specimens from the same ore or the way they differ in specimens cut in different orientations.

Thus,

$$N_v = \frac{3K_1 V_v}{\pi K_4} \mu_{-3}^{1}(D_c)$$

where N_V is the number of grains per unit volume;

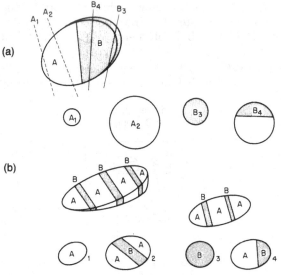

Fig. 8.5 Diagram to illustrate that sections, and linear probes across those sections, invariably provide simplified information regarding particle textures: (a) simple two-phase particle; (b) complex banded particle.

V_v is the volumetric proportion of the specified mineral; D_C is the equivalent screen size of the mineral grains, and μ^1 is the moment – measured from the origin.

2.8. Values that cannot be determined by measurement on a section surface

Some specimens contain mineralogical features that cannot possibly be measured by the examination of individual plane sections cut through them. Among these unknown details is the *connectivity* of any of the features seen on that surface (Fig. 8.6), i.e. it is not possible to determine whether or not two features that appear to be separate from each other when viewed on a section are joined together in the three-dimensional bulk specimen. Only serial-sectioning procedures will answer such a question.

3. Summary

Image analysis techniques can only be used to measure real mineralogical features when the number of dimensions exhibited by those features in three-dimensional space (D_f) plus the number of dimensions actually measured during an analysis (D_m) exceeds 2 (see Table 8.1). For example, a real two-dimensional feature, such as the contact area between two minerals, cannot be measured by a zero-dimensional point analysis method, since $D_f + D_m$ in this instance does not exceed 2. Table 8.1 shows what can, and what cannot, be measured – thus point analysis can only provide global volumetric data, while a line analysis, on the other hand, can provide data both about areal and volumetric features: area analysis can be used to measure the linear, areal

and volumetric features that exist in three-dimensional space.

There are some types of mineralogical feature that cannot possibly be measured on any two-dimensional section through an ore, or through a population of particles. For instance, if a rock consists of two or more intermixed, continuous phases (like long strips of Plasticine which have been inter-twined and moulded together to form a block), then no section through that rock can show that both phases are, in fact, continuous. Instead, one of the phases will appear to be distributed as discrete grains through a single continuous matrix. No image-analysing procedure based on the examination of random sections cut through such a material will establish whether or not the apparently discrete grains are in fact continuous within the original rock (Fig. 8.6).

As mentioned above, image analysis often produces biased results and, although the *direction* of the bias is often known, its *amount* is difficult to determine. These biases are the inevitable consequences of the sectioning and measuring procedures and cannot be overcome merely by increasing the numbers of measurements. Although mineralogists and mineral engineers may be well aware of the existence of these biases, nevertheless they often ignore them and use uncorrected measured values to compare the characteristics of

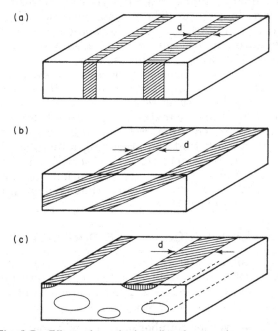

Fig. 8.7 Effect of sectioning direction on the apparent widths of lamellar inclusions: (a) section cut at right angles to inclusions; (b) section cut at oblique angle to inclusions; (c) section cut parallel to long axes of inclusions (d = apparent width).

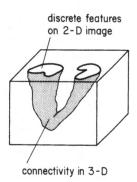

discrete features on 2-D image

connectivity in 3-D

Fig. 8.6 The two features seen on the section surface are discrete and no information regarding their possible connectivity in the body of the specimen can be obtained by studying that surface.

different specimens. However, these uncorrected results should, strictly speaking, only be used for comparing closely similar specimens and they should never be treated as if they were the true, three-dimensional values.

Other somewhat confusing results can also be obtained during the analysis of mineral sections. For example, during an area-measuring analysis lamellar inclusions of uniform width may appear to vary in width from grain to grain. These variations in thickness may, however, be caused by differences in the orientations of the different grains (Fig. 8.7). Furthermore, it is possible for grain-boundary precipitates to appear to lie slightly *within* one of the bounding grains instead of being *along* the junction between these grains.

The experienced mineralogist is well aware of the difficulties of interpreting the results of an image analysis and subconsciously tries to make the necessary qualitative adjustments. In many instances stereological transformation methods enable quantitative adjustments to be made to the measured data, but the skill of the experienced mineralogist is still of inestimable value when interpreting the results obtained from complex images.

Chapter 9

X-rays, electron beams and miscellaneous methods of mineralogical analysis

1. Introduction

X-radiation, or X-rays, make up that part of the electromagnetic spectrum that lies between the comparatively low-energy ultraviolet rays and the very high-energy gamma-rays (Fig. 9.1). The X-rays consist of individual photons of energy and the energy E associated with each photon is inversely related to its wavelength:

$$E = h\nu = \frac{hc}{\lambda}$$

where h = Planck's constant,
ν = frequency of the radiation,
c = velocity of electromagnetic radiation in vacuum, and
λ = wavelength of the radiation.

These photon energies range from about 120 kV to 0.25 kV. The wavelengths associated with these energies range from about 1.0 nm to about 500 nm.

X-rays are formed whenever matter is bombarded by photons of sufficiently high energy. In the mineralogy laboratory, X-rays are generated either by bombarding matter with high-energy electrons (and in this way producing *primary* X-rays) or by irradiating matter with high-energy X-rays so as to produce what are termed *secondary* or *fluorescent* X-rays of lower photon energies. The dominant effect of either form of bombardment is the production of heat but, in

addition, the bombardments produce small amounts of other forms of radiation: these radiations include X-rays, light and various forms of low-energy electron beams (see below). X-rays can be used, in appropriate circumstances, to determine (1) the chemical compositions of mineralogical specimens, and (2) the crystallographic structures of the components of mineralogical materials.

2. The laboratory production of X-rays

To produce primary X-rays it is necessary to provide

(1) a source of suitable "missiles" that can be used to bombard matter;
(2) a method of imparting large amounts of kinetic energy to these missiles, and
(3) a method of converting this kinetic energy into radiation energy within the X-ray wavelength range.

The "missiles" (or energy-carriers) are usually electrons and these electrons are produced in large numbers by thermal emission from a metal filament heated within a vacuum. This filament is made to act as a cathode whilst a nearby metal *target* forms an earthed anode. A large potential difference between the cathode and the anode

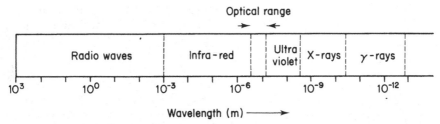

Fig. 9.1 Electromagnetic spectrum showing relative position and range of wavelengths of X-radiation.

provides the energy needed to accelerate the electrons towards the target. A metal cup held at the same potential as the cathode serves to focus the electron beam towards the target (Fig. 9.2).

When an electron impinges on to the target its kinetic energy is dissipated, either by a single collision with a target atom or by a series of ricochets with the atoms of that target. Some of the electron collisions give rise to X-rays but the energy of an X-ray photon is always lower than that of the electron that produced it.

Any increase in the kinetic energy of the impinging electrons – achieved by increasing the electrical potential V – will result in the production of some X-rays of shorter wavelengths and the increased energy will also increase the *total* amount of X-ray energy produced (Fig. 9.3). Conversely, any reduction in the kinetic energies of the bombarding electrons (brought about either by reducing the accelerating voltage V or by increased loss of energy due to successive

ricochets from the target atoms) will result in the production of X-rays of longer wavelengths and will decrease the total amount of X-ray energy being produced.

2.1. Continuous X-ray spectrum

When electrons of sufficient energy strike any matter X-rays of many wavelengths are produced, and these X-rays show a wide, *continuous* spectrum of energies (and wavelengths) (see Fig. 9.3). These energy levels range from the maximum energy of the electrons in the incident electron beam (often called the *short-wavelength limit*) down to energies which approach those of ultraviolet radiation. The short-wavelength limit is not a function of the material that is emitting the X-rays: when the kinetic energy of the electron beam is increased, the short-wavelength limit decreases accordingly. In addition, if the

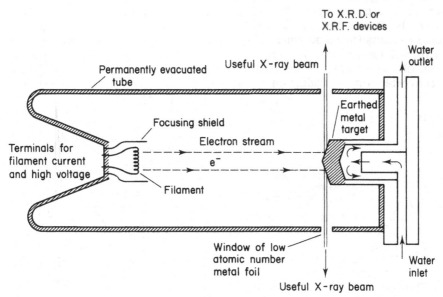

Fig. 9.2 Diagram of an X-ray generator. This consists of a thick metal tube, evacuated continuously by a vacuum pump, in which electrons from the heated filament (cathode) are accelerated towards the target (anode) and roughly focused by the focusing shield.

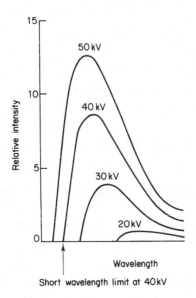

Fig. 9.3 X-ray spectra showing the distribution of intensity with wavelength (nm) in the continuous X-ray spectrum of tungsten at various accelerating voltages. The area under each curve indicates the total amount of X-ray energy produced at the stated accelerating voltage.

where Z is the atomic number of the target element and V is the accelerating potential of the electrons (expressed in volts). The conversion efficiency of electron energy into X-ray energy is thus very small; with a tungsten target ($Z = 74$) and an accelerating potential of 50 kV only about 0.4% of the energy association with the incident electrons is converted into X-radiation. Most of the electron energy is converted into heat within the target.

2.2. Critical excitation potential

There is a marked change in the *nature* of the X-ray distribution curve once the electron energy exceeds a critical value called the *critical excitation potential* (*CEP*). At potentials higher than the CEP, very intense X-rays are produced at certain very closely defined wavelengths. These critical wavelengths are found to depend on the nature of the target material: in fact, these wavelengths are so distinctive that they form the *characteristic* line spectra of the target materials (see Fig. 9.5).

2.3. Characteristic spectra

The sharply peaked "line" spectra, described above, are produced when the electrons have sufficient energy to penetrate to the inner levels

total amount of energy in a beam of electrons is increased (by increasing the beam current) then the area under the X-ray intensity curve will also increase, i.e. the amount of X-ray energy produced is a function of the amount of electron energy that is available (Fig. 9.3).

Figure 9.4 shows the effect of changing the atomic number of the target material. The higher the atomic number, the greater the amount of X-rays produced; however, the SWL (short-wavelength limit) remains the same since this is a function of beam voltage and not the target.

The proportion of the energy in the electron beam that is converted into X-radiation (ϵ) is given by:

$$\epsilon = 1.1 = 10^{-9} ZV$$

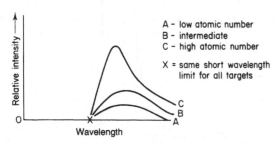

Fig. 9.4 Relative intensities of the continuous X-rays produced from different target materials using an electron beam of fixed energy.

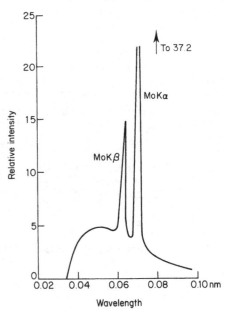

Fig. 9.5 Curve showing characteristic wavelengths superimposed on the continuous X-ray spectrum of molybdenum.

of the target atom and displace other electrons from their normal orbiting shells. When the displaced electrons return to their original orbits, the atom emits quanta (or photons) of X-ray energy. The energy of an X-ray "packet" is specific to, and characteristic of, a particular type of atom (Fig. 9.6). In other words, the energy associated with an orbiting electron (E_i) whilst it is in an outer shell of an atom is higher than its energy when in an inner shell (E_f). Therefore, when the displaced electron returns to an inner shell from an outer shell, an X-ray photon is emitted and this photon has an energy equal to ($E_i - E_f$). This difference can only take a limited number of values and these values are characteristic for each element.

The full characteristic spectrum of any element contains many X-ray peaks: the K, L and M series of peaks are derived from vacancies in the K, L and M orbital shells. The energy difference ($E_i - E_f$) is greatest when K shell electrons are concerned: the energies of L lines are lower than those of K lines, and those of M lines are lower still. For example, if an electron is initially displaced from the K shell of a copper atom to its L orbital shell then a quantum of X-radiation of specific energy will be released when another electron returns from the L shell to fill the K shell vacancy. This radiation is called the Cu K_α radiation. An electron displaced from the K shell to the M shell will ultimately give rise to a quantum of Cu K_β radiation. The probability that the K-to-L-to-K transformation will take place is

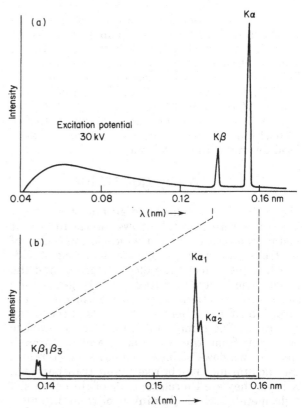

Fig. 9.7 Electron energies involved in various K-shell transitions: (a) X-ray spectrum of pure copper; (b) expanded version of the spectrum shown in (a).

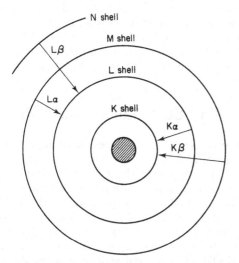

Fig. 9.6 Orbital shells of an atom: electrons displaced from the K shell to the L shell give rise to $K\alpha$ characteristic X-rays on returning to their original shell. Similarly, displacement from the K shell to the M shell gives rise to the $K\beta$ X-rays, and so on.

much greater than the probability of the K-to-M-to-K transition occurring; the K_α lines are therefore always much stronger than K_β lines (Fig. 9.7). Electrons are more readily displaced from the L shells and from the M shells than they are from the K shells, and therefore the M and L characteristic radiations are produced at lower critical excitation potentials than those corresponding to the K lines.

Once the critical excitation potential of a target material has been exceeded then the characteristic line spectrum of that material becomes superimposed on the continuous line spectrum which, as described above, is produced as a result of primary electron deceleration effects. The width of a characteristic line in an X-ray spectrum is extremely small – only about 0.01 nm at half the peak height (see Fig. 9.5). Furthermore, the peak intensity (i.e. peak height) of a characteristic line is usually very high compared with the intensity levels of the adjacent continuous radiation. Thus, when a copper target is bombarded by electrons having kinetic energies of 30 keV the copper gives rise to a Cu K_α signal that is almost 100 times as high as the adjacent "background", continuous signal. An increase in excitation potential of the bombarding electrons results in an increase in the intensity of the characteristic radiation but it does *not* change the wavelength of that radiation. When this sharply defined characteristic radiation is isolated from the accompanying continuous X-ray signals (see section 4.1.1 below) it provides a convenient high-intensity, highly monochromatic source of energy that can be used for many analytical purposes.

2.4. X-ray generator

Figure 9.2 is a schematic diagram of a typical X-ray generating unit. It shows the vacuum-tight casing, the metal filament which is the source of electrons, the anti-cathode (or focusing shield) which tends to focus the electron beam, and the target. The X-rays produced in the target radiate out in *all* directions; some are absorbed in the target itself and others by the walls of the X-ray generator. The remaining X-rays are allowed to pass out from the vacuum chamber through special "windows". These windows must retain the vacuum and must be reasonably transparent to X-rays; they are therefore made of thin sheets of light metals such as aluminium or from thin films of organic compounds such as mylar. The high vacuum is needed to prevent the dissipation of the electron energy by collision with air molecules. The vacuum is achieved either by continuously evacuating the generator unit, or by using generators that are evacuated and then permanently sealed by the manufacturer.

Most of the energy in the bombarding electron beam is converted into heat, and so much heat is sometimes generated that provision must be made to cool the target. This can be done (1) by rotating the target about an axis which is *not* parallel with the axis of the primary electron beam, (2) by a water-cooling system whereby water is circulated through the target (Fig. 9.2) or (3) by making special provisions for conducting away the heat through thin, conducting films (the method used in X-ray microanalysers is described in section 9.2 below).

3. Detection of X-rays

3.1. Photographic films

These were the first devices used to detect X-rays and, even now, photographic films with especially thick emulsions are used to detect X-ray beams and to estimate their intensities. The film is blackened by exposure to X-rays: the degree of blackening at any point on the film varies (up to its saturation level) with the number of X-ray quanta that strike that point. Of course, the number of X-ray quanta will vary both with the intensity of the X-ray source and with the length of time of the exposure.

Photographic films of this kind are especially useful for the integrating over very long periods the results produced by very weak X-ray signals. In fact, the exposure times can vary from a fraction of a second (for example, in a chest clinic where the intensities and the locations of the X-rays that pass through a patient's body are recorded on film) up to many weeks for the films used as radiation "badges" by people who work with X-rays (these people do not expect to be exposed to the dangers of X-radiation, but the "badges" are periodically developed to check whether any hazardous exposure has occurred). The photographic films provide useful permanent records, especially of very weak beams, but they provide no information regarding the wavelengths or the energies of the X-rays in those beams.

3.2. Fluorescent screens

These are usually glass plates coated with a compound, such as zinc sulphide, that fluoresces when irradiated with X-rays. Such screens are often used to establish the *locations* of strong X-ray beams (and also strong electron beams) and are sometimes incorporated within an X-ray analytical device to align the X-ray (and also the electron) beams. These screens tend to saturate readily and they therefore provide little information about beam intensities, nor do they provide any information on the wavelengths of the detected X-rays.

3.3. Scintillation counters

These counters are, in effect, sophisticated versions of the fluorescent screens in which the screens are coupled with photomultiplier devices. The photomultiplier counts the individual light photons (*scintillations*) produced when X-ray quanta strike the screen. The screens are generally much more complex than the fluorescent screens described above, and they may be coated with thallium-activated sodium iodide phosphor. Scintillation counters of this kind have high counting efficiencies and can be used to detect weak X-ray beams; however, they supply no information regarding the wavelengths of the

X-ray beams that cause the scintillations. These counters have been found especially useful for detecting high-energy X-rays.

3.4. Ionisation devices

These devices rely on the ionisation effect produced in a gas by the presence of X-rays. Electrodes of opposite polarity collect the ions which are formed during ionisation and the change in potential difference between the electrodes is recorded in an electrical circuit; for example, this is the procedure used in a Geiger tube (or Geiger–Müller counter). Each absorbed X-ray photon ionises a number of gas atoms within the counter and produces positive ions and negative ions. These products give rise to a cascade of many ions. In this way an ionisation device increases (i.e. amplifies) the original X-ray signals. They are often called *proportional counters*: the current produced by the ionisation process being proportional to the intensity of the X-ray signals.

The gas within a counter can be selected so that it is readily ionised by low-energy X-rays (compare scintillation counters, which work best with high-energy photons). Arrangements must be made so that the cascade of ionisation when an X-ray is detected takes as little time as possible, so that the next, incoming, X-ray quantum can be detected as a separate event – i.e. each electrical discharge pulse must be quenched very quickly. In this way, the *dead-time* between successive detectable pulses can be minimised. On modern counters the dead-time is about 5×10^{-6} seconds.

X-rays are produced within a chamber held at a very good vacuum: this vacuum chamber is separated from the gas-filled ionisation detector by a thin "window". These counter "windows" are made as thin as possible to minimise absorption of ingoing X-rays; however, the gas within the counter tends to diffuse through the thin windows and is slowly lost from the counter. Consequently, a small, continuous flow of gas is maintained through the counter in order that the correct gas concentration can be ensured.

Like the scintillation counters and the photographic films, the ionisation devices are unable to measure the wavelengths of the X-rays.

3.5. Solid-state X-ray detectors

In these modern, solid-state detectors an X-ray quantum generates an electrical response which is proportional to its *energy* and it is, therefore, possible to determine the energy spectrum of a beam of polychromatic X-rays with these devices (Fig. 9.8). The solid-state detectors (*energy-dispersive detectors*) have, therefore, made it possible to determine both the location of an X-ray and its energy (or its wavelength).

The detector is made from a specially produced crystal of lithium-drifted silicon. The very minute signals detected by this device must be greatly amplified by suitable electronic circuitry and the random electrical "noise" in these circuits is minimised by operating the detector at liquid-nitrogen temperatures.

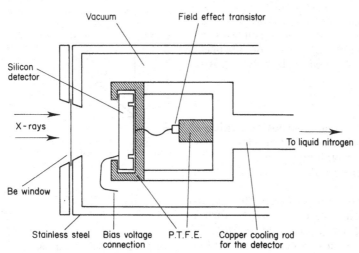

Fig. 9.8 Diagrammatic cross-section of a lithium-drifted silicon detector (after Long in Zussman, 1977).

4. Interaction of X-rays with matter

4.1. Absorption

X-rays are partly absorbed, partly diffracted, and partly scattered by matter of any kind. Absorption is often the dominant effect and, for a given material, the amount of absorption generally increases as the energy of the incident X-ray decreases. Similarly, the absorption of an X-ray quantum of specified energy is generally greater in "heavy" (high-atomic-number) elements than in matter composed of "lighter" elements.

The proportion of a beam of X-rays that will be absorbed in passing through matter can be calculated:

$$I = I_0 e^{-\mu x}$$

where I_0 is the initial intensity of the beam,

I is the final intensity,

x is the thickness of matter that has been penetrated,

μ is the linear absorption coefficient of that material.

However, it is more usual to use the mass absorption coefficient μ/ρ instead of μ, where ρ is the density of the absorbing material.

The mass absorption coefficients of individual elements for given X-ray energies are known. Therefore, if the mass proportions of all the elements in an absorbing medium are known, the corresponding mass absorption coefficient for the whole absorber can be calculated. In general, the mass absorption coefficient of a material increases with increase in the wavelengths of the absorbed rays, except for marked discontinuities that occur at particular wavelengths and which give rise to *absorption edges* (Fig. 9.9).

The relationship between mass absorption coefficients and the wavelengths of the absorbed X-rays *between* absorption edges is given approximately by:

$$\mu/\rho = K\gamma^{\mu}Z^{v}$$

where K, μ and v are constants (but the value of K changes at each absorption edge),

γ is the X-ray wavelength, and

Z is the atomic number of the absorbing element.

The dramatic change of absorption that takes place at an absorption edge is due to the formation of fluorescent X-radiation (see section 6.4 below), and this radiation is a characteristic feature of the absorbing material.

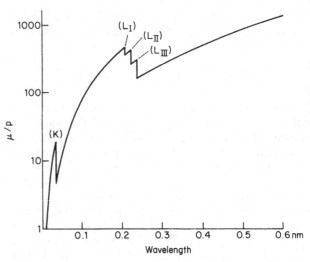

Fig. 9.9 Variation of the mass absorption coefficient (relative intensity, μ/p) of barium carbonate as a function of X-ray wavelength (after Norrish and Chappell in Zussman, 1977).

4.1.1. X-ray filters

For many analytical purposes it is an advantage if all the X-rays in a beam are of the same wavelength, i.e. some analyses require beams of *monochromatic* X-radiation. One way of producing such radiation is by using *absorption filters*. Polychromatic radiation from a pure metal target is passed through a sheet of metal foil that has an absorption edge between the K_α and the K_β wavelengths of the target metal (Fig. 9.10).

For example, before being passed through a filter the ratio of Cu K_α to the Cu K_β signal inten-

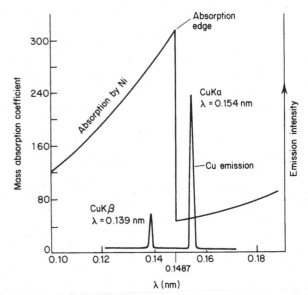

Fig. 9.10 Relative positions of the absorption edge of nickel and the $K\alpha$ and the $K\beta$ emission lines of copper.

TABLE 9.1

Suitable filter materials for X-rays produced from different target materials

Target element (Z)	Filter element (Z − 1)
Copper	Nickel
Cobalt	Iron
Iron	Manganese
Chromium	Vanadium
Molybdenum	Niobium

sity would normally be about 5:1; but after passage through a nickel filter (of the optimum thickness) the $Cu\,K_\alpha$ to $Cu\,K_\beta$ ratio can be as high as 500:1. The $Cu\,K_\beta$ signal can, of course, be completely removed by increasing the thickness of the filter but, in doing so, the *amount* of available $Cu\,K_\alpha$ radiation would also be significantly reduced. The most suitable filter for producing essentially monochromatic radiation from the characteristic radiation of an element of atomic number Z is a material composed of atomic number $Z - 1$ (see Table 9.1).

5. Potential hazards in the use of X-rays

All X-ray equipment uses high electrical voltages. The manufacturers of such equipment take great care to ensure that both the anode and the cathode are inaccessible to the operator whilst the power is switched on. No attempt should ever be made to short-circuit the safety devices which are built into the equipment; furthermore, it is important never to use any inflammable liquid in the vicinity of the high-voltage contacts – the liquids can ignite explosively.

X-rays are dangerous and they can kill or seriously damage human tissue. Once again, the manufacturers take great pains to make it virtually impossible for the X-rays to escape from the equipment. The manufacturer's safety switches must *not* be tampered with and the working environment surrounding any X-ray equipment must be checked at frequent intervals to ensure that no X-ray "leaks" are occurring.

6. Application of X-rays in mineralogy

Most minerals are crystalline bodies that have regular internal atomic structures and chemical compositions that fall within well-defined limits (see Chapter 1, section 2.2). The range of wavelengths in the X-ray spectrum are of the same order of magnitude as the distances between the various planes of atoms in crystalline minerals. X-rays are, therefore, diffracted by minerals in much the same way as light beams are diffracted by suitably sized repetitive patterns – for example, the lines of a diffraction grating spaced apart by about the wavelength of light. The X-ray diffraction effects can be used to determine the inter-planar distances in crystals and from this information it is possible to determine the structure of a crystalline substance.

Furthermore, when any matter (including, of course, minerals) is bombarded by a beam of sufficiently high-energy photons (which include X-ray beams) it gives off its own characteristic X-radiation. The characteristic wavelengths in the radiation will show which elements exist in the bombarded material and the intensities of these radiations indicate the concentrations of the elements (see section 6.4 below).

Thus it is possible to use X-rays to determine both the structure and the chemical composition of a mineral. With these two major pieces of information, it is a comparatively simple matter to identify virtually any mineral. As we shall show later, it is also possible to use X-ray based measuring techniques to determine important values such as mineral proportions, mineral grain sizes and particle compositions.

6.1. X-ray diffraction analysis

As mentioned above, the wavelengths of X-rays are similar to the atomic spacings in minerals, and an X-ray beam which falls on a crystal will be diffracted in the same manner as a light beam is diffracted by an optical diffraction grating. The distance between the "lines" of a diffraction grating (a few micrometres) is approximately the same as the wavelength of light (0.3 to 1.0 μm); the distance between the atomic planes in crystalline minerals (a few nanometres) is within the wavelength range of X-rays.

Laue showed in 1912 that crystals do, in fact, diffract X-rays, but the first person to use X-rays for crystallographic analysis was Bragg in 1913. Figure 9.11 is a two-dimensional representation of a regular crystal lattice. It shows, diagrammatically, the effect of bombarding a perfect crystal with a beam of parallel monochromatic X-rays. Most of the X-rays will pass through a thin specimen and will form an unaltered, and easily detected, transmitted beam of high intensity.

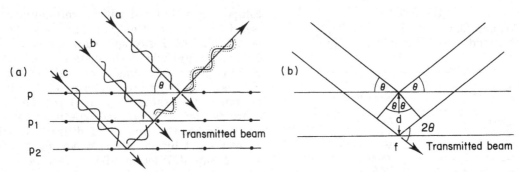

Fig. 9.11 Condition for diffraction of X-rays from a regular crystal lattice: (a) Bragg angle of diffraction (P, P_1 and P_2 are successively deeper atomic planes uniformly spaced within the crystal); (b) path-length difference in condition for diffraction, i.e. when the incidence angle is θ and d is the interplanar distance.

Other X-ray photons are scattered by successively deeper atomic planes within the specimen. At most values of the angle of incidence (i.e. the angle between the electron beam and the set of diffracting atomic planes) the X-rays scattered from the different atomic planes are out of phase with each other. Consequently, because of their varying path lengths through the crystal, most of these rays interfere destructively with each other. However, there is one particular angle, θ, at which the path lengths of the scattered rays differ from each other by exact multiples of the incident monochromatic X-ray wavelength. In this case (and in this case only) the scattered rays are all in phase: this gives rise to a reinforcement, or a constructive interference, of the X-rays, and a comparatively strong diffracted beam is formed. The critical angle, θ, at which this diffraction effect occurs is called the *Bragg angle*, after W. L. Bragg who, as mentioned above, pioneered the use of X-rays to analyse the internal structures of crystals. Bragg showed that a strong diffracted beam is formed only when

$$n\lambda = 2d \sin \theta$$

where n is a small integer (frequently 1),

λ is the wavelength of the incident beam, (and also, of course, the wavelength of the emerging undiffracted beam),

d is the distance between adjacent atomic planes, and

θ is the angle between the incident beam and the reflecting crystal plane.

The incident beam, the normal to the reflecting atomic plane, and the diffracted beam are always in the same plane.

It is almost invariably very difficult to measure θ. However, simple geometrical considerations show that the angle between the diffracted beam and the *transmitted* beam is 2θ: this angle is comparatively simple to measure. The Bragg equation shows that diffraction of X-rays in crystals *only* occurs when the X-ray wavelength (λ) is of the same order as the distance which separates the crystallographic planes. Thus:

$$n\lambda = 2d \sin \theta$$

and

$$\lambda = (2d \sin \theta)/n$$

But $\sin \theta$ cannot exceed unity; therefore

$$\lambda \leqslant 2d/n$$

and, since $n_{\min} = 1$, the smallest value of $\lambda = 2d$, and λ must be less than d.

The average d value for minerals is roughly 0.3 nm and, therefore, the wavelengths of X-rays used for diffraction purposes in mineralogy should be of this order. It is, of course, possible in theory to use wavelengths which are very much shorter than 0.3 nm to determine crystal structures, but in these instances 2θ becomes very small and very inconvenient to measure accurately.

The diffracted beam is, at all times, very weak compared with the original, incident beam (because even in the most favourable cases only a small fraction of the incident energy is diffracted). The incident, the transmitted and the diffracted beams can nevertheless all be detected readily and their intensities can be measured by a variety of X-ray detecting systems (see section 3 above).

Monochromatic rays are needed for many diffraction analyses and they are produced by "filtering" polychromatic (white) rays obtained from a typical X-ray generator; however, some small amounts of other wavelengths will almost inevitably remain in a "monochromatic" beam produced by filters (see section 4.1.1 above).

6.1.1. The flat-plate (Laue) method

In this method a parallel beam of *polychromatic* radiation is directed at a single crystal of the unknown material (see Fig. 9.12). (*Note*: X-rays cannot readily be focused in the same way as optical beams are focused. Consequently, their direction and position are controlled by *collimators*: these are narrow, parallel metal tubes or plates that permit X-rays to pass only when they are travelling virtually along the axis of the collimator.) Whatever the specimen, the requirements of the Bragg equation ($n\lambda = 2d \sin \theta$) will always be met by *some* wavelengths within the incident beam and these rays will be diffracted by the specimen. The positions and the relative intensities of these diffracted rays are recorded on a flat photographic plate set at right angles to the incident (and, of course, to the transmitted) beam. This plate (Fig. 9.12) will show a pattern of spots: these spots form the outlines of complex ellipses and hyperbolas. A comparison of these shapes, their sizes and their positions with similar values obtained from standard materials can, with some difficulty, be used to establish the identity of the unknown mineral. However, since each of the diffracting planes diffracts X-rays of different wavelengths, it is very difficult to derive quantitative structural information by this method.

This flat-plate system does not (indeed cannot) produce the complete diffraction pattern of the irradiated specimen because some of the crystal planes – for instance, those parallel to, or nearly parallel to, the incident beam – cannot possibly diffract the X-rays.

6.1.2. The rotating-crystal method

This method was developed in order to overcome some of the problems encountered with the Laue

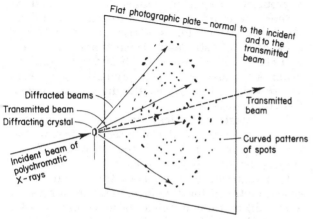

Fig. 9.12 Diagram of a Laue plate method of determining crystal structure, by recording the X-ray diffraction pattern.

method. A single-crystal specimen is set with one of its crystallographic axes normal to the direction of a parallel beam of *monochromatic* X-rays. The crystal is partly surrounded by a concentric, cylindrical photographic film (this film exists in the form of a "strip" which is then formed into a cylinder). The crystal is rotated about its selected axis whilst being bombarded by the X-ray beam. During this rotation, some atomic planes within the crystal will, momentarily, satisfy the requirements of the Bragg equation for the monochromatic X-ray beam being used. The positions of the diffracted beams on the photographic film are marked by a pattern of bright spots and these diffraction patterns can be analysed to reveal the structure of the single, analysed crystal. Once more, the diffraction pattern is incomplete because some of the atomic planes in the crystal cannot be properly presented to the incident beam. The value of such a diffraction pattern would clearly be improved if it were possible to rotate the crystal about more than one of its crystallographic axes.

6.1.3. The powder method

It is often difficult for the mineralogist working in the mineral industry to obtain mineral specimens that consist of single crystals. Even on those rare occasions when a single crystal is available, it is not possible (as shown above) to obtain complete diffraction patterns unless that crystal is rotated about *all possible* axes.

The *powder method* is an attempt to overcome two problems: that of obtaining good specimens, and that of the difficulty of rotating those specimens about all possible axes. A pure (or nearly pure) specimen (which need not be in the form of a single crystal) is ground into particles of less than about $50 \, \mu m$ in size. By using a binder such as glue or resin, this powder is then formed into a thin cylindrical shape about 10 mm long. It is assumed that the mineral fragments in this cylinder are randomly orientated, and rotation about its long axis is equivalent to rotating the original specimen about all possible crystallographic axes. Consequently, when a suitably prepared powdered specimen is rotated in a parallel beam of monochromatic X-radiation, each set of crystallographic planes will, momentarily, produce a diffracted ray. This system of presenting the specimen produces so many diffraction "spots" that they coalesce into "lines" which are recorded, as before, on a surrounding strip of photographic film. These line patterns (Fig. 9.13) are comparatively easy to measure and to analyse, since the linear *distance* from the centre of the transmitted beam to the position of the diffracted beam can be

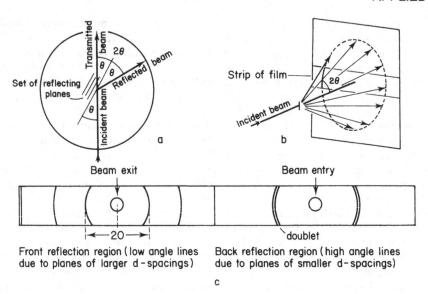

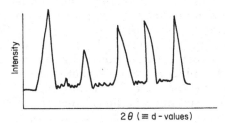

Fig. 9.13 Illustration of the formation of powder diffraction lines: (a) angle between transmitted and reflected beam is 2θ – this value can be measured easily and the value θ between the incident beam and the reflecting plane can thus be obtained; (b) "cone" of diffracted X-ray beams; (c) X-ray powder film laid flat.

measured and hence the angle 2θ can be calculated readily.

One of the most modern versions of the powder diffraction method uses an *X-ray diffractometer* instead of a photographic film to locate and measure the diffracted rays. The specimen is prepared as a flat layer of powder and is then set on the circumference of an X-ray *focusing circle*. A stationary, linear source of X-rays also lies on the circumference of that circle (i.e. the collimator is shaped so as to produce an apparently linear source for the X-rays). On the other side of the circumference of this circle is a narrow slit behind which lies a suitable X-ray detector (usually a proportional counter). When an analysis is carried out, the specimen holder and the detector slowly rotate about the centre of the circle; in order to maintain the requirements of the Bragg equation, the angular velocity of the detector is arranged to be twice that of the specimen. In this way, when a crystal plane within the specimen meets the demands of the Bragg equation the detector will be correctly placed to detect the diffracted beam. If the orientations of the powder grains have been correctly randomised then all the diffracting planes in the specimen will be detected sequentially by this method. The signal intensities recorded by the X-ray detector as it slowly traverses around the circle can be plotted on a chart (see Fig. 9.14 for an example). The values obtained from this chart, with its abscissa showing 2θ values and its ordinate showing signal intensities, can be used to calculate the

Fig. 9.14 Diffractometer trace, showing diffracted-ray intensity values plotted against 2θ values.

detailed atomic structure of the specimen. More commonly, the chart is used to identify minerals by comparing the signal intensities and the appropriate d-spacing values against standard values contained in powder diffraction data files. (*Note*: 2θ values for a particular mineral are dependent on instrumental factors and, moreover, will change if the wavelength of the irradiating beam is changed; it is therefore better to convert the 2θ values to d values, the distances between the atomic layers.)

The latest analytical systems use stepper motors to drive the X-ray detectors around the focusing circle and, since the diffracted X-ray signals are digital in character, it is possible to obtain digitised data on both the signal intensities and the 2θ (or d) values. In the most modern X-ray diffraction systems these values can be compared against standard data held on computer files and, in many instances, an analysed mineral can be identified automatically.

6.1.4. The energy-dispersive method

In the powder methods of X-ray diffraction described above the incident beam of X-rays must always be *monochromatic*: each set of atomic planes that has the appropriate *d* spacing will then diffract the beam at the appropriate angle.

A more recently developed *energy-dispersive method* of determining diffraction effects uses solid-state energy-dispersive X-ray detectors (section 3.5 above). In this method the incident X-ray radiation is *polychromatic* (cf. the original, Laue method of X-ray diffraction), and the 2θ values are fixed – i.e. the positions of the effective X-ray source, the specimen and the detector are fixed. Each set of planes within the specimen will have different inter-planar (*d*) values and will diffract only those X-rays of appropriate wavelength (or energy) into the energy-dispersive X-ray detector. This device records the intensities associated with each energy level (i.e. with each set of atomic planes in the specimen). In this way, all the diffraction data can be obtained without the need for producing monochromatic illumination; furthermore, the information can be obtained without any mechanical movements of the specimen, and without moving the detector.

6.2. Identifying an unknown material

Every mineral has its own unique X-ray diffraction pattern. When a mixture of minerals is analysed by an X-ray diffraction method, each mineral will simultaneously produce its own characteristic pattern and the patterns will be superimposed on one another. If the mixture contains only two or three minerals the individual patterns can be disentangled and the minerals responsible for them can be identified easily. This identification procedure becomes more and more difficult as the number of minerals in the specimen increases, because the different patterns overlap one another. The lower limit of detection for any single mineral in a mixture of minerals is in the range 1–5% by mass, and this factor alone ultimately restricts the number of different minerals that can be identified in one specimen. The modern X-ray diffraction procedure for identifying an unknown substance is as follows.

The specimen is made up in the form of a thin layer of fine-grained powder, from 1 to 10 μm in particle size, on a glass (i.e. non-crystalline) plate. The easiest, and hence the most commonly used, method is merely to smear the powder on to the

---

24 – 177   24-178

d	2.65	1.61	2.96	2.96	$Ca_3Al_2Si_3O_{12}$	$3CaO \cdot Al_2O_3 \cdot 3SiO_2$	★
$I I_1$	100	50	25	25	Calcium Aluminum Silicate	Grossularite, Garnet	

Rad. CuKa$_1$ λ 1.54051 Filter Ni Dia.
Cut off $I I_1$ Diffractometer $I I_{cor}$ = 2.0
Ref. Smith and Auh, JCPDS Report, 1971

Sys. Cubic   S.G. Ia3D (230)
a_0 11.850 b_0 c_0 A   C
a β γ Z 8 Dx 3.596
Ref Ibid.

$\epsilon\alpha$ $\eta\omega\beta$ 1.737 $\epsilon\gamma$ Sign
2V   D 3.6   mp   Color
Ref. Pabst, Am. Min., *21* 1–10 (1936)

Crystals from Georgetown, California supplied by A. Pabst.
Analysis given in Deer, Howie, Zussman Rock Forming Minerals (1962) Vol. 1, page 95, no. 1
Composition = $Gr_{96.8}An_{2.2}Al_{0.6}Ur_{0.4}$

d A	I/I$_1$	hkl	d A	I/I$_1$	hkl
2.959	25	400	1.263	7	664
2.647	100	420	1.249	2	851
2.524	11	332	1.222	1	931
2.417	20	422	1.197	5	941
2.321	18	431	1.173	1	1011
2.162	17	521	1.162	2	1020
2.093	4	440	1.130	8	1031
1.921	25	611	1.100	15	1040
1.872	2	620	1.082	11	1042
1.710	17	444	1.073	1	873
1.643	25	640	1.047	10	880
1.612	2	721	0.9875	3	1200
1.607	50	642	.9740	4	1220
1.504	2	651	.9611	10	1222
1.481	10	800	.9197	2	992
1.458	2	741	.8931	3	1244
1.417	1	653	.8832	9	1260
1.324	10	840	.8735	3	1262
1.292	18	842	.8551	3	888
1.278	1	921	(Plus 4 more lines)		

FORM M-2

B

Fig. 9.15 Powder diffraction information card (after Zussman in Zussman, 1977).

plate. If the glass slide has previously been coated with a thin layer of grease the particles will stick to the grease and, furthermore, will tend to assume random orientations. (In the absence of grease, platy minerals, such as clay, tend to lie on their stable, and therefore non-random, basal planes.)

An alternative method of mounting the specimen is to use a "cavity" mount. A slight depression is etched in a glass plate; the cavity is filled with mineral powder and the surface of the powder is smoothed off. This method generally produces better random orientation of mineral particles than the smear mount.

If the prepared specimen is truly randomised then all possible crystallographic directions will be represented in the plane of the specimen surface. A monochromatic X-ray beam is then directed at this surface: the angle of incidence is continuously varied during the analysis by slowly rotating the specimen, and the diffracted beams are detected and measured by a suitably placed, movable X-ray detector. Any range of θ values (i.e. Bragg angles) can be scanned.

The diffraction results obtained from a substance are, in the first instance, described by the angular positions of the reflections and by their *relative* intensities. As we have seen (section 6.1.3 above) a substance is best characterised by its individual set of d values (i.e. by its interplanar spacing values). A complete diffraction pattern contains a large number of reflections: the d values and the relative intensities of these values

are unique to a particular substance (rather in the way that a fingerprint is unique to a particular person). So, if there is an up-to-date reference catalogue of the patterns of all the known minerals, it should be possible to match one of them with the diffraction pattern of the unknown mineral.

However, there are over 3500 different minerals and it is not always easy to locate a matching pair of diffraction patterns. Consequently, separate information cards for each mineral have been provided by the Joint Committee on Powder Diffraction Standards. These cards list details such as mineral name, chemical composition, major d spacing values, relative intensities of these spacings, crystal system and so on (see Fig. 9.15). Instructions in the use of a suitable identification procedure are provided along with the cards: for example, the d values of the strongest reflections for each mineral are listed in decreasing order. However, variations in the intensities of the reflections can be caused by non-random specimen presentation and as a result the four most intense reflections are permuted and each mineral can be found in at least four locations of this determinative scheme.

6.3. Determining mineral proportions

Although X-ray diffraction patterns are primarily used by the applied mineralogist as a means of establishing which minerals occur in a specimen,

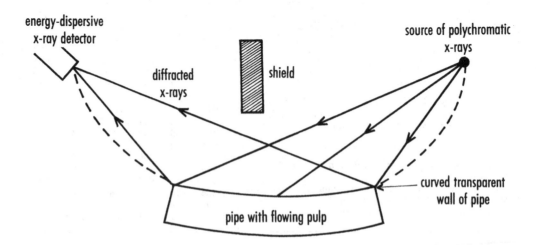

Fig. 9.16 Diagrammatic representation of an on-stream X-ray diffraction system. The dotted line shows the focusing circle: the detector, curved pipe and X-ray source lie on the circumference of that circle.

these patterns can also be used to provide estimates of the proportions of each mineral (provided that the minerals are few in number and that they occur in mass proportions of greater than about 5%). Quantification is achieved by comparing the results obtained from an unknown specimen with values obtained from carefully prepared, standard specimens of compositions similar to that of the unknown material. However, this procedure is not easy to carry out, even with two-component mixtures, and in more complex specimens it may not be possible to get highly accurate results because of the variable degrees of absorption of the diffracted X-rays by the various (and often unknown) constituents of the analysed specimens.

Many attempts have been made to provide instantaneous, *on-stream* estimates of mineral proportions by using the X-ray diffraction technique. These attempts have been based on the use of energy-dispersive X-ray detectors (see above). A flowing stream of fine-grained mineral pulp is passed through a specially shaped pipe which is transparent to X-rays. The top of this pipe (Fig. 9.16) is curved to conform with the circumference of the X-ray focusing circle. The specimen surface is irradiated by a divergent beam of polychromatic X-rays originating from a "line" source situated on the circumference of the focusing circle. The detector is also placed on this circle and measures the energies of the various diffracted rays which are "focused" on to it. The diffraction peaks are measured very rapidly by a pulse height detector and selected peak heights can (in theory, at least) be used to estimate the proportions of selected mineral components in the flowing pulp.

6.4. X-ray fluorescence spectroscopy

The Bragg equation ($n\lambda = 2d \sin \theta$) can be used in two distinct ways to provide information on mineralogical materials:

(1) As shown in section 6.3 above, if λ in the equation is known (or fixed) and θ can be measured, then d can be calculated for the various atomic planes that occur in a crystalline mineral: i.e. X-rays can be used for determining the crystal structure of a mineral.

(2) If, however, the value of d is fixed or is already known and θ is measured then λ can be calculated, i.e. if a crystal of known d spacings is used to diffract the X-rays obtained from a specimen then the measured values of θ will allow values of λ to be calculated. The wavelength (λ) of

a characteristic X-ray is uniquely related to a specific element (as shown in section 2.3). Therefore, the wavelengths of the various characteristic X-rays obtained from a composite material will show which elements are contained in that material. The intensity of a characteristic wavelength is a complex function of the proportion of the specific element that gave rise to it and of the types and amounts of the accompanying elements. These relationships can be established, and the X-rays derived from a substance can be used to determine the proportions of the various elements it contains.

6.4.1. Methods of analysis

As long as the appropriate techniques of excitation are used, all elements can be made to emit characteristic radiation in the X-ray wavelength region. These radiations can then be used to identify and to estimate the concentrations of the elements in a specimen. This technique is called *X-ray fluorescence (XRF) analysis*. The procedure produces *chemical* information rather than specifically *mineralogical* information, i.e. an XRF analysis does not directly provide any information about the identities, natures or proportions of the *minerals* in a complex specimen. Instead, it determines the overall elemental composition, and this chemical information can sometimes be used to derive estimates of the mineral proportions (see below). However, an XRF analysis does not, and cannot, provide any information on mineral grain sizes, mineral associations or rock textures. But although the chemical results seldom provide direct and unambiguous mineralogical information, such an analysis can still provide the mineral engineer with very useful, comparative data that he can use for process control purposes; for example, when the mineralogy of a rock, an ore, or a plant product is known to be essentially uniform, then a series of chemical analyses of a number of similar specimens will provide a rapid, convenient and valuable check on possible *changes* in the proportions of the mineral components: if the feed to a section of a treatment plant is known to contain chalcopyrite as the only copper-bearing mineral, then a quick check on the copper content of the feed will show any changes in chalcopyrite content. But if the mineralogy of that feed material alters significantly (say, by the inclusion of an unknown amount of the mineral chalcocite) then the chemical results will of course have little mineralogical value. The major advantage of the chemical analysis is that the information can be obtained very swiftly (almost instantaneously with on-stream XRF analyses; see section 7).

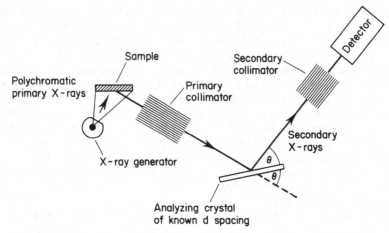

Fig. 9.17 Principles of operation of an X-ray spectrometer, showing path of fluorescent X-rays.

In X-ray fluorescence spectroscopy an appropriately prepared specimen (see section 6.5 below) is irradiated by a beam of polychromatic X-rays. The fluorescent X-rays produced from the specimen are passed into an X-ray spectrometer (see Fig. 9.17) where specified characteristic wavelengths are detected and measured.

In order to obtain a *qualitative* analysis of the specimen, it is only necessary to identify the elements responsible for these major characteristic peaks and then to observe their heights. A much more complex procedure, however, must be carried out before a fully *quantitative* chemical analysis can be obtained.

These quantitative analyses are carried out by comparing the results from the unknown specimen with the results obtained from a known (and, if possible, closely similar) standard material:

$$I_{sp}/I_{st} = K(C_{sp}/C_{st})$$

where I_{sp} is the intensity of a characteristic line from the unknown specimen,

I_{st} is the intensity of the same characteristic line from the standard material,

C_{sp} is the concentration of the specified element of the unknown specimen,

C_{st} is the concentration of this element in the standard material, and

K is a constant (see section 9.4 below).

The ratio of the X-ray intensities in the unknown specimen and the intensities in the standard are approximately equal to the ratio of the concentration of a specified element in the unknown specimen to its concentration in the standard. This ratio is affected by the effects of matrix absorption, multiple excitations, X-ray background signals, etc. It is quite difficult to calculate

these effects and, therefore, all quantitative XRF analyses are carried out by comparing the spectrum of the unknown specimen with that of a closely defined standard of composition similar to that of the unknown substance.

(i) *Matrix absorption*
If the chemical compositions of two specimens are different from one another there will inevitably be corresponding differences in the nature of the matrices that surround the X-ray-producing elements. As the matrices differ, so also will the absorptions of both the primary (incoming) and the secondary (outgoing) X-radiations within those matrices. (Suitable correction factors can be employed to offset these variations in absorption effects during X-ray microanalysis, see section 9.1 below.)

(iii) *Multiple excitations*
If an unknown specimen contains both element A and B, then the primary X-radiation (if of sufficiently high energy) will cause both elements to emit their own characteristic radiations. If, say, the energy of the characteristic radiation of element B is greater than that of the K_α radiation of element A, then some of the radiation from element B will also excite element A – which then gives off *additional* A radiation. As a result, the amount of element A in the specimen will appear to be larger than its true value. (Appropriate correction factors must, therefore, be applied to counter this effect during X-ray fluorescent analysis.)

(iii) *Background radiations*
Any sensitive X-ray detector always picks up "background" noise. A small part of this noise is

derived from cosmic rays but the bulk is derived from scattered primary radiation from within the instrument. This effect can be reduced (but not entirely eliminated) by careful attention to the design of the equipment.

(iv) *Dead-time*

An X-ray detector takes a small, but finite, time to register the presence of an X-ray photon. During this time the counter is inoperative and cannot register or count any other photon. This period of inaction is called the *dead-time*, and any photons which arrive at the counter during this time are lost. The observed count rates must, therefore, be corrected for these losses.

(v) *Absorption of electron beams and X-rays within the measuring equipment*

Low-energy X-rays (and electron beams) are easily absorbed and scattered by gas molecules. If the fluorescent X-rays within a spectrometer are required to pass through air then the characteristic radiations from elements of atomic number less than about $Z = 21$ are seriously absorbed. However, many modern spectrometers operate under vacuum, and these spectrometers can measure X-radiation from elements down to $Z = 11$ without serious loss of signal due to absorption effects. (In the electron probe X-ray microanalyser (see below), it is possible, although quite difficult, to measure all elements of atomic number greater than $Z = 4$: only the elements hydrogen, helium, lithium and beryllium cannot be measured.)

6.5. Specimen preparation for X-ray fluorescence analysis

Many kinds of specimen can be analysed by X-ray fluorescence. Solid rock specimens need only be cut to a suitable size with a fine diamond saw: this produces flat, but not necessarily polished, surfaces. Specimens of mineral powders are mixed thoroughly and are then pressed into pellet form. Liquid specimens, or specimens of mineral pulps, are held in special containers which are transparent to X-rays – these containers must, of course, be very carefully isolated from any vacuum chambers in the instrument.

Only a comparatively thin layer (about 1 mm deep) of the specimen is actually examined during an X-ray fluorescence analysis. The surface geometry of the specimen must, therefore, be reasonably uniform and representative of the total mass. Adequate uniformity is generally achieved in solid specimens by careful slicing, and

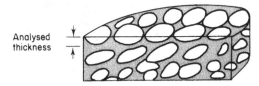

Fig. 9.18 Effect of particle size on specimen preparations: the fine-grained (shaded) component must fit in between the coarse-grained mineral component (white), and an inaccurate proportion of the former may collect in the analysed volume of the specimen.

uniformity is achieved in particulate specimens by grinding the specimen to particle sizes of $50\,\mu m$, or less. Figure 9.18 shows the danger of using a material of wide size range: one component has been preferentially ground and, as a result, an unrepresentative proportion of this component may collect in the shallow, analysed layer.

7. On-stream XRF analysis

Representative portions of the many streams of mineral pulp in a typical mineral processing operation can be passed continuously through specially designed XRF analysing devices. Parts of the pipes used to transport the different pulp streams must be made transparent to X-rays (see Fig. 9.19) in order that these pulps can be analysed.

Such a system provides an essentially continuous measure of the proportion of a given element in a pulp. If necessary, the system can be arranged to determine first one and then another of a range of elements every few minutes. This method of analysis is excellent for routinely monitoring the elemental compositions of plant products that are intended to remain at set values. It does *not*, ordinarily, provide any information about the mineralogical content of the analysed material; however, this method of analysis is especially useful for checking the elemental content of a product which is being sold on the basis of its chemical composition (for example, a copper concentrate).

8. X-ray fluorescence analysis: general

(1) XRF analyses employ intense polychromatic primary beams, e.g. the beam obtained from a copper target at 50 kV.

(2) The intense primary beam gives rise to a

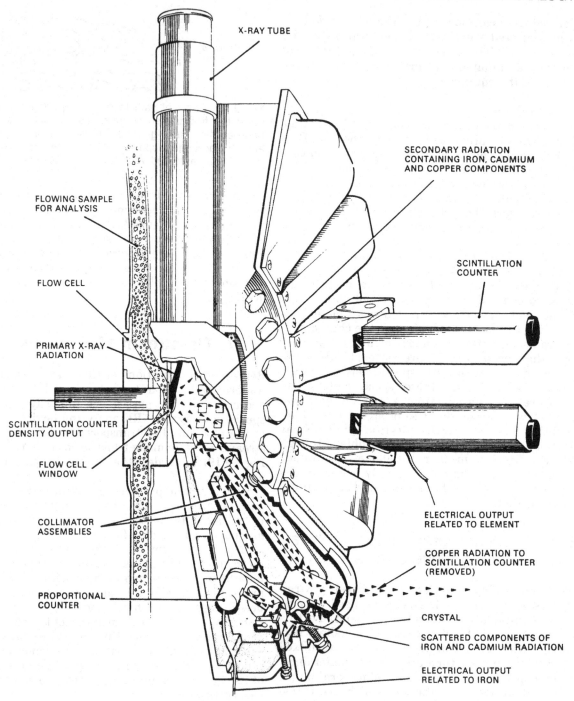

Fig. 9.19 On stream X-ray fluorescence.

comparatively strong beam of secondary (fluorescent) X-rays. A strong secondary beam either can reduce the time needed to collect an adequate number of X-ray photons, or it will reduce the statistical counting error on the count rates achieved per unit time.

(3) The intensity of the characteristic radiation from the specimen will vary with the concentration of a given element in the specimen.

(4) The *intensity* of the characteristic radiation will be affected by (*a*) matrix absorption, (*b*) secondary fluorescence, (*c*) background, and (*d*) counter dead-time.

(5) Elements greater than $Z = 5$ can be

detected and qualitatively measured in some instruments; elements above $Z = 12$ can be accurately measured. The lower limit of measured Z value is due to:

(a) absorption of the low-energy secondary radiation within the spectrometer by stray gas molecules,

(b) absorption of the secondary X-rays by the counter window (the counter contains ionisable gas and must, therefore, be separated from the vacuum in the spectrometer by a suitable window),

(c) absorption of the secondary X-radiation by the specimen itself, i.e. low-energy characteristic radiation produced within a specimen is strongly absorbed by the specimen itself and, therefore, may never reach the detector,

(d) difficulties in providing analysing crystals with the required, very large d spacings needed for the long-wavelength X-rays which are produced by low-atomic-number elements.

(6) Because of the self-absorption effects mentioned above, it is not practicable to detect very small proportions of a "light" element (i.e., one of low Z value) in a "heavy" matrix. Very small proportions of a "heavy" element can, however, often be detected in a "light" matrix (for example, the radiation emitted by carbon within a uranium matrix is easily absorbed whilst very small amounts of uranium can be detected in a carbon matrix).

(7) The accuracy of an analysis depends, in part, on the counting times used during that analysis, and on the signal-to-noise ratio (or peak-to-background ratio) of the measured characteristic line.

9. Electron probe X-ray microanalysis

9.1. Introduction

The electron probe X-ray microanalyser (XMA) combines some of the features of electron microscopy (described in section 10 below) and X-ray fluorescence spectrometry. In a microanalyser (see Fig. 9.20) a beam of high-energy electrons is fosused on to an area of about $1-2\,\mu m^2$ on the surface of a suitably prepared specimen which, in this instrument, acts as the "target". A variety of signals is produced from the surface layers of the specimen and these can be used to derive information about that specimen. These signals vary widely in their photon energies and are generated from slightly different depths within the specimen (see Fig. 7.13).

On striking the specimen some of the bombarding electrons are reflected or *back-scattered* from a thin surface layer, about $1\,\mu m$ deep, in a manner roughly analogous to the reflection of light from a similar surface. These back-scattered electrons (BSE) are easily detected and can be used to provide information about (1) the topography of the specimen surface, and (2) the mean atomic number ($\bar{Z}$) of the material in the small irradiated volume of the specimen. Other electrons in the primary beam penetrate to depths of 1 or $2\,\mu m$, and excite the atoms in the specimen to give off their characteristic X-rays (cf. section 2.3 above). These characteristic X-rays are detected and measured, either in one of a number of X-ray spectrometers or, increasingly nowadays, in an energy-dispersive solid-state detection device.

The wavelengths of the X-rays produced from the small volume of the specimen irradiated by the electron beam indicate which elements occur in that volume. The intensities of these characteristic radiations are proportional to the amounts of these elements present (cf. section 6.4.1 above).

On striking the specimen some of the energy from the original electron beam may be converted into visible light – this effect is called *cathodoluminescence* and it can be used, on occasion, to provide useful information regarding the trace-element composition of the irradiated volume of specimen. For example, cassiterite (SnO_2) often glows with a blue-green colour – this colour varies with the titanium and the niobium contents of the mineral.

The bulk of the energy contained in the electron beam is, however, converted within the specimen into heat – so much heat, in fact, that particular care has to be taken to ensure that delicate minerals are not harmed by the local build-up of temperature. A build-up of electrical charges will also take place on non-conductive specimens unless these charges are conducted away to earth (i.e. to ground) – see section 9.2 below.

Electron probe microanalysers are, nowadays, invariably equipped with vacuum-path spectrometers and with diffracting "crystals" of appropriate d spacings so that they can analyse all elements above atomic number $Z = 4$. Man-made, crystal-like, materials called *pseudo-crystals* are used for measuring X-rays produced by the lower atomic number elements. These microanalysers are extremely useful since they can be used to carry out rapid and accurate analyses of extremely small amounts of material (typically about $1\,\mu m^3$ in volume and weighing only about 10^{-12} grams). Such small volumes of material

Electron – optical column

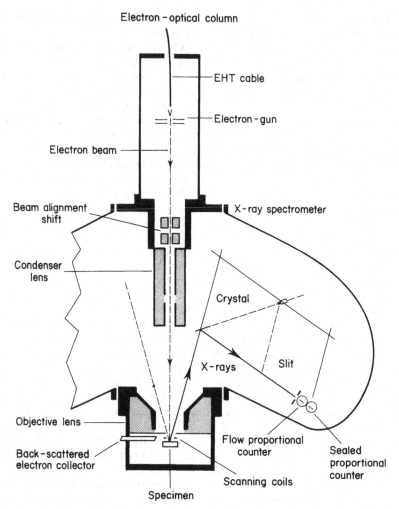

EHT cable

Electron – gun

Electron beam

Beam alignment shift

X-ray spectrometer

Condenser lens

Crystal

X-rays   Slit

Objective lens

Flow proportional counter

Sealed proportional counter

Back-scattered electron collector

Scanning coils

Specimen

Fig. 9.20 Diagrammatic plan view of the "Geoscan" microanalyser showing arrangement of electron-optical column and the two crystal spectrometers. Although no longer in manufacture, the diagram illustrates the arrangement of the electron-optical column and crystal spectrometer (after Long in Zussman, 1977).

very often consist of only one mineral phase and, therefore, it is possible with this equipment to analyse single, small mineral grains without first having to remove them from their rocky matrix or having to produce purified fractions of the mineral.

In the electron probe X-ray microanalyser a stream of high-energy electrons is focused into a fine beam by a series of electromagnetic lenses (see Fig. 9.21). This beam can be made to scan electronically over small areas of specimen: these areas range from $2\,\mu m \times 2\,\mu m$ in size to about $1.5\,mm \times 1.5\,mm$. Whilst the electron beam is moving over the specimen the BSE signal produced from each irradiated point on the specimen is used to modulate the brightness of a cathode-ray tube (or oscilloscope) which is being synch-

ronously scanned. In this way, an image is built up on the tube and this shows the variations of the BSE current values from different parts of the specimen surface (Fig. 9.22).

The back-scattered electron signal is a function of both compositional and topographical details of the specimen surface. However, the topographical detail can, if necessary, be reduced almost to zero by suitable polishing techniques or by using electronic "tricks". As a result, the contrasts in the BSE image generally mirror the $\bar{Z}$ variations within the specimen; i.e. variations in the back-scattered electron signal indicate variations in the minerals in different parts of the specimen surface. Unfortunately, as shown in Appendix 5, it is possible for several minerals to have the same mean atomic number, and mineral dis-

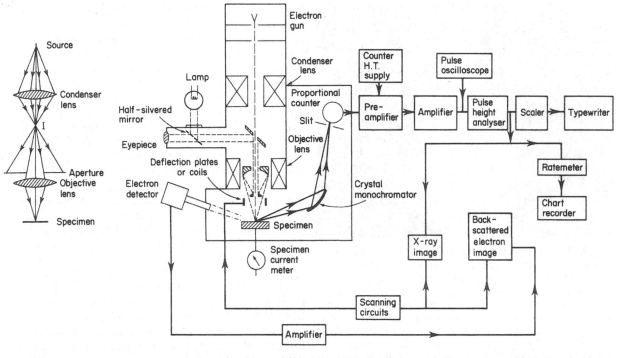

Fig. 9.21 Schematic diagram showing the components of a typical electron probe microanalyser including the light ray paths in the electron-optical system which allow the specimen to be viewed whilst it is being bombarded by the electron beam (after Long in Zussman, 1977).

crimination by BSE signal alone is therefore not always possible.

Even better contrast between minerals can be achieved by using the characteristic X-ray signals from these minerals to modulate the brightness of the oscilloscope (see Fig. 9.24). The oscilloscope will then show, in a semi-quantitative manner, the distribution and the concentration of specified elements on the area of specimen scanned by the electron beam. An "element" image can be photographed with an oscilloscope camera, or the signals can be fed directly into a computer for further analysis.

More detailed information concerning, say, the nature and the composition of the boundary zone between two minerals can be obtained either by electronically traversing the beam across that boundary, or by mechanically traversing the specimen beneath a stationary beam. The results of an electronic line scan can be shown on the oscilloscope and, if necessary, it can be superimposed onto the BSE image of the surrounding area (Fig. 9.23).

If the focused beam and the specimen are held stationary it is possible to obtain a fully quantitative chemical (elemental) analysis of the minute volume of specimen (about $1 \mu m^3$) being irradiated by the beam. This is done in much the same way as in X-ray fluorescence analysis except that the *primary* X-rays are measured. The intensities of the characteristic X-rays derived from the specimen are compared with those obtained from standard materials. As with XRF, the effects of matrix absorption, secondary fluorescence, background and dead-time must be taken into account (see section 6.4.1 above). With XMA, an additional effect must be taken into account; the proportion of the incident electrons lost in the back-

Fig. 9.22 Back-scattered electron photograph (cf. Fig. 9.24).

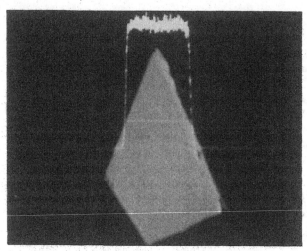

Fig. 9.23 Line scan photograph showing variation in X-ray signal intensity across a field of view in an electron probe X-ray microanalyser. The bright horizontal line across the particle is the scan line. The other line shows the signal strength (and, therefore, the elemental composition) along the scan line. The particle is a grain of quartz. The variation in signal strength shows the variation in silicon concentration along the line scan which, in this instance, is about 100 μm from left to right (with acknowledgements to Mrs E. A. Lewis).

scattered electron signal (as mentioned above, this proportion varies with the mean atomic number value of the analysed material).

Only very small volumes of the specimen are analysed at a time, and because such small volumes are likely to be homogeneous, it is only necessary to use pure element (or pure oxide, or pure sulphide) standards (cf. quantitative XRF, which demands large numbers of standards of widely varying compositions).

9.2. Specimen preparation

Thin sections, polished sections and even fine-grained particulate materials can be examined in the microanalyser. However, because the surface topography of a specimen affects the results, it is usual to use flat, polished specimens.

Delicate specimens, such as hydrated minerals, must be protected from damage by the heat generated by the electron beam and from the effects of the high vacuum used in the instrument. In addition, the free electrons introduced into the specimen by the beam must be conducted away to earth. Very few mineral specimens are sufficiently conductive to do this and they must be coated with thin conducting films. These films must be essentially transparent to the incident

electron beam, they must be transparent to the X-rays produced in the specimen, and they must also be transparent to the back-scattered electrons. Furthermore, the conducting films must be capable of being earthed easily. Vacuum deposition of low-atomic-number metals produces films which are only about 5 nm thick and these provide good electrical conductivity and are also transparent to X-rays and to the electron beams. Unfortunately, these metal films strongly reflect light so that metal-coated specimens act as mirrors and cannot readily be studied in the reflected-light microscope fitted in the microanalyser. Consequently, mineralogical specimens are usually coated with a layer of deposited carbon about 10–15 nm thick. This coating has adequate electrical properties and, at the same time, is largely transparent to light (although some unusual interference colours are assumed by the coated mineral surface: for example, pyrite grains can appear distinctly pink in reflected light rather than their usual yellow).

A great deal of heat can build up within a mineral of low thermal conductivity whilst it is being bombarded by an electron beam. The accompanying rise in temperature can seriously alter some minerals by causing them to dehydrate or even to decompose. In the more usual kind of X-ray generator the heat produced in the target is conducted away by water-cooling that target, but this is not possible in the XMA because the mineral specimens themselves form the "target".

These undesirable thermal effects can be minimised either by reducing the amount of energy in the electron beam (i.e. by reducing the beam voltage and/or the beam current) or by coating the specimen with a thin film of a good thermal conductor. Although carbon is not the ideal material for this task, it is adequate for most purposes and, as shown above, it also provides the required electrical conductivity whilst still allowing the specimen to be viewed in vertically reflected light.

In the microanalyser, the specimen, the beam and the X-ray spectrometers are, necessarily, enclosed within a high vacuum (approximately 10^{-6} mmHg) and the vacuum can occasionally dehydrate minerals such as chrysocolla or hydrated iron oxides. This effect cannot easily be avoided.

Particulate specimens are normally mounted in a cold-setting epoxy resin. In this way, any undesirable thermal effects that might occur during solidification of the resin are avoided. Epoxy resin does not readily conduct either heat or electricity and it may melt or even evaporate under the electron beam. If the resin is mixed with 10–20%

of powdered graphite, however, the thermal conductivity and the electrical conductivity are greatly improved. In any case, the resin-based specimen mounts are still coated with a carbon film to maximise their electrical and thermal conducting properties.

9.3. Use of the XMA (X-ray microanalysis) in mineralogy

The mineralogist desires to know the identities, the compositions, the sizes and the textures of the minerals in rocks and in granular products (Chapter 1, section 6). This type of information can be produced by manual microscopy methods or by automated, optical methods of image analysis (Chapters 6 and 7). However, the applicability of these methods of analysis is often limited because of poor optical contrast between the components of a specimen. The microanalyser, on the other hand, can provide *chemical* contrast, and this is often better than the optical contrast seen in the microscope. Furthermore, the X-ray spatial resolution of the microanalyser is probably better than the best optical image analysers and the microanalyser can often be used to provide a great deal of useful mineralogical information. Much of this information can be produced automatically, with virtually no direct involvement of the mineralogist.

For example, one of the major functions of the microanalyser within the mineral industry has been to examine, identify, and then quantify the mineralogical details of mineral deposits and ores. Figure 9.24 shows a group of oscilloscope pictures obtained from a complex material containing sulphide minerals. These pictures (and Fig. 9.22) show the distributions of copper, lead, zinc, iron, bismuth and sulphur, and also the variations of BSE values, in a specimen of this rock. From these pictures (and, if necessary, from a few confirmatory "spot" analyses carried out in the microanalyser) the identities and the locations of chalcopyrite, galena, sphalerite, pyrite and bismuth over a small area of the specimen can be observed and, if necessary, the mineral proportions and the grain sizes can be measured. If required, the overall proportions of the various minerals in the complete specimen can be determined by detailed measurements of a mosaic of such pictures (or by linear analysis methods – see Chapter 6, section 5.2).

The microanalyser is especially useful for determining the detailed composition (and also the variability of composition) of an individual mineral species, even when that mineral occurs as very small, individual grains. Selected areas a few square micrometres in size within a zoned mineral can be accurately analysed and this information can be useful for determining the genetic history of a rock.

The microanalyser is also extremely useful for providing the mineralogical information needed to solve some of the design and production problems encountered in the mineral industry. For example, it can be used to establish the manner in which the valuable elements exist in the tailing (discard) product from a treatment plant. An examination of such a product from a tin-recovery operation might show that 30% of the tin was being lost in the form of unliberated cassiterite: some of this cassiterite (say, 50%) might be coarse-grained but the rest of the lost mineral might be very fine-grained. With this information the mineral engineer would be able to take the remedial action needed to improve the recovery of tin (see Chapter 10).

9.4. X-ray microanalysis: summary

The XMA provides a means of determining the chemical compositions of small, selected regions, about 1 to $2 \mu m^2$ in area, on the surfaces of polished mineral specimens or polished grain mounts. It does this by making the mineral specimen the "target" in a novel kind of X-ray generator unit. In this instance, the wavelengths of the primary X-rays produced by electron beam bombardment are used to determine the elements in the specimen. The intensities of these characteristic radiations are proportional to the concentrations of the elements in the small irradiated volume of that specimen.

A system of magnetic lenses focuses an electron beam on to an area of about $1 \mu m^2$ on the specimen surface. The specimen and a group of standard specimens (which in the microanalyser can be pure elements) are mounted together on a motorised stage: a co-axial microscope allows the area of interest to be selected optically before being analysed by the electron beam. Multiple spectrometers (commonly three or four) are used to detect X-rays of selected wavelengths and, as in the XRF system, the intensities of the X-rays are measured with proportional flow counters. Energy-sensitive solid-state X-ray detectors (see Fig. 9.8 and section 3.5 above) can also be used and these detectors permit the whole of the emitted X-ray spectrum to be recorded in one measurement. This "energy-dispersive" technique is quick and simple to operate and is being in-

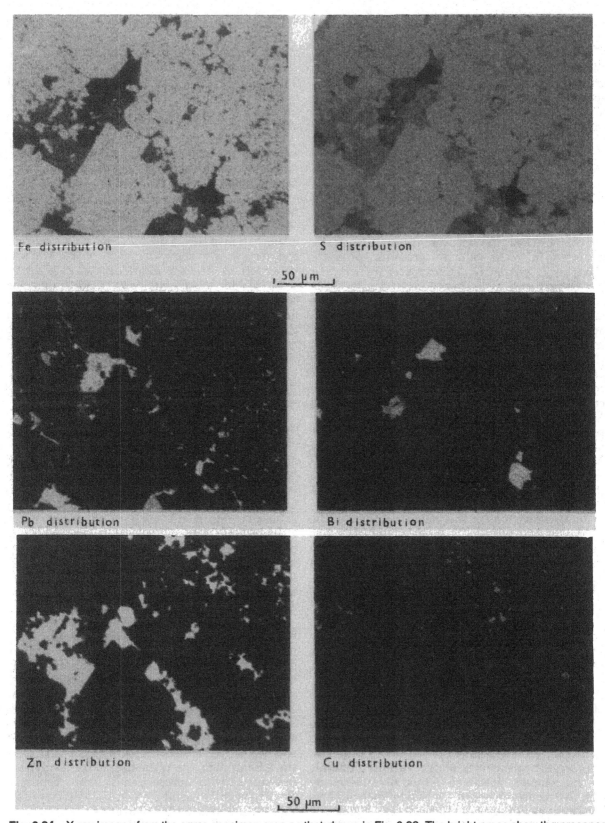

Fe distribution

S distribution

50 µm

Pb distribution

Bi distribution

Zn distribution

Cu distribution

50 µm

Fig. 9.24 X-ray images from the same specimen area as that shown in Fig. 9.22. The bright areas show the presence of the stated element – the brighter the element the higher the concentration of that element. Pyrite (FeS$_2$) is shown by the presence of iron and sulphur; sphalerite (Fe.Zn)S is shown by low concentrations of iron and sulphur with high proportions of zinc; etc.

creasingly used for rapid identification of unknown minerals: it can also be used to provide quantitative analyses of these phases.

The characteristic X-ray spectra generated by the electron beam are similar to those produced in a common X-ray generator. In addition to the characteristic peaks, the bombarding electron beam produces a continuous spectrum and this continuous spectrum forms a background signal for the peaks. A quantitative analysis at a "point" involves a comparison of the net intensity of a characteristic line (i.e. peak value minus background level) of each element with the corresponding value generated under identical conditions in a standard specimen of known composition. Peak intensities (I_{sp} and I_{st} for the specimen and standard respectively) and element concentrations (C_{sp} and C_{st}) are then related according to:

$$I_{sp}/I_{st} = K(C_{sp}/C_{st})$$

where K is a correction factor that takes into account the differences in the matrices of specimen and standard. Values of K can be *calculated* from readily available data and these calculated values are used instead of the carefully matched standards required with XRF analysis.

The XMA does not (and cannot) directly provide any mineralogical data, but its chemically based data are often readily interpreted in mineralogical terms. Quantitative analyses for elements above sodium ($Z = 11$) in the Periodic Table can be performed with relative accuracies of 1–2% for concentrations greater than about 1%. Quantitative analyses for elements of atomic numbers 5 to 10 (inclusive) are more difficult to carry out and relative accuracies of 10% are usually all that can be attained. The limit of detection for an element varies inversely with its atomic number: it is typically about 50 to 100 parts per million in the analysed volume, but it can be lower in favourable instances.

10. The scanning electron microscope

10.1. General

The electron microscope was first used in the 1930s: the development of the *scanning* electron microscope (SEM) began in the early 1960s and has proceeded almost in parallel with that of the microanalyser. In mineralogy, the SEM is generally used in the reflecting mode because most minerals strongly absorb electron beams and the electrons can only pass through *very* thin specimens (about 1 μm thick, or less). Such exceedingly thin mineral specimens are very difficult to prepare.

Today, the scanning electron microscope and the X-ray microanalyser are sometimes difficult to distinguish from one another. The electron-optical resolution of the true microanalyser has been improved until it now rivals (but does not equal) that of the true scanning electron microscope. Furthermore, most scanning electron microscopes are now either fitted with at least one X-ray-wavelength spectrometer or, more commonly, with energy-dispersive X-ray detectors. However, the true SEM is designed to achieve the best possible spatial resolution, whilst the microanalyser is designed to achieve the best possible chemical resolution.

The electron beam in the SEM is of the order of 10 nm in diameter and the probe current is of the order of 1–10 pA (much smaller than in the XMA). As with the microanalyser, the beam is scanned over a suitably prepared specimen by sets of scanning coils. The beam produces much the same effects in the specimen as does the beam of the microanalyser. However, the main discrimination feature used in the SEM is the electron signal emitted from the point of impact of the beam. By using appropriate electron filtering and detection systems it is possible to show either the compositional differences or the topographical effects on the specimen surface. As described in section 9.1 above, the back-scattered (reflected) electrons are electrons from the original beam which, after multiple ricochets within the specimen, re-emerge from its surface. The energies of these back-scattered electrons range from the beam energy (a few kilo electron volts) to zero. Secondary electrons are also produced: these are the electrons that were initially bound into the atoms of the specimen but have received sufficient additional energy from the incident beam to escape from the specimen. Secondary electron energies range from about 50 eV to zero (see Fig. 9.25) and these electrons are characteristic of the elements in the irradiated specimen.

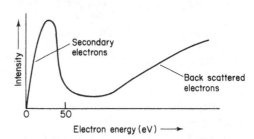

Fig. 9.25   Differences in energies of secondary electrons and back-scattered electrons.

10.2. Back-scattered electrons

The proportion of the incident electrons back-scattered from a specimen is a function of the mean atomic number of the small part of that specimen being irradiated by the electron beam (for carbon the proportion is less than 10%, for uranium it is over 50%) (Fig. 9.26).

A topographic feature on such a specimen surface will increase the back-scattering in some directions and decrease it in others. If two back-scattered electron detectors are positioned on diametrically opposite sides of the specimen then, by subtracting one signal from another, it is possible to produce an image that ignores the variations in atomic number effect and which consists almost entirely of topographical detail. On the other hand, if the signals from a pair of diametrically opposed electron detectors are added together, the topographic detail is more or less cancelled out and an image based largely on differences in atomic numbers can be produced.

Greater spatial resolution (about 10 nm) and greater depth of focus are possible by using the "secondary" electron image. The secondary electrons are emitted, almost entirely, from the point of impact of the electron beam with the specimen and the spatial resolution is approximately the width of that beam. On the other hand, although

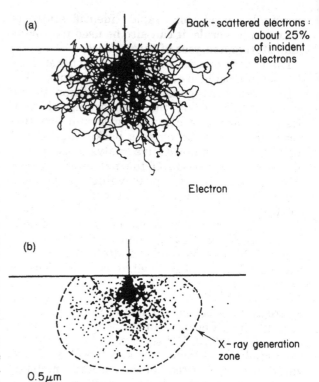

Fig. 9.26 Computer plot of (a) electron trajectories, and (b) X-ray production from a copper target at 20 kV (after Long in Zussman, 1977).

TABLE 9.2
Signals generated within a specimen by a scanning electron beam

Signal	Energy range of signal	Depth of emission	Kind of information obtained
Characteristic X-rays	Depends on Z values of the elements in the specimen	1 to 3 μm	X-ray spectra, quantitative analyses, element distribution images
Continuous X-rays	From zero to the energy of the beam electrons	1 to 3 μm	"Noise" or interfering signal
Back-scattered electrons	From zero to the energy of the beam electrons	about 10 nm	High-resolution topographic images; channelling pattern images
Secondary electrons	0 to 50 eV	10 to 1 nm	High-resolution topographic images
Absorbed electrons (specimen current)			Material contrast, i.e. contrast, low-resolution topographic contrast
Cathodoluminescence	1.0–3.1 eV	Various	Phase discrimination, trace-element contents of minerals
Transmitted electrons	Energy of primary electrons	Very thin specimens	Transmission images

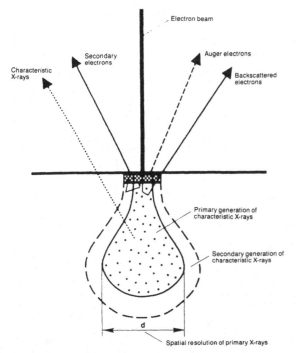

Fig. 9.27 Point of emission of secondary electrons from a small near surface zone (other radiations are also shown).

the electron beam in the SEM is very small, the spatial resolution obtained by using the emitted X-ray signals is still only 1 to 2 μm (see Fig. 9.27).

The SEM is used in a manner analogous to the ordinary optical microscope and the images produced in the SEM are often similar to those obtained with a good, conventional microscope. However, the SEM can also provide images with a very wide range of useful magnification, and at the same time it provides much greater depth of focus than any optical system.

Table 9.2 lists the various signals that are generated when a specimen is bombarded by a beam of high energy electrons: it also shows the depths within the specimen where these signals originate (and hence provides a good indication of the spatial resolution that can be obtained by using that signal).

11. Miscellaneous methods of mineralogical analysis

Introduction

There are many other analytical methods that can be used to establish mineral identities and to determine mineralogically useful parameters. This section describes very briefly some of the more important of these methods.

11.1. Radiographic techniques

A few minerals are distinctly radioactive. These spontaneously emit a variety of high-energy radiations, which can be used either to identify the minerals or to locate their positions within a specimen.

11.1.1. Autoradiography

This method was first used in mineralogical work to show the distribution of pitchblende (which contains uranium and thorium) in uranium ores. Specimens of these ores were placed on unexposed photographic plates and it was found that the radioactive mineral(s) produced black images when the plates were developed. The degrees of blackening provided clues to the nature of the minerals whilst the locations of the blackened areas indicated the mineral locations. (Positive prints of these *autoradiographs* would, of course, show up white instead of black.)

This procedure can also be used to identify the radioactive minerals in a group of mineral particles. The particles are sprinkled over and then pressed into the emulsion of an unexposed photographic plate; after an exposure period that may be measured in hours or even days, the developed film will show a distinct blackening around each radioactive particle. Specially thick photographic emulsions are available for this work and small mineral particles (100–200 μm in size) can be firmly embedded in the emulsions and will not fall out during the development stages. These photographic emulsions do not, however, differentiate between the various kinds of radiation produced by different radioactive minerals.

(i) *Alpha-particle radiography*
Alpha-particles are produced during the decay of the uranium atom and these particles have a range of about 20–25 μm in photographic emulsion (Fig. 9.28). Therefore, the spatial resolution

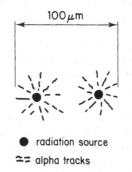

Fig. 9.28 Diagram of alpha-particle tracks derived from two adjacent uranium sources.

of this technique is usually no better than about 30 μm, i.e. two radioactive particles which are closer together than 30 μm will appear as only one, larger image on the radiograph. Resolution is at its best when the emulsion thickness is reduced and when the mineral specimen is in intimate contact with the detector (i.e. it is pressed against the photographic emulsion). Solid specimens are therefore usually polished and then placed in contact with an emulsion 5–10 μm thick.

The alpha-particles produce visible tracks in the emulsion and the lengths of these tracks are indicative of the type of radioactive nuclei (thorium or uranium) which give rise to them.

(ii) Beta-particle radiography

Fast emulsions produce records of the paths of low-energy electrons (beta-particles) as well as the high-energy alpha-particle tracks. Beta-tracks are usually curved (the alpha-tracks are always straight) and it is often difficult to trace them to their source. However, careful study of the radiographs can differentiate between the alpha-tracks derived from uranium nuclei and the beta-tracks which have originated in other nuclei.

11.2. Infrared spectroscopy

11.2.1. Introduction

The first studies on infrared spectroscopy took place in the early 1900s, but it is only comparatively recently that any substantial progress has been made in the use of this method for mineral identification purposes.

When a mineral is irradiated by radiation within the infrared range (wavelengths ranging from about 1 μm to 1 mm) the mineral produces a characteristic spectrum that yields information concerning its interatomic bonding. The absorption, or reflection, of infrared radiation by the mineral gives rise to changes in the vibrational energy of the constituent molecules: these changes are specific and can be used to identify the mineral.

It is possible to obtain an infrared spectrum by the (1) absorption, (2) reflection or (3) emission of suitable radiation. However, of these methods, the absorption method is the most commonly used in mineralogy.

The absorption spectrum of a specimen is a plot of the amount of radiation absorbed (or, conversely, transmitted) by a specimen against the frequency (or the energy) of that radiation (Fig. 9.29). An infrared absorption spectrum yields information on both the structure and the bond-

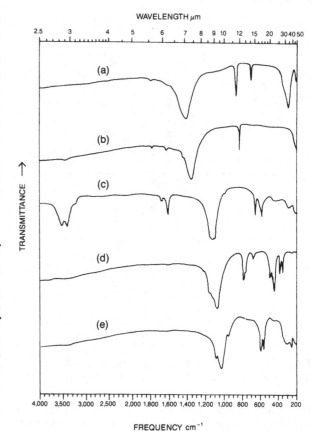

Fig. 9.29 Infrared absorption spectra of some commonly studied minerals: (a) Calcite (CaCO$_3$); (b) Soda-nitre (NaNO$_3$); (c) Gypsum (CaSO$_4$.2H$_2$O); (d) Quartz (SiO$_2$); (e) Fluorapatite (Ca$_5$(PO$_4$)$_3$F) (after Estep-Barnes in Zussman, 1977).

ing of substances and, although this kind of information is of greatest value for characterising organic materials, it can also be used to identify certain kinds of unknown mineral specimens.

The most dominant bands (or peaks) in the infrared absorption spectra of minerals are due to the fundamental vibrations of isolated, and tightly bound, molecular groups such as CO$_3^{2-}$ and SO$_4^{2-}$. Weaker bands are often caused by lattice vibrations. Figure 9.29 shows the absorption spectra of a few typical minerals.

Further details of this analytical technique can be found in modern textbooks, such as Zussman 1977.

11.2.2. Sample preparation

The infrared absorption spectra obtained from a mineral will be of poor quality unless the specimen has been properly prepared. For example, particle size is very important. Theoretical considerations dictate that the particle size should be less than the wavelength of the incident infrared radiation in order to minimise losses due to scattering and reflection effects. The optimum particle size is, in fact, less than about $2\,\mu m$; however, the large amount of grinding needed to produce such small particles can result in the generation of local high temperatures and unexpected changes in the structure of the mineral being studied. All material should, therefore, be wet-ground whenever possible to minimise the formation of local "hot-spots". If water-soluble minerals are thought to be present then the specimens should be ground in alcohol.

To obtain high-quality infrared absorption spectra it is also important, whenever possible, to match the refractive index of the specimen with that of its mounting matrix (see below). Significant differences in the refractive indices of the specimen and its dispersion medium can distort and shift the absorption bands in an unpredictable manner.

11.2.3. Measuring methods

A sample of the pulverised mineral is suspended in a material of similar refractive index. This matrix should be transparent to infrared radiation in the spectral range of interest, it should be chemically stable, and it should need only a low sintering pressure to form a pellet. Potassium bromide and caesium iodide are widely used for this purpose.

Ordinarily, the pellets are prepared using about 1 mg of ground mineral in about 300–400 mg of the alkali halide. The mixture is pressed to form a firm pellet. Modern developments, however, allow the use of small samples that weigh only about $10\,\mu g$. These small samples permit comparisons to be made between infrared data and complementary data which have been obtained from other microanalytical techniques such as electron probe X-ray microanalysis, X-ray diffraction analysis and petrographic analysis.

Infrared spectroscopy has been used to show that the mineralogically complex, manganese-rich, nodules found on the floors of the deep oceans consist mainly of hydrated iron and manganese oxides. Similarly, very fine-grained marine phosphate deposits have been shown to contain significant amounts of the minerals quartz and calcite.

11.3. Thermal techniques

Thermal analysis is a general term used for a group of related techniques, all of which measure the variation with temperature of some physical property – e.g. specimen mass, or linear dimension, or magnetic permeability, or electrical conductivity. In addition, it is often possible, at the same time, to analyse any gases that are evolved or to study fluorescent radiation that may be produced on heating a specimen.

These effects provide information regarding the absorption and desorption of moisture and other substances from the specimen. They also indicate chemical reactions, such as decomposition and oxidation, that take place. Polymorphic transformations from one crystal structure to another are reflected in exothermic or endothermic effects. Recrystallisation effects and changes of phase (from solid to liquid or gas) can also be detected. These thermal effects are sometimes so obvious and so characteristic of one particular mineral that they can be used to identify that mineral within a composite material and even provide a quantitative estimate of its mineralogical composition. The most commonly used techniques of thermal analysis are briefly described below.

11.3.1. Differential thermal analysis (DTA)

In this method of analysis the differences in temperature between an unknown specimen and an inert reference material are established when the two substances are subjected to identical heating or cooling rates. These temperature differences are caused by the preferential evolution or absorption of heat by the substance which is being examined.

Specimens of the unknown material and of an essentially inert material (such as alumina) are heated in the same environment at a controlled rate. The temperature of the inert material will steadily increase whilst the rate of temperature increase of the unknown material is unlikely to remain steady. The *differences* in temperature (Δt) between the two materials are plotted against time or against the temperatures of the inert material (Fig. 9.30). Exothermic reactions in the unknown specimen will show as positive Δt peaks, whilst endothermic events will show as negative peaks. The positions and the sizes of these peaks can be interpreted in terms of the chemical reactions or crystallographic transformations that are taking place in the specimen; for example, effects such as desorption of a substance, the loss of a structural hydroxyl group, the decomposition of carbonates and sulphates, and changes of phase all produce endothermic effects. On the other hand, the oxidation of sulphide minerals, the collapse of defect structures and the recrystallisation of a mineral all produce exothermic peaks.

Figure 9.30 shows the DTA curve for the min-

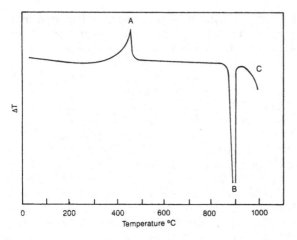

Fig. 9.30 DTA curve of anglesite with approximately 5% pyrite contamination (after Neumann in Zussman, 1977).

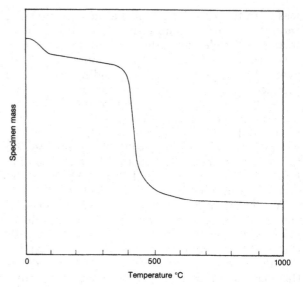

Fig. 9.31 Thermogravimetric curve of kaolinite (after Neumann in Zussman, 1977).

eral anglesite ($PbSO_4$). The left-hand, exothermic peak is due to the oxidation of a small amount of pyrite which occurs as a contaminant in the specimen; this reaction can be seen to begin slowly at about 300°C and the rate of oxidation increases as the temperature rises to about 450°C when the reaction ceases. The middle, large, endothermic peak represents the change of the anglesite from the orthorhombic crystal system to the monoclinic system: this change occurs very rapidly and is completed over a small temperature range. The incomplete curve at the right-hand end of the trace shows the decomposition of the anglesite: the decomposition starts slowly and was incomplete when the experiment was concluded.

The base-line values of such a graph are not always as easy to establish as those shown on Fig. 9.30, and the interpretation of the results is then more difficult.

Use of DTA in mineralogy
DTA is mainly used for identifying minerals, especially clay and clay-like minerals, which undergo dehydration, oxidation, decomposition or crystallographic inversion on being heated. The clays occur as very small primary particles (i.e. they need not be ground) and because the particles are so small (around 1–2 μm in size) they are very difficult to characterise by the traditional, optically based methods of examination.

11.3.2. Thermogravimetric analysis (TGA)
In a thermogravimetric analysis the changes in *mass* of a specimen are measured when it is

heated (or, rarely, cooled) at a constant rate in a controlled atmosphere (Fig. 9.31). The specimen is attached to a suitable balance and is placed in a small furnace. The furnace is heated at a constant rate and the mass of the specimen is recorded on a chart (cf. DTA method).

The typical thermogravimetric curve starts with a level plateau, indicating that the weight remains unchanged for some time. Rapid mass losses then show up as marked steps in the curve. Ideally, these steps should be sharp and well-separated from each other, showing the temperatures at which a series of separate effects take place. Unfortunately, in practice the steps become blurred because the reactions and the removal (or accretion) of the reaction products (which, after all, cause the loss of mass) tend to occur over ranges of temperatures.

Mass *losses* can be caused by (1) the removal of adsorbed moisture, (2) loss of hydroxyl groups, (3) removal of volatile substances other than water (e.g. carbon dioxide from the decomposition of carbonate minerals, sulphur trioxide from sulphates, the loss of sulphur dioxide from sulphides in an oxidising atmosphere). Mass *gains* can occur by the oxidation of specimen components.

Use of thermogravimetric data
The range of minerals that can be studied by thermogravimetric analysis is limited to those which release volatile substances when they are heated by themselves or with selected reactants, and those which gain weight when reacted with reagents such as air. These minerals include the

carbonates, the clays and the clay-like minerals, organic "minerals" such as coals and resins, sulphates such as gypsum, sulphides, and so on. The minerals in a simple specimen can be identified by comparison with the results obtained from standard materials but the interpretation of thermogravimetric curves can become difficult when the specimen contains a number of mineral components and when two or more reactions overlap.

Chapter 10

The role of mineralogy in mineral processing

1. Introduction

As described in Chapter 1, most of the valuable minerals of commerce occur in nature as dilute rocky mixtures, and these mixtures are of little value unless they can be upgraded into saleable concentrates.

Whether a mineral deposit can ever be exploited commercially depends on several factors. Some of these factors, in turn, depend on strategic, economic, political or social conditions which are beyond the control (but not the concern) of the mineralogist and the mineral engineer. However, even when economic and political conditions are at their most favourable, it still frequently happens that the major factor which finally decides if a mineral deposit will ever be exploited is the technical ability of the engineer to produce saleable products at a socially acceptable cost.

In order to establish whether or not such saleable products can be produced, the engineer must have (among other things) large amounts of accurate, readily available mineralogical data concerning the deposit. The enormous value of detailed mineralogical data becomes apparent at every stage in the exploitation of a mineral deposit – from its location by a geological exploration team to its final exhaustion by mining. For example, large amounts of mineralogical information are first needed for the evaluation of the quality and the extent of a recently located deposit. Questions must then be answered concerning the nature and the amount of its "contained values", in terms both of minerals and of elements. In addition, the engineer will be asked to establish both the compositional variability of the deposit and whether this variability will affect the mining procedures or the treatment processes. To provide realistic answers to these questions, sampling tests and small-scale amenability tests must be carried out as soon as suitable material becomes available. These tests will establish first of all whether or not it is technically feasible to prepare potentially saleable products from that deposit.

If these amenability tests are successful, then larger amounts of material will be put through extended test programmes to determine whether the saleable products can be produced at economic cost. Later, even larger amounts of material will be passed through a continuously operating, pilot-scale plant in order to design the optimum treatment process(es) for that deposit and also to assess the effects of continuous operation on the likely build-up of deleterious products in the plant.

Later still, large amounts of quantitative mineralogical data will be required by the mineral engineer when designing, commissioning and operating the full-scale mineral treatment operation and, finally, similar information is needed for designing safe, long-term facilities for storing the waste products produced by the plant. These will require continuous monitoring for decades, or even centuries, in order to ensure that

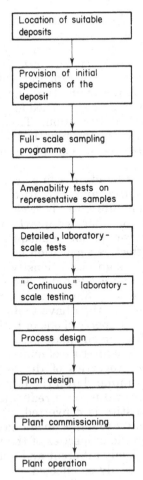

Fig. 10.1 Flowchart showing types of investigation undertaken by the applied mineralogist.

they do not become health or safety hazards (see Fig. 10.1).

The mineralogical data required by the mineral engineer for these eminently practical purposes differ in many respects from the information usually collected by the more academically orientated, "pure" mineralogist. Thus, the academic mineralogist may require only qualitative or descriptive information regarding rock textures, or may seek only rough estimates of mineral grain sizes. The mineral engineer, on the other hand, would prefer to have quantitative data on textures and on grain sizes whenever possible. The other main difference between the information required by the engineer and mineralogist is that the latter is primarily concerned with the features and the characteristics of unbroken rocks, whilst the mineral processing engineer is, *in addition*, greatly concerned with the manner in which rocks break and with the nature, and the behaviour, of the broken fragments under various plant conditions. Furthermore, the mineral en-

gineer is always anxious to obtain information as soon as possible, whereas the mineralogist may often be working to a longer time-scale.

Before looking in more detail at the type of data which are specially required by the engineer, let us look at the operation of a typical mineral treatment plant.

1.1. Typical mineral treatment operations

1.1.1. Liberation

On some occasions the mineral deposits exploited by the mineral engineer exist in nature in the form of unconsolidated materials such as beach sands, alluvial deposits and soils. The individual minerals in these deposits already exist in the form of discrete, single-phase particles (see Fig. 10.2 in colour section), and are usually completely liberated from each other.

More commonly, the material being exploited consists of a hard, competent, consolidated rock in which the individual mineral grains are firmly bound together (see Fig. 1.1 in colour section). The valuable minerals in such rocks can, at present, only be released (i.e. liberated) from the surrounding rocky matrices in a commercial operation by expending large amounts of mechanical energy in crushers and grinders to break the rocks into small fragments.

From the mineral processing engineer's point of view, the ideal kind of liberation would be achieved if a hard rock could be broken *along* the boundaries between the valuable mineral grains and the worthless matrix or *gangue* (Fig. 10.3). This kind of breakage is very difficult, but not impossible, to achieve – after all, it is continually being achieved by the natural agencies which produce the alluvial deposits and the beach sands mentioned above. Much more commonly, liberation is achieved by breaking a rock "at random" into progressively smaller fragments (Fig. 10.4). Liberation commences when the fragment size becomes smaller than the original grain size of either the gangue material or the valuable mineral. Since the gangue usually occurs in much greater proportion than the mineral, it may be possible to liberate some of the rocky matrix at a coarser size than the valuable mineral. This kind of liberation by random breakage will, unfortunately, only achieve 100% liberation when the rock is broken to zero particle size. Figure 10.5 compares broken and unbroken ore samples.

1.1.2. Separation

It is necessary to limit the amount of size reduction to that at which *adequate* liberation is

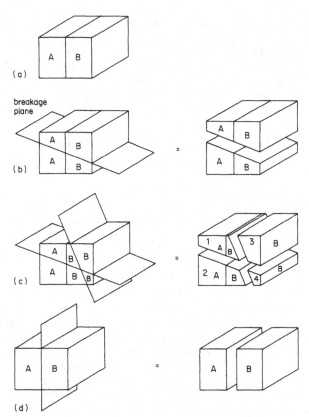

Fig. 10.3 Diagram showing breakage in a rock fragment composed of two mineral grains: (a) The unbroken fragment containing minerals A and B; (b) A single random break achieves no liberation; (c) Multiple random breaks liberate some of mineral B (particles 3 and 4 are liberated B; particles 1 and 2 are composite particles containing both A and B); (d) A single non-random break along a grain boundary produces perfect liberation of both A and B (after T352 O.U. publication).

could be made the basis of a separation technique that would provide mineral fractions consisting largely of clean quartz or clean magnetite. Unfortunately, the composite quartz–magnetite particles have a range of property values between those for clean quartz and those for clean magnetite and these must be returned for further grinding and further liberation. This sequence of events is shown diagrammatically in Fig. 10.7, which is a flowsheet of a simple magnetic separation process.

On the other hand, in a fine-grained rock consisting of quartz and hematite (rather than magnetite) the differences between the magnetic properties of the liberated quartz and the liberated hematite would not be adequate to form the basis of an effective separation procedure. There is, of course, a significant density difference between the two minerals but this property cannot be exploited at very fine particle sizes. The mineral engineer may then have to devise a method for *changing* a property of one or both minerals in order to produce an adequate property difference. For example, in a mixture of quartz and hematite the magnetic properties of the quartz and the hematite are similar, but if the mixture is heated to a high temperature in a reducing atmosphere then the hematite is converted into magnetite, whilst the quartz remains unchanged; after heating, the magnetic properties of the two minerals would be markedly different and a magnetic separation could easily be effected.

2. Mineralogical information required

Most mineral assemblages are much more complex than the examples given above, and the design and control of suitable processing operations for separating a multicomponent mixture of minerals requires the provision of a great deal of high-quality mineralogical information. Let us look, therefore, in detail at the *particular* mineralogical data which are required by the mineral processing engineer for these purposes. These data are summarised in Table 10.1, along with a brief indication of their importance in a mineral processing context.

2.1. Mineral identities

It is clearly important that the mineralogist should know the identities of the minerals in the rocks which are to be studied. It is equally impor-

achieved ("adequacy" being related to the cost of grinding and to the performance of the concentration methods that follow). The liberated, valuable mineral grains are then separated, along with the richer of the composite particles, from the unwanted waste material. This is done by separation techniques which rely on differences in the properties of the various particles.

Different minerals often show some naturally occurring exploitable property differences; for example, a ground-up quartz–magnetite ore would consist of liberated quartz particles, liberated magnetite particles and some composite particles consisting of quartz and magnetite (see Fig. 10.6). The quartz particles differ from the magnetite particles in (among other things) density, magnetic response and surface chemistry, and any one (or any combination) of these properties

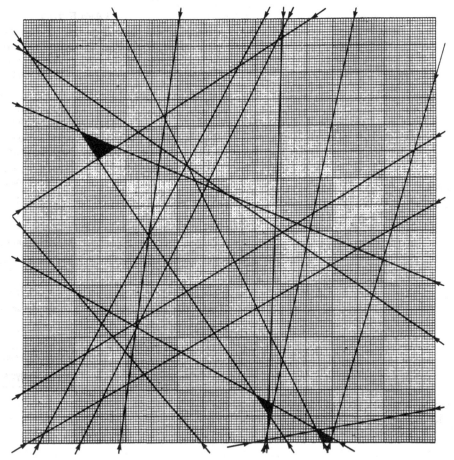

Fig. 10.4 Superposition of random breakage lines (indicated by arrows) over a two-dimensional, fine-grained, regular pattern. Mineral A is shaded, mineral B is plain. The amount of liberation achieved is very small (shaded areas) (after T352 O.U. publication).

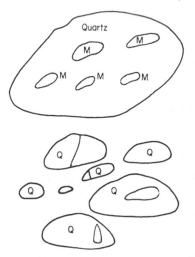

Fig. 10.5 Comparison of broken and unbroken ore samples. Above: unbroken ore fragment consisting of magnetic grains (M) in a quartz matrix; below: broken ore consisting of particles of liberated quartz, some particles of liberated magnetite and many composite particles that still contain both quartz and magnetite.

tant that the mineral engineer should know which minerals occur in the deposits that are to be treated. Once the identity of a mineral is known the engineer will have good indication of some of its more useful "industrial" properties. For example, cassiterite always has a high density, ilmenite is always moderately magnetic, galena is always dense and brittle and so on. This knowledge of the likely properties of a mineral can also provide the engineer with advance warning of the problems that might be encountered with certain minerals during various types of processing operations. These problems occasionally arise even when these minerals only occur in very minor proportions.

For example, the mineral engineer will know that particular care is needed when small amounts of the minerals graphite and pyrophyllite occur in a gold ore. The presence of these minerals will, almost certainly, result in unacceptable losses of gold during a cyanide-leaching process for recovering that gold because, in the presence of graphite and/or pyrophyllite, gold that

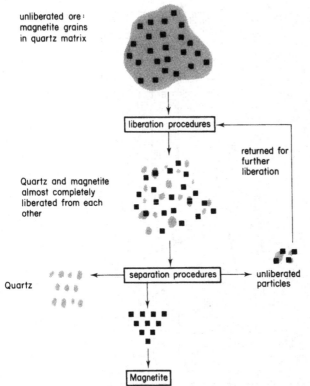

unliberated ore:
magnetite grains
in quartz matrix

liberation procedures

returned for
further
liberation

Quartz and magnetite
almost completely
liberated from each
other

separation procedures → unliberated
particles

Quartz

Magnetite

TABLE 10.1
Mineralogical data required for the design and the control of mineral treatment operations

Mineralogical characteristic	How normally obtained
(1) Mineral identity	Determinative test schemes; detailed chemical analyses; XRD analyses
(2) Mineral composition	XMA or chemical analyses of carefully cleaned material
(3) Mineral proportion	Mineral separations; point, line or area measurements
(4) Element distribution	Calculation from (2) and (3) above
(5) Grain-size distribution	Chemical liberation followed by screening; image analysis
(6) Mineral properties	Experiment (not by reference to the literature)
(7) Ore texture	Image analysis
(8) Particle-size distribution	Screening and/or classification
(9) Particle shape	Image analysis
(10) Particle composition	Laboratory separations or image analysis
(11) Particle texture	Image analysis

Fig. 10.6 Schematic representation of a simple mineral treatment operation for separating quartz and magnetite from a quartz-magnetite ore.

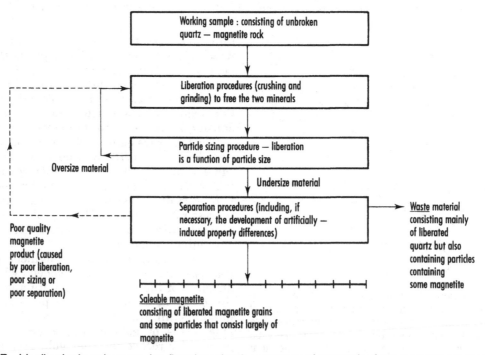

Working sample : consisting of unbroken quartz — magnetite rock

Liberation procedures (crushing and grinding) to free the two minerals

Particle sizing procedure — liberation is a function of particle size

Oversize material

Undersize material

Separation procedures (including, if necessary, the development of artificially — induced property differences)

Poor quality magnetite product (caused by poor liberation, poor sizing or poor separation)

Waste material consisting mainly of liberated quartz but also containing particles containing some magnetite

Saleable magnetite consisting of liberated magnetite grains and some particles that consist largely of magnetite

Fig. 10.7 Idealised mineral processing flowsheet for the recovery of magnetite from a quartz-magnetite rock.

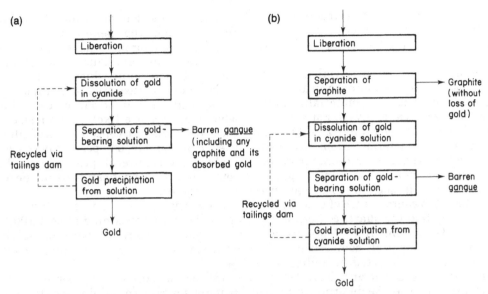

Fig. 10.8 Two alternative outline flowsheets for the recovery of gold by cyanidation: (a) may result in unacceptable losses of gold by absorption of gold-bearing solution by graphite; (b) the graphite is removed *before* the gold is taken into solution.

has been dissolved in the cyanide solution is selectively re-adsorbed by these minerals, and is permanently lost when they are discarded along with the other plant tailings. However, if the engineer is aware in advance of the presence of such gold-adsorbing minerals, he can make provision for these minerals to be separated from the remainder of the ore at an early stage in the process, before any attempt is made to recover the gold by cyanidation (Fig. 10.8).

It is often impossible to predict accurately the "useful" properties of a mineral and, when this is so, provisions must be made to *measure* those properties most likely to affect the probable separation procedures.

2.2. Mineral compositions

A detailed knowledge of the chemical composition of a mineral will often enable the mineralogist to identify it. In fact, this procedure may well be the only way of identifying amorphous or metamict minerals (see Chapter 1, section 2.2.5). In addition, a detailed knowledge of the chemical compositions of the minerals in an ore will allow the mineral engineer to predict the qualities (or grades) of the best products that he can hope to produce from the ore. Thus, if the only tin-bearing mineral in a tin ore is cassiterite (nominally SnO_2) then there is a good chance that it will be possible to design a process which will produce at least some high-grade, cassiterite-rich concentrate containing upwards of 75% tin. If, on the other hand,

the tin in an ore occurs as the mineral cylindrite ($Pb_3Sn_3Sb_2S_{14}$) then the maximum possible tin content in any concentrate will be limited to 21%. Furthermore, this tin-rich concentrate will *always* contain up to 37% lead. Mixed tin–lead concentrates of this kind are of no great value; they are difficult to sell to a smelter because it is hard to separate the two elements during the usual smelting procedures.

A detailed knowledge of the compositions of minerals will often explain otherwise unaccountable losses of valuable elements into the waste products produced during mineral treatment operations. For example, the major gangue mineral in some kinds of copper ore is pyroxene. An analysis of the pyroxene may show that it contains about 0.5% copper. The waste products from the treatment of such an ore would inevitably contain large amounts of copper – copper that cannot economically be recovered.

2.3. Mineral proportions

The economic value of a mineral deposit is almost invariably controlled by the mass proportions of the *minerals* it contains, rather than by its overall chemical composition. Thus, in a deposit that contains 0.5% copper, all this copper may occur as coarse grains of the mineral chalcopyrite ($CuFeS_2$); such a deposit may well become a useful economic source of copper. Another deposit may also contain 0.5% copper, but if all this copper occurs as a minor constituent in pyroxene (i.e. if the copper forms part of the structure of the pyroxene) then that copper is economically un-recoverable and the deposit is of no commercial value.

There are only a few, often trivial instances, when it is possible to estimate accurately the *mineralogical* composition of a material from the results of a bulk chemical analysis. For example, if a material contains *only* quartz (SiO_2) and hematite (Fe_2O_3), then determinations of the silicon and/or iron contents of that material will provide accurate information on mineral proportions. For instance, if such a specimen were assayed and found to contain 14% of silicon, then it is possible to say that the quartz content is 30% and the hematite content 70%. More usually, however, the mineralogical composition of a specimen cannot be calculated unambiguously from its chemical composition, and the mineralogical proportions must then be determined by one of the methods described in Chapter 2, section 7.

In practice, however, the mineralogist often uses bulk chemical analyses to provide rough estimates of the proportions of the various minerals in much more complex rocks and particulate specimens. These estimates usually demand that the mineralogist has a good pre-knowledge of the minerals concerned. For example, if a specimen is known to contain *only* calcite ($CaCO_3$) and rutile (TiO_2), then assays for either calcium or titanium will provide accurate estimates of both the calcite and the rutile contents. Similarly, if a specimen is known to contain only calcite ($CaCO_3$) and sphene ($CaTiO_3$), then a titanium assay will provide the basis for an accurate deduction about the amounts of sphene and calcite. If, however, a specimen contains all three minerals – calcite, rutile and sphene – then it is *not* possible to deduce the proportions of these minerals by simple analyses for calcium and titanium. Can the proportions of these three minerals be obtained by a more extended chemical analysis? Yes: they can be differentiated either by dissolving away the calcite in dilute acid and then analysing the residue for calcium, or by analysing the bulk specimen for carbonate, calcium and titanium. It is apparent from examples such as this that as specimens become mineralogically more complex it becomes more and more difficult to deduce their mineralogical compositions from sets of bulk chemical analyses. The value of bulk chemical analysis in this context is further limited by the fact that the compositions of many minerals are variable and are seldom known with any accuracy.

Information on mineral proportions, coupled with data on mineral compositions, enables the mineral engineer to determine how the valuable elements and the penalty-incurring elements are distributed (or partitioned) among the various minerals in an ore. This information makes it possible to determine which, if any, of these minerals should be recovered in a mineral processing plant, and such information can also indicate what processes are likely to be needed to collect these minerals into concentrates of different qualities. For example, a deposit may contain a seemingly attractive overall proportion of the element copper but the copper can exist in one or all of the following forms:

(1) native copper (i.e. a naturally occurring metallic phase containing almost 100% copper),

(2) one or more copper-rich sulphides (e.g. chalcopyrite, $CuFeS_2$, containing about 34% copper),

(3) one or more copper-poor sulphides (e.g. cupriferous pyrite, containing less than 5% copper),

(4) copper carbonates (e.g. malachite, $Cu_2CO_3(OH)_2$, 58% copper),

(5) copper-bearing phosphates (e.g. turquoise, $CuAl_6(PO_4)_4(OH)_8.4H_2O$),

(6) copper oxides (e.g. cuprite, Cu_2O, 89% copper),

(7) copper-bearing silicates (e.g. shattuckite, $2CuSiO_3.H_2O$),

(8) copper arsenates (e.g. olivenite, $Cu_2AsO_4(OH)$, 26 % copper).

It may well be possible to recover the copper sulphides from such an ore by using froth-flotation methods (see Chapter 3, section 3.5), but it is likely that the copper carbonates and the native metal would have to be recovered by a chemical leaching process. The copper phosphates, arsenates and silicates, with their comparatively low copper contents, would probably not be worth recovering.

2.4. Mineral grain-size distributions

A great deal of useful information about the geological history of a rock can be deduced from its mineralogical characteristics – features such as mineral content, mineral composition and textures. In addition, the grain-size distributions of the various minerals can also provide useful clues to the manner in which the rock was formed. More important, from a mineral engineer's point of view, is the fact that the grain-size distribution of a mineral determines, to a large extent, the ease with which that mineral can be liberated from its rocky matrix. Minerals that occur as large grains are liberated at comparatively large particle sizes (see Chapter 1, section 2.6) whilst small grains can only be liberated at very fine particle sizes. If a mineral can be freed at coarse particle sizes then the liberation process will be comparatively cheap and, furthermore, at these coarse particle sizes the minerals are easy to concentrate into saleable products. Consequently, the mineral engineer is most anxious to obtain good estimates of the grain-size distributions of the various minerals in a deposit, and this information is needed as early as possible during an investigation into that deposit's likely value.

Mathematical methods have been developed whereby grain-size distribution data obtained from an unbroken rock are used to predict the mineralogical compositions (and, hence, the amount of liberation) of the particles likely to be produced when the rock is broken in a random manner to different particle-size ranges (see section 1.1.1 above). From this information the engineer can estimate the likely costs of the liberation procedure(s) required and can also gain some insight into the nature and the likely cost of the concentration processes that will subsequently be needed.

2.5. Mineral properties

Some of the properties of a mineral, such as its hardness, density and refractive index, do not vary significantly from one mineral grain to another within a mineral deposit or, even, from one deposit to another. As a result, these properties are particularly useful for mineral *identification* purposes. The mineral engineer, on the other hand, is greatly interested in those mineral properties that can be used to effect a *separation* of one mineral from another. Some of these mineral properties vary markedly from one mineral grain to another, and there can be, for example, large differences in mineral grain size from one part of a deposit to another. It is important, therefore, that the engineer knows the details of these variations, otherwise his or her separation procedures may not produce the desired products.

Within a separation process, each mineral particle reacts as an individual unit. It is therefore important to measure not only the properties of the pure minerals but also the properties of individual *particles* produced in a plant – it is not enough to establish the mean value of a property that is exhibited in various degrees by all the individuals in a large population of particles. Each composite particle has its own distinct mineralogical composition: it is not enough to establish the *mean* composition of a group of particles – the mean value of the particle compo-

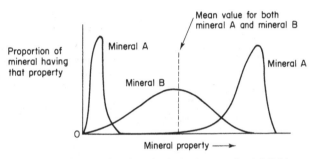

Fig. 10.9 Graph showing the importance of establishing the *distribution* of a mineral property: both minerals have the same mean value, but it is nevertheless still possible to separate these two minerals on the basis of the measured property.

sitions shown in Fig. 10.9 is about 50% of mineral A and 50% of mineral B, but this single value conceals the fact that most of these particles are either "liberated A" or "liberated B".

2.6. Ore textures

The problems of liberating the minerals from an ore are closely related to the textural complexities of that ore. Intimately intergrown or felted aggregates of minerals are difficult to liberate, whilst a more open-textured material is comparatively easy to treat. Consequently, the mineral engineer seeks quantitative (i.e. numerical) data concerning ore textures (and the variations of texture within an ore) so that first, the most suitable liberation process for that ore can be designed, and then the quality of the feed into the working plant can be monitored and, if possible, controlled.

The textural details of an ore can often be assessed, and then *described*, by the skilled mineralogist after only a brief examination of a typical specimen. It is much more difficult to provide the *quantitative* description required by the mineral engineer. However, it is now possible (thanks to the availability of automatic image-analysing techniques) to measure quantitatively some important texture-related values – for example, values such as the distribution of distances between the grains of a selected mineral along a random traverse line. Variations of these readily measured values in different specimens (or in different parts of a single specimen) provide accurate, and often very effective, measures of the textures of these specimens (Fig. 10.10).

Variations of other texture-dependent values, such as the mean distance (in three dimensions)

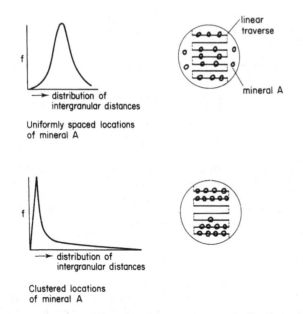

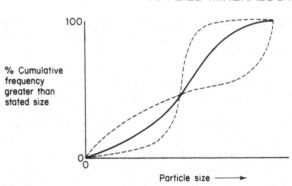

Fig. 10.11 Cumulative size-distribution graphs, showing three different size distributions which have the same mean values and the same ranges.

Fig. 10.10 Graphical illustration of the quantification of textural differences in ores from the distribution of the distances between adjacent grains.

between the nearest-neighbour grains of a single mineral species, are also measurable (see the discussion of stereology in Chapter 8).

2.7. Particle-size distributions

Almost every mineral processing operation works best when treating a limited range of particle sizes. Some treatment operations can be used only for coarser sizes, other processes only work efficiently on very fine-grained particles, whilst a very few treatment processes work well with a wide range of particle sizes. It is therefore necessary to carry out frequent, routine measurements of particle-size distributions on a large number of different plant products in order to check process performance. The mineral engineer is not only interested in the size of particles deliberately broken in a plant – he is also interested in the size distributions of particles found in unconsolidated mineral deposits such as soils and sands.

In a size analysis it is not enough merely to determine, say, the size of the largest particle (viz. "all the particles are less than size x") nor is it sufficient to determine only the top and bottom values of a range of particle sizes ("the particles range from x to y in size"). It is essential to establish the complete size distribution (Fig. 10.11). This can be done in a number of different ways, e.g. screening or classification in a fluid (see

Chapter 5, section 3.2), but some care is needed when comparing the results obtained by different methods because, generally speaking, each sizing method measures a different value of particle "size".

2.8. Particle shape

Although particle shape is difficult to define, it is fortunate for the mineralogist that mineral particles occur in a variety of generally simple shapes. For example, some are acicular (needle-like), others are platy; most are convex, others are mildly concave; and so on. The behaviour of these particles in fluids depends in large measure on the specific surface area between the solid and the fluid, and this in turn depends on factors such as (1) the proportion of solids in the fluids, (2) the size distribution of these solid particles, (3) the compositions, and therefore the densities, of the particles and (4) the shapes of the particles. For example, particles of fine-grained ferrosilicon (Fe_xSi_y) are frequently used to form the "dense media" employed in industry to separate minerals according to their densities. Changes in the shape of the ferrosilicon particles can markedly alter the viscosity of a "dense medium" and can thus seriously affect the efficiency of the separation procedure. A detailed knowledge of the actual shape may not always be necessary – all that may be required is a shape coefficient that can be used to compare the shapes of different populations of particles.

Similarly, the natural shapes of the mineral particles in alluvial and in beach-sand deposits greatly affect their flow characteristics in fluids and a study of these shapes provides useful information regarding the transportation conditions that existed during their period of formation.

Furthermore, particle shape exerts a marked control on the packing characteristics of mineral powders in bins and hoppers and greatly affects the "covering power" of the pigments in paints.

2.9. Particle composition

As explained above, the initial aim in any mineral treatment operation is to liberate the various minerals in an ore, and to do this as well as possible so that the subsequent concentrating processes can be used to best advantage. In other words, the aim is to produce particles that are essentially monominerallic, and to produce these particles at as coarse a size as possible.

The efficiency of a liberation process is generally determined *indirectly* by measuring the size distribution of the particles which are produced; it is then assumed (with justification) that the finer the particles the greater the amount of liberation. It is sometimes possible to measure liberation *directly* by determining the compositions of the particles from each size fraction of the broken material, i.e. attempts are made to determine how many of the grains are liberated, how many grains are nearly liberated, and how many grains are nowhere near to being liberated in each of a number of size ranges (Table 10.2). From this information the mineral engineer can then predict the likely response of the particles in treatment processes such as tabling, jigging and dense-medium separation, and can also estimate the compositions of the various products likely to be produced from these separation processes.

Very occasionally it is possible (and even worthwhile) to use chemical methods of analysis to determine the proportion of a specified mineral within *individual* mineral particles. In this procedure, a number of particles are separately analysed (the neutron activation method is used for this kind of work since comparatively large particles can be analysed, and the particles are retained intact for further examination if necessary). If the mineralogy of the particles is simple, then the analytical results can be converted into mineralogical compositions. Figure 10.12 shows a typical set of results obtained by determining the uranium contents of a few hundred individual mineral particles. If *all* the uranium had existed as the mineral uraninite (UO_2), then the chemical analyses could have been converted readily into their equivalent uraninite contents. Unfortunately, the uranium in this material was shown to occur both as uraninite and as brannerite (a mineral of very variable composition but, nominally $(U,Y)Ti_2O_6$). In these circumstances it was not possible to determine accurately the relative proportions of the two minerals from the results of the uranium assays.

As mentioned in section 2.4, attempts are being made to develop methods of *predicting* the compositions of the particles likely to be produced by various comminution processes. If these attempts

TABLE 10.2
Table showing the "apparent" compositions of a population of mineral particles (as seen by linear analysis of a sectioned specimen)

Particle composition (% of stated mineral)	Proportion of mineral A	Proportion of mineral B	
0–10	0.1	0.4	
10–20	0.4	0.7	
20–30	0.5	2.2	
30–40	0.3	1.8	
40–50	0.7	2.0	
50–60	0.4	1.4	
60–70	1.1	2.8	
70–80	0.9	13.1	
80–90	1.3*	13.8	unliberated
90–100	2.1	37.5	B-rich particles
Liberated	92.2	24.3	
	100.0	100.0	

*That is, 1.3% of mineral A occurs in the form of particles that show between 80% and 90% of A.

Mineral A is seen to be almost completely liberated; only 24% of mineral B appears to be liberated, although a further 50% of the B-rich particles are nearly liberated.

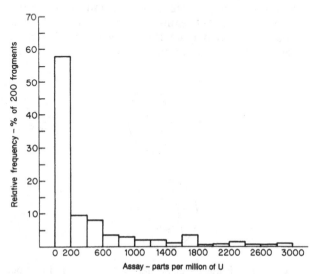

Fig. 10.12 Uranium ore sample – histogram of relative frequencies of assay values for 200 fragments 8 mm in size (10 fragments analysed greater than 3000 ppm), analysed by neutron activation analysis.

become successful, the engineer will be able to make initial design calculations and obtain estimates of treatment costs evaluations well ahead of the completion of the usual liberation test programme.

2.10. Particle textures

The behaviour of mineral particles in some mineral treatment operations is controlled not only by particle size, shape and composition, but also by particle texture. That is, the behaviour of a particle is controlled, in part, by its structure – the spatial relationships between its various mineralogical components. For example, both particle 1 and particle 2 in Fig. 10.13 are the same shape, the same size and the same composition, but they are of different texture. Particle 1 would easily float in a flotation circuit designed to recover mineral B, whilst particle 2 would report to the "sink" product. Similarly, particle 1 would react very differently to particle 2 in an electrodynamic separator – or in any other separating system that depends on the reaction of the surface of the particle to external forces.

Again, particles 3 and 4 are the same size and shape and each has the same proportions of minerals A and B. However, particle 4 is approaching the size at which virtually complete liberation of minerals A and B will be achieved, whilst particle 3 must be greatly reduced in size before any significant amount of liberation of either mineral will take place.

Mineral B can be swiftly and completely recovered from particle 6 by a chemical leaching action, whilst the same mineral in particle 5 is totally screened from the effect of the leaching action. Good quantitative indications of the

TABLE 10.3

Specific surface area values of minerals A and B in an epoxy matrix, C (interfacial surface area per unit volume of mineral, μm^2 per μm^3)

Mineral	A	B	C	Total
A	—	0.109	0.005	0.114
B	0.016	—	0.001	0.017
C	0.029	0.040	—	0.069
				0.200

Mineral A has a total interfacial area of 0.114 μm^2 per μm^3; 96% of this area is in contact with mineral B. Only 4% of mineral A is in contact with the matrix, and therefore available for reaction with, for example, flotation reagents, or reagents used in cyanidation of gold ores.

availabilities of mineral grains to attack by reagents can be obtained by measuring their specific surface areas (Table 10.3): high values of interfacial surface area between a mineral and the epoxy resin matrix shows a high degree of availability.

3. Summary of information required

The mineral engineer needs large amounts of detailed, quantitative information on mineral grains, mineral particles, and ores when designing suitable treatment methods for recovering the various minerals, and for the subsequent monitoring of plant performance.

Significant variations inevitably occur in the mineralogical characteristics of specimens – even in specimens which have been derived from a

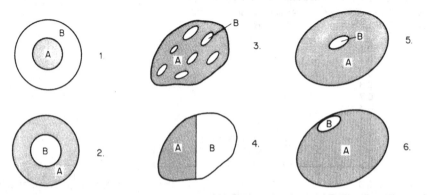

Fig. 10.13 Particles of similar size, shape and composition but differing texture: particles 1 and 2 will react very differently in a flotation process; particles 3 and 4 are the same size and shape and they contain the same proportions of minerals A and B. In particle 4 minerals A and B are nearly liberated, but particle 3 will have to be greatly reduced in size before any liberation takes place. In particle 5, the grain of mineral B is not available for attack by any leaching agents, but a similar grain in particle 6 is readily available.

single deposit. Many of these variations reflect differences in the mode of formation of the minerals in various parts of a deposit. For example, the minerals and textures in the weathered cappings of ore bodies are generally very different from those in the underlying unweathered ore. The mineral engineer is anxious to know the details of these variations, in order to plan ahead to counter their effects on treatment operations.

Mineral particles from the various parts of a treatment plant invariably differ in size, composition and texture. The mineral engineer needs information concerning these variations and, furthermore, must have this information quickly and on a routine basis, so that the operation of that plant can be efficiently controlled.

Some mineralogical features, such as particle size and particle composition, may change with time. Many of these changes are of great importance to the engineer and they can take place whilst an ore is temporarily stored in a mine stope or in stock-piles or bins awaiting treatment. Oxidation of sulphide minerals under these circumstances can seriously affect the character of their surfaces and, as a result, their responses to flotation processes may change drastically. The mineral industry also permanently stores large quantities of very finely ground waste minerals in ponds, tips, dams and so on. The mineralogical changes that take place in these materials over extended periods of time can affect their chemical and their physical stabilities and may give rise to serious hazards, and these changes must be monitored so that their effects can be countered.

If the mineralogical information discussed in this chapter is to be of maximum value within an industrial context, then it must be produced as swiftly as possible – and it must be produced in numerical form. Speed of data collection is often so important on the operational side of the mineral industry that it may be permissible to sacrifice some accuracy in order to obtain the analytical results as quickly as possible. On the other hand, if the information is needed for process design purposes or for research work then time may be less important and, in these circumstances, it may be possible to insist upon results of the highest possible accuracy even if the analytical procedures take a very long time.

The techniques used for collecting mineralogical data fall into two main categories: (1) indirect, *chemical* methods, and (2) direct, physical methods of *mineralogical* analysis. The chemical methods are frequently used by mineral engineers and mineralogists because these methods are comparatively cheap, and also quick and easy to carry out. The chemical results must be used with care, however, since they seldom provide unambiguous mineralogical data. In addition, the value of a chemical analysis for, say, mineral identification purposes, often depends in large measure on the purity or otherwise of the analysed materials. The problem of obtaining sufficient high-purity material is especially difficult when comparatively large amounts of material are needed for an analytical procedure. It is now possible, however, by using the right kind of equipment, to analyse very small individual mineral grains, even whilst those grains are embedded in a rocky matrix (see Chapter 9, section 9.1).

However, there are many mineralogical features that cannot possibly be measured or inferred from chemical analyses – features such as particle sizes, rock textures, particle textures and so on. Consequently, various physical methods of mineralogical analysis are always required to complement the information provided by the comparatively cheap and swift (and well-established) chemical methods of analysis.

Many physical methods of analysis (e.g. the use of specific gravity measurements for identifying a mineral) require that the measured specimen is as pure as possible. There are many practical advantages to be gained by using methods that produce unambiguous results from a small number of carefully selected, clean fragments, or which can produce good results from the analysis of mixed, impure particles. Thus, the crystal structure (and hence the identity) of a mineral can be established by X-ray diffraction analysis without, necessarily, producing clean, pure mineral grains for the analysis.

TABLE 10.4

Natural grain size of cassiterite in (1) a pyritic Cornish tin ore and (2) a tin-bearing granite from Thailand

Screen size (μm)	Weight %		Cumulative % coarser	
	Cornish	Thai	Cornish	Thai
>420	39	8	39	8
300–420	6	4	45	12
210–300	6	10	51	22
150–210	12	15	63	37
105–150	14	16	77	53
75–105	11	16	88	69
< 75	12	31	100	100
	100	100		

(Note: grain sizes are given in some parts of the mineral industry in the form +420 μm, −420 + 300 μm etc. . . . . −75 μm.)

Mineral proportions in rocks and mineral products are often determined by first grinding the specimen to liberate the minerals of interest and then dividing the liberated grains as cleanly as possible into different groups using the appropriate methods of separation (see Chapter 3). However, it is sometimes extremely difficult to achieve an acceptably high degree of liberation at particle sizes which are large enough to allow efficient separation of the minerals. In these circumstances, the results obtained from a particle-separation analysis are more likely to reflect the efficiency of the separating process than to measure the mineralogical composition of the original specimen.

The mineral engineer is frequently interested

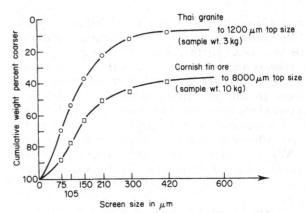

Fig. 10.14 Natural grain-size distributions (by weight) of cassiterite from a Cornish sulphidic tin ore, and tin-bearing granite from Thailand.

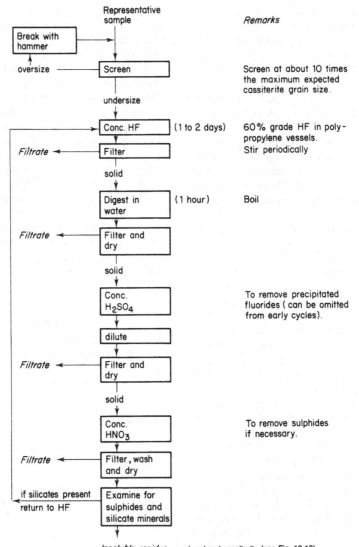

Fig. 10.15 Procedure for disintegration of a silicate-rich tin ore in order to recover the cassiterite at its natural grain size.

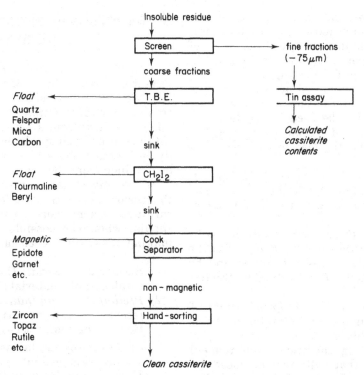

Fig. 10.16   Procedure for examining the acid-insoluble residue from a granitic tin ore (see Fig. 10.15).

in the natural grain sizes of a mineral. This information can sometimes be determined by selectively dissolving the rocky matrix to release the grains; these grains can then be sized by ordinary screening or sedimentation methods. Table 10.4 and Fig. 10.14 show the results obtained by dissolving first the silicate minerals and then the various sulphide minerals that surrounded cassiterite in a Cornish tin ore and in a Thai granite. Figure 10.15 gives the flowsheet for the dissolution procedure and Fig. 10.16 the procedure for examining the insoluble residue. More commonly, estimates of the natural grain-size distribution of a mineral can only be established by image analysis procedures (see Chapter 6).

Particle compositions can sometimes (but only very rarely) be accurately determined by heavy liquid tests. Occasionally, it is even possible to deduce particle compositions from the chemical analyses or from magnetic permeability values of the individual particles. Both these methods are only applicable when there is a clearly defined, unique relationship between the measured property (density or magnetic permeability) and the composition of the particle. Such a relationship is restricted to simple, two-component particles.

Estimates of particle compositions and particle textures can be obtained by image analysis.

4. General conclusions

Value of quantitative mineralogical information to the mineral industry

Detailed quantitative mineralogical information is essential for the exploitation of low-grade and of complex mineral deposits. This information is used for the following purposes:

(1) *Locating suitable deposits.* Mineralogical data are used, in conjunction with geochemical and geophysical data, to pinpoint the location of economically promising deposits. It is important to establish *how* the potentially valuable minerals (or elements) occur within a deposit in order to assess the likelihood of recovering these minerals.

(2) *Determining the extent of a deposit.* The limits of an ore body are sometimes controlled by geological factors – for instance, an ore-bearing vein is often confined by barren country rock. At other times the limits of an ore body are controlled by economic factors: thus, the boundaries of a porphyry copper deposit may be very indistinct and the extent of the ore may be controlled by the price of copper rather than by mineralogical changes.

(3) *Controlling the planning of a mine.* Mineralogical information is especially useful in the planning and in the control of operations in open-pit mines. Drilling within the mine (perhaps for blasting purposes) can produce large numbers of ore samples. The mineralogical information obtained from these samples can be used to decide which parts of the exposed ore should be mined in order to optimise the feed to the treatment plant.

(4) *Designing the preliminary, amenability testing stages* to establish whether or not saleable products can be produced at an acceptable cost.

(5) *Controlling the process testing operations* by enabling the process engineer to predict the most favourable processing routes and by establishing the results obtained by the test procedures.

(6) *Process development and optimisation.* A great deal of mineralogical information is needed to establish the combination of processes, separating conditions and reagent concentrations that will give the "best" result (where "best" may refer to maximisation of profits, maximum recovery of the valuable mineral(s), and so on).

(7) *Plant design.* Mineralogical information on, for example, particle sizes and shapes, can be of great value when specifying the size of equipment, etc. needed for a working plant.

(8) *Process control.* Each part of the working plant must be kept in near-optimal conditions, even though the amount of feed and the quality of that feed may change: advance mineralogical knowledge is vital for this purpose.

(9) *Product control.* Products of saleable (or discardable) qualities must be maintained; the product is often sold on a strictly *mineralogical* (rather than chemical) basis, e.g. gravels, clays, mica, precious minerals, etc.

(10) *Monitoring the physical and chemical stability of tailing dumps.* Long-term mineralogical changes can release undesirable elements into the natural drainage, or they can seriously affect the stability of a dump or tip. These changes must be regularly monitored.

(11) *Research* into the properties and behaviour of minerals must, of course, be based on mineralogical knowledge.

(12) *Research into new treatment methods* must also rely on the ready availability of quantitative information on particle sizes, shapes, textures, mineral properties, etc.

5. Mineralogical examination of a specimen

The following procedure is recommended for the receipt and examination of a specimen.

In the early stages of an investigation it is only possible to obtain *specimens* (rather than properly taken *samples*) from a potential ore. The results obtained from such specimens may not, necessarily, be representative of the results that will later be obtained from the ore.

As the investigation proceeds, it is essential that good *samples* are used for all the mineralogical (and other) work. The mineral engineer should, whenever possible, be involved in the making of sampling decisions.

(1) *Receipt of material.* Carefully number and catalogue all materials received.

(2) *Preliminary examination of the material "as received"* – always carry out this examination on the coarsest possible material

 a) determine (as far as possible) the major minerals present,

 b) study any large-scale textures or segregations of these minerals,

 c) determine the possibility of hand-sorting the coarser fractions into mineralogically-different groups of particles,

 d) determine whether any of these "groups" contain
   (i) barren gangue,
   (ii) pure, valuable component,

 e) is it possible to establish a preliminary liberation size from the results of (d)?

(3) *Determine mineral identities*, proportions, grain sizes, associations, etc. Is it now possible to predict the liberation size of the test material?

(4) *Crush and/or grind* the specimen to the predicted liberation size.

(5) *Check the amount of liberation* actually achieved by the selected comminution method: i.e. determine the mineralogical compositions and the textures of the broken particles.

(6) *Establish the "useful" properties* of all the minerals (and not only the valuable minerals).

(7) *Design a test programme* that can be carried out on representative samples from different parts of the deposit.

(8) *Establish the mineralogical performances:* i.e. where have the various types of particles reported? What accounts for any losses of valuable minerals in the discarded products?

Why is the concentrate grade so low? Is the material adequately liberated? etc.

(9) *Adjust the test conditions* in the light of the information obtained during the detailed mineralogical analyses.

(10) *Repeat* Sections 8 and 9 until adequate performance is achieved.

(11) *Carry out "cyclic tests"* in which "middling" products are re-circulated into the test: in this way, it is possible to determine the likely effects of the build-up of middling (composite) particles and/or the build-up of deleterious reagents in an operating plant.

(12) *Carry out continuous "pilot" tests* in which a few tonnes of representative material is treated.

(13) *Carry out detailed mineralogical examinations* of the products of these continuous tests and use the results to adjust the separating conditions.

Appendix 1

Practical hints and laboratory exercises

Introduction

In the mineral industry a great deal of mineralogical information is collected by the mine geologist – both during the initial period when a deposit is being evaluated and, later, whilst that deposit is being exploited. However, large amounts of additional mineralogical information are needed to design and to control the associated mineral treatment operation and this information must often be collected by the mineral engineer. Both the geologist and the mineral engineer must know what information is needed and how to collect the necessary data. They must also realise that, under industrial conditions, the data must be gathered swiftly and often with a limited amount of analytical equipment.

The following selection of practical exercises was first developed for undergraduate students in the Department of Mineral Resources Engineering at Imperial College, London. The aims of that course were to teach the mineralogical skills needed (1) to *identify* the more common minerals encountered in the mineral industry, and (2) to obtain *quantitative estimates* of the mineralogically important characteristics of minerals, ores and plant products.

The purpose of these practical exercises is to enable the student (or any other newcomer to the mineralogical field) to become familiar with the methods which are available for sampling, handling, examining and analysing the minerals and features which occur in solid rocks and in particle aggregates. These exercises are purposely *not* concerned with "mineral-spotting" procedures in which sets of mineral specimens are "recognised" by repeated handling. Instead, the exercises are designed to teach the student how to *identify* minerals that he or she may never have seen before and measure their important properties. The exercises are also intended to teach the student how to make best use of the basic tools of mineralogy – his/her eyes and the simple microscope – and how to make mineralogical observations and how to record and present analytical results.

The first practical exercise gives a list of the special equipment and materials needed for the test; each subsequent exercise gives a list of additional equipment and materials that are required.

Laboratory notebooks

Notebooks must be used to record all observations and all measurements made during every investigation. A hard-cover, A4-size notebook with numbered pages provides a very convenient and durable record of all the work done during a test programme. It is also the only kind of record that conveniently preserves the *sequence* in which the work was carried out. Alternate pages in the book should, preferably, be lined and plain (or "squared") and the following layout is recommended when using the book:

(1) the lined pages should contain full experimental details, including the date and time;
(2) the lined pages should also contain infor-

mation on the equipment used and details of all the observations, measurements and results;

(3) the plain (or squared) pages should contain all the calculations, and all the manipulations that have been carried out on the measured data;

(4) the plain pages should also contain any preliminary interpretations of the results, together with any annotations and references from the technical literature.

In this way the observations and measurements are kept quite separate from any deductions made on the basis of these measurements; after all, the measurements are (or should be) inviolate, whilst the deductions can be (and often are) changed as more information becomes available during the course of an investigation.

Each day's report should begin with the date; each new project, or exercise, must start with a statement of the aims of the work. Each project record should end with a brief summary of the results which have been achieved and a comparison of the achievements with the original aims of the project. Cross-references to earlier projects and any annotations regarding earlier work should be made by reference to the page numbers in the notebook.

Many students (commendably) try to protect their notebooks from the "rough and tumble" of the mineral laboratory environment with its dust and moisture, and its ubiquitous, fine-grained mineral pulps. It is important, however, to record all adjustments made to the test equipment and, what is more important, to record these adjustments *as they are carried out*. These records must be immediately incorporated in the appropriate place in the notebook: they must *not* be jotted down initially on odd scraps of paper (so that the laboratory notebook can be kept clean), and later transferred to the book at some convenient time in a cleaner, quieter atmosphere. Scraps of paper are easily lost, and a lost scrap of paper may contain some vital piece of information.

It is preferable to have the *complete* record of an experiment in a slightly dirty notebook than it is to have a spotlessly clean but incomplete record. (A spotlessly clean, complete record is, of course, even better.)

Perhaps the most important task undertaken by the mineralogist is the production of an accurate, concise report of the work done during a mineralogical investigation and a properly kept laboratory notebook is absolutely essential for this task.

Practical no. 1: Sampling

Aim: to illustrate the variability of the products obtained by various sampling procedures.

Special equipment and materials needed: laboratory-sized "micro"-riffler 5–8 cm in size; 100 g of sized quartz grains, 10–20 g of magnetite grains; bar magnet; spatula; assortment of brushes and beakers.

Riffle sampling

Weigh out approximately 90 g of quartz grains that are about 100 μm in size, and 10 g of magnetite grains of the same size. Mix the two minerals thoroughly to obtain a "bulk material" containing about 10% magnetite. Use the riffling method to split this into two equal fractions. Weigh the two fractions and comment on the results obtained. Separate the quartz and magnetite from both fractions by using a bar magnet covered with a polythene bag, and again comment on the results you obtain.

Coning and quartering

Recover all the bulk material used in the riffling experiment and re-sample it by using the cone-and-quartering method (see Chapter 2, section

6.1.1). Repeat the magnetic separation on both fractions to determine their quartz and magnetite proportions and comment on the results obtained.

Nine-point sampling

Again recover all the products from the bulk specimen and now sample it by using the nine-point method (see Chapter 2, section 6.1.2); determine the quartz and magnetite proportions of this sample as before. Then take another nine-point sample from the remaining material and determine its mineral proportions. Comment on your results.

Tabulate your final results thus:

Test procedure	Percentage (by mass)	Percentage by mass:	
		Quartz	Magnetite
Bulk specimen	100	90	10
Riffling product 1 Riffling product 2			
Cone-and-quartering product 1 Cone-and-quartering product 2			
Nine-point sample 1 Nine-point sample 2			

Practical no 2: Mineral identification, particle sizing and particle separation methods

Mineral identification

Aim: to provide an opportunity for the student to identify small fragments of unknown minerals.

The mineralogist and the mineral engineer frequently need to identify unknown minerals. These minerals often occur as rare, fine-grained particles in ores and in mineral products. It is always difficult to identify such minerals and it is, therefore, important to use a *systematic* approach.

Appendix 2 gives full details of a simple, systematic mineral identification scheme that has been designed for use by students. The scheme does not use expensive, sophisticated equipment and it requires only modest mineralogical skills.

Minerals A, B, and C occur in the same ore body. Use the determinative procedure to identify these minerals. Make sure that you follow each step and carefully record the results of your investigations in your laboratory notebook.

Particle sizing

Aim: to establish the size distributions of particulate specimens by (i) screening, (ii) classification, and (iii) diffraction of a laser beam.

Screening

Special equipment: set of 20 cm diameter laboratory screens; screen shaker; micro-mesh sieves.

Square-aperture screen surfaces made of woven wire cloth are available with aperture sizes from about 4 mm to 38 μm.

The ratios between the aperture sizes of successive sieves in a British Standard Sieve Series is $\sqrt[4]{2}$, but it is common practice to use only half of the available number of sieves in any single laboratory investigation, i.e. a screen series having aperture ratios of $\sqrt{2}$ is commonly used (see Table A.1.1).

Accurate woven-wire screens are difficult to make and considerable tolerances are allowed in their manufacture. These tolerances increase as the aperture sizes decrease and the inaccuracies generally become unacceptable at sizes below about 75 μm. Particles less than 75 μm in size must then be screened through "micro-mesh" sieves (see below), or they must be "sized" by using a method that does not rely on the passage of a particle through an aperture (see "Classification" – below).

TABLE A.1.1
Nominal aperture sizes of woven-wire screens (B.S.S. 410)

Mesh number	Aperture size (μm)
4	4000
5	3350
6	2800
7	2360
8	2000
10	1700
12	1400
14	1180
16	1000
18	850
22	710
25	600
30	500
36	425
44	355
52	300
60	250
72	212
85	180
100	150
120	125
150	106
170	90
200	75
240	63
300	53
350	45
400	38

The underlined $\sqrt{2}$ series of screens are the ones most commonly used during a laboratory screen analysis.

Standardised screening procedures are essential if comparisons of particle sizes are to be possible from day-to-day or from one laboratory and another. Details of these procedures along with the total weights of materials that should be screened at any one time on a particular screen are given in B.S.S. 1796, as well as the types of shaking actions that should be employed. However, for the purposes of this practical exercise, it is sufficient to limit the overall sample weight for any single test to 100 g and to use any reasonably repeatable, manual shaking action (see below).

Coarse screening
Use a "nest" of $\sqrt{2}$ screens that cover the size range 300 μm to 53 μm (Fig. A.1.1). Place about 100 g of your dry material on the top screen and shake the whole nest of screens gently from side to side. Occasionally, include a slight "bumping" action by striking the screens with the ball of the hand.

Continue the screening action for 30 seconds; weigh each size fraction and return it to its

(a)

(b)

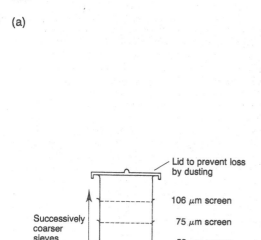

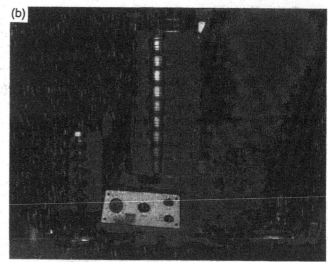

Lid to prevent loss
by dusting

Successively
coarser
sieves

106 μm screen

75 μm screen

53 μm screen

Cannister to receive
−53 μm fraction

Fig. A.1.1 (a) Illustration of nest of sieves for screening size ranges from 106 μm to 53 μm; (b) Photograph of Fritsch Analysette sieve shaker with dry and wet sieving and precision microsieving accessories (courtesy of Christison Scientific Equipment, Gateshead).

appropriate screen. Resume the screening action for a further 30 seconds and re-weigh all the size fractions. Continue to screen and weigh the fractions until the amount passing through from sieve to sieve after 30 seconds approaches zero. Plot your results on a graph showing "cumulative weights held on the various screens against screening time" (Fig. A.1.2).

The times at which the graphs flatten out are the minimum times that should be taken for a screening test with this type of specimen. The minimum amount of specimen needed to achieve a given precision during a screening experiment depends on particle size and can be calculated

from the following equation:

$$\theta^2 = \frac{fpd^3}{P}\left(\frac{1}{m} - \frac{1}{M}\right)$$

$$= \frac{fpd^3}{P} \quad \text{if} \quad m \leqslant M$$

where: θ is the coefficient of variation (standard deviation/mean); f is the shape factor (normally 0.5); P is particle density; d is mean diameter of particles in the largest measured fraction; p is the mass fraction of the largest measured size fraction; M is the original mass of material; and m is the mass of sample taken for analysis.

For the usual confidence limits of 95% and a precision of 5%, the value of $\theta = 0.02555$ (see Table A.1.2).

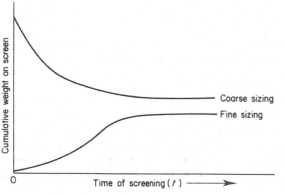

Fig. A.1.2 A typical graph for cumulative weight held on screen plotted against screening time.

TABLE A.1.2
Minimum sample weights needed during a size analysis to achieve results within 95% confidence limits. (Particle density 3000 kg m⁻³.)

Mean particle diameter of largest measured size fraction (μm)	Minimum weight of sample needed for:	
	5% precision	2% precision
10 000	23 kg	150 kg
1 000	23 g	150 g
100	0.023 g	0.15 g
10	0.000023 g	0.00015 g

(a)

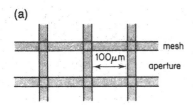

The area proportion of holes
to woven wire varies with wire
thickness but this proportion is
comparatively large

(b)

The area occupied by
apertures is always small (10 μm)

Fig. A.1.3 Illustration showing the difference in aperture area in a woven wire sieve (a), and on electro-formed micro-mesh screen (b).

Fine screening
Special, micro-mesh, electro-formed laboratory screens have accurately controlled apertures that range in size from 150 μm to 5 μm. The apertures are either circular or square in shape and the tolerance on any aperture size is ±2 μm. These screens are very delicate, and therefore, expensive. The proportion of screen surface occupied by "holes" is always small and the fractionating procedure is inevitably very slow. Consequently, these sieves handle only a gram, or so, of material at a time (Fig. A.1.3).

Dry screening is possible with particles larger than about 25 μm, but it is usual for a micro-mesh sieving operation to be carried out (one screen at a time) with the screen and its canister immersed in an ultra-sonic water bath. The vibration in this bath disperses the particles and also increases the speed of the screening action.

Take about 1 g of the 53 μm fraction from the previous test and screen on a 50 μm micro-mesh screen. Comment on the result you obtain.

Particle sizing with a stereoscopic microscope
Full details of this procedure are given in B.S.S. 3406. The method is tedious and time-consuming but, since it is one of the few methods that measure individual, discrete particles, it is often used to check the results obtained by other methods of

"size" determination. However, it does not produce sized particle fractions for further study.

Only a few hundred particles are normally measured by this method (compared with the few millions that are measured during a screening operation), and it is very important that these particles are truly representative of the material being studied.

True diameters
Take some ballatini (glass beads), nominally, 100 μm in diameter. Sprinkle them onto a "sticky" glass slide. The perfectly spherical balls will, of course, present their true diameters for measurement. Measure 100 of these diameters and plot your results as a histogram or as a cumulative plot of "number of balls versus diameter".

Feret diameters
Take a small sample of the −75 to +53 μm fraction obtained during the coarse screening practical (above). Sprinkle the particles on to a glass slide. The particles will assume their most stable positions and each will present its largest projected diameter for measurement. (Alternatively, the particles can be sprinkled on to a sticky surface such as a piece of double-sized adhesive tape stuck on to a glass slide: these particles will tend to assume random orientations and will present random projected areas.) Measure the Feret diameters of about 200 particles from each kind of specimen: the Feret diameter (d_f) is the distance between two vertical tangents on the opposite sides of a particle. (B.S.S. 3406 specifies that not less than 625 particles should be measured – but this would take too long for the purpose of this experiment.)

Compare your results with those you obtained by screening. How did you try to ensure that the particles that were measured were representative of the particles on the slide?

Particle sizing by multiple decantation
This is a particle sizing method based on the settling rates of the particles in a fluid.

Beaker decantation method
A dilute suspension of particles containing less than 5% solids (by volume) is carefully stirred in a 1-litre beaker to produce as near a homogeneous, dispersed suspension as possible. The time (t) that it takes a particle of size (d) and density (δ) to fall from the water surface to the bottom of the beaker is calculated from Stokes Law (see page 30 for nomogram showing the relationships be-

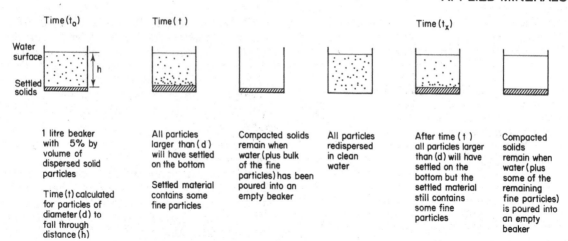

Fig. A.1.4   Beaker decantation procedure.

tween particle size, particle density and particle velocity).

After time (t) *all* particles that are larger than (d) will have reached the bottom, along with those smaller particles that started their descent from suitable intermediate positions. At time (t) you should compact the sediment which has collected at the bottom of the beaker by firmly (but not too firmly!) tapping the beaker with a glass rod encased in a thick rubber tube. The liquid, along with its dispersed small particles, should then be poured into a second large beaker. The coarse-grained solids left in the first beaker are separated from some of the entrapped fine particles by being re-dispersed in a litre of *clean* water, and being allowed to settle for a further period (t): any particles that have not settled are poured into a third beaker. This procedure, consisting of dispersion, timing and decantation, is repeated a further 3 times (i.e. 5 times, in all). At the end of this procedure the sedimented layer in the final beaker will consist, almost entirely, of particles greater than size (d). This product forms the first, and coarsest, size fraction; it is removed from the beaker, dried, and weighed.

The complete procedure (Fig. A.1.4) is then repeated for a smaller size fraction (d_1) with settling time (t_1) and using the earlier "wash" liquid, and so on until the required number of size fractions is obtained.

Take about 50 g of the $-75\,\mu$m size material obtained during the screening test. Carry out a fractionation procedure using the "beaker decantation" method and produce "sized" products that follow a $\sqrt{2}$ progression. Use ordinary tap water at 15°C but add a few drops of a detergent solution to help disperse the mineral particles. (The complete fractionation takes 5 to 6 hours but the experi-

menter need not be present, or actively involved, during much of this time.) Compare your results with the results obtained by screening and by the microscope sizing method.

Cyclosizer method
The Beaker Decantation method of particle sizing has been replaced in many laboratories by the "Cyclosizer". This consists of a series of small, inverted hydrocyclones mounted in series. Closed boxes are fitted over the apex nozzles and, therefore, solid particles that would normally leave the system through these nozzles cannot do so and continually fall back into the hydrocyclone. Small particles leave the system through the vortex finder until, eventually, only the coarsest particles remain (Fig. A.1.5).

A series of 5 hydrocyclones is arranged in order of decreasing diameters (and increasing fluid flow rate) and the "fines from one hydrocyclone become the feed to the next. A specimen can, therefore, be fractionated into five size fractions (plus a final "overflow" product which contains material smaller than about 8 μm and is often discarded).

Samples weighing a few tens of grams, and containing particles in the 50 to 10 μm size range, can be fractionated without supervision in 30 to 40 minutes.

Particle sizing by diffraction of a light beam
When a parallel beam of monochromatic radiation (e.g. the light from a laser) falls on to a particle then the light is diffracted by an amount which depends on the size of that particle. A suitable lens system can be used to concentrate the diffracted beams from a particle into a narrow annulus, the diameter of which is a function of the particle size.

(a) 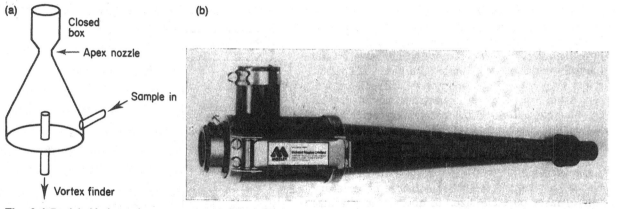 (b)

Closed box
Apex nozzle
Sample in
Vortex finder

Fig. A.1.5 (a) Hydrocyclone arrangement in "Cyclosizer" apparatus; (b) Photograph of Mozley hydrocyclone (courtesy of Christison Scientific Equipment Ltd, Gateshead).

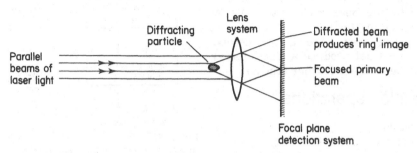

Diffracting particle
Lens system
Diffracted beam produces 'ring' image
Parallel beams of laser light
Focused primary beam
Focal plane detection system

Fig. A.1.6 Principle of particle sizing using a light beam.

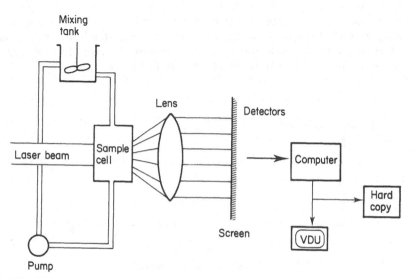

Mixing tank
Lens
Detectors
Laser beam
Sample cell
Computer
Hard copy
Pump
Screen
VDU

Fig. A.1.7 Diagram of laser diffraction sizing apparatus, such as the "Malvern" system.

A population of different size particles will produce its own, unique, set (or pattern) of diffracted light rings, where the diameter of any particular ring is inversely proportional to the size of the particle which is responsible for that diffraction effect (Fig. A.1.6). The complete diffraction pattern from a population of particles can be detected electronically and the pattern converted by a small dedicated computer into a size distribution (by weight) of the particles that are involved.

Laser sizing

Use a "Malvern"® (or other suitable proprietary make) laser particle sizer to determine the size distribution of the material previously used for the Beaker Decantation exercise.

Take about 1 g of specimen and disperse the particles in 1 litre of water. This suspension will be pumped through a glass cell held in a parallel, monochromatic beam of light from a low-power helium–neon laser (Fig. A.1.7). Comment on the results obtained by the Beaker Decantation and the "Malvern" sizing methods.

(*Note:* the Malvern Sizer – like the microscope sizing method – does not produce a fractionated product that can be used for further study.)

Particle separations

Magnetic separations

These separations are carried out in the laboratory in order to: (1) produce magnetic fractions from a mixture of mineral particles; and (2) determine the distribution of magnetic permeabilities of a population of mineral particles.

Magnetic fractionation

Take 10 g of a 50:50 quartz–magnetite mixture (−150 to +100 μm in size) and separate into two fractions with a simple hand magnet. To do this, spread out the mixture on to a clean sheet of paper and use a plastic bag over the magnet to

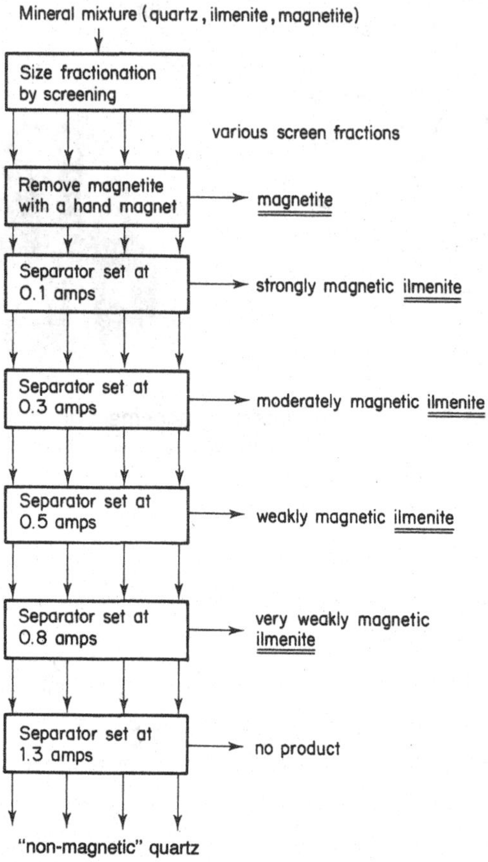

Fig. A.1.8 Flowsheet for magnetic permeability measurements using the Frantz or Cook isodynamic separator.

prevent the magnetite grains from sticking (the grains will easily fall off the magnet when the bag is removed).

Weigh the magnetic fraction and compare your results with the mineral proportions you used to make up the feed material. Comment on any differences between the expected value and the measured value.

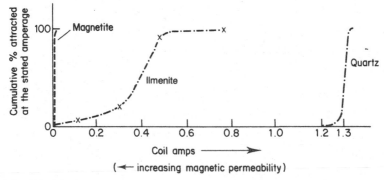

Fig. A.1.9 Graphical presentation of results from magnetic permeability measurements.

Magnetic permeability of minerals

Take a mixture of quartz, ilmenite and magnetite and screen it into a number of "size" fractions. Pass each size fraction through an isodynamic magnetic separator at the following settings:

longitudinal slope – 20° to 30° depending on particle size: the smaller the particle, the steeper the slope.

back slope – 8°.

feed rate – arranged to provide a continuous, single file of particles through the separator.

magnetic field – coil amperage to be set, sequentially, at 0.1, 0.3, 0.5, 0.8, and 1.3 (see flowsheet, Fig. A.1.8).

Weigh each product and use your results to calculate the overall mineralogical composition of your feed material (see Table A.1.3). Draw a graph showing the variations of magnetic permeabilities for the different minerals (Fig. A.1.9). What coil amps would you use to obtain a clean ilmenite fraction from a mixture of liberated particles consisting of magnetite, ilmenite and quartz?

TABLE A.1.3
Mineralogical details (example only)

Size fraction (−μm)	Mass (g)	Mass (%)	Percent of minerals in each size fraction		
			Magnetite	Ilmenite	Quartz
−400 + 300	500	50	20	20	60
−300 + 200	300	30	30	30	40
−200 + 100	100	10	40	40	20
−100	100	10	40	60	NIL
Totals	1000	100	27*	29*	44*

*To calculate the "totals" for each mineral multiply its "percent weight" by the "percent weight" in each size fraction. Thus, the "total percent magnetite" is calculated as follows:

Size (μm)	Wt. %	% magnetite	% Wt. × % magnetite
−400 + 300	50	20	(50 × 20 = 10.00)
−300 + 200	30	30	(30 × 30 = 9.00)
−200 + 100	10	40	(10 × 40 = 4.00)
−100	10	40	(10 × 40 = 4.00)
Totals	100	27	27.00

Practical no 3: The stereoscopic microscope

Aims: to gain familiarity with the stereoscopic microscope and to practise its use in the study of particulate mineral specimens.

Special equipment and materials needed: low-power stereoscopic microscope; microscope lamp; tweezers; sheets of white and black paper; ultra-violet lamp; glass slides; flat-bottomed Petri dishes; mounted needle; small spatula; fine brush; soap solution; gloves; safety glasses; selection of mineral grains; blue-glass filter; Leitz–Jelley refractometer; refractive-index liquids.

General information: It cannot be too strongly emphasised that practice in looking at and in measuring minerals is of vital importance in the development of mineralogical skills. You can read a detailed description of a mineral, or you can study coloured photographs of unusually good examples of that mineral, but you will not necessarily thereby develop any of the skills required for identifying that mineral when it occurs as fine-grained broken particles enclosed in a variety of natural (or even unnatural) matrices.

The stereoscopic microscope is a cheap and simple instrument that allows the student and the engineer to study fine-grained mineral specimens. The aim of this exercise is to provide practice in identifying minerals, and in quantifying mineralogical features such as grain sizes and mineral associations.

A typical stereoscopic microscope is shown in Fig. A.1.10. It consists of two separate monocular microscopes which are locked together in such a way that they provide a stereoscopic view of any object placed on the glass stage. Changes of magnification are generally achieved by changing both the objective lenses, and these are housed in a rotatable turret (a "zoom" facility for changing magnification is fitted to many modern instruments, however).

The stereoscopic microscope is quite robust, but it should nevertheless be treated carefully. It has a long focal length so that there is sufficient space between the objective lenses and the specimen for the specimen to be manipulated with a pair of tweezers. This facility allows the mineralogist to turn over loose mineral particles so that they may be examined from a variety of angles.

This stereoscopic microscope provides a range of magnifications up to about ×100, and this magnification range is adequate for much of the routine mineralogical work that is carried out

Fig. A.1.10 Typical modern stereoscopic microscope, fitted with zoom lenses for changing the magnification.

in the mineral industry. The microscope also conveniently provides a non-inverted, three-dimensional image of the specimen (cf. the petrological microscope which inverts the image and makes manipulation of the specimen a difficult operation).

Illumination is provided by a lamp of adjustable intensity which can either be attached to the microscope frame or stand entirely separately. Coloured filters provide approximately monochromatic light beams which may be useful for certain purposes: for example, it is often easier to distinguish chalcopyrite from pyrite in blue light than in white light.

Use of ordinary white light

The light should be directed on to the specimen so as to provide the maximum *contrast* between the various minerals. For example, the light can be directed across the specimen from almost any angle or direction. A strong light may not always be desirable: the angle and the intensity of the light and the colour of the background should be arranged so as to give the best contrast between selected features. A dark background is useful for studying light-coloured mineral grains and such a background can be obtained very simply by inserting a black cloth or sheet of black paper beneath the microscope stage. Similarly, a light background can be useful for studying dark mineral grains and a light-coloured background can be obtained by placing a white tile or a sheet of white paper beneath the stage. Transmitted light can be obtained by shielding the specimen with the hand from all "top" (reflected) light and by using *only* the light reflected from a sheet of white paper placed beneath the stage. The mineralogist must experiment with the light conditions to obtain the best contrast for each specific task.

Ultraviolet light

Ultraviolet light can sometimes be useful for illuminating a specimen, but *it must only be used with the appropriate safeguards* to prevent excessive exposure of the eyes and skin tissue to this light (e.g. by wearing protective glasses and gloves). Some mineral grains, such as scheelite (calcium tungstate) and zircon, fluoresce and the fluorescent grains can be seen readily in a specimen – even when they occur in only very small proportions. These minerals can be distinguished readily from optically similar (but non-fluorescent) minerals such as calcite or garnet.

Accessory equipment

The specimens being studied under a stereoscopic microscope should not be placed directly on to the glass stage but should always be placed on a

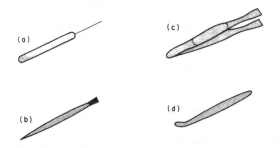

Fig. A.1.11 Useful accessories for use with the stereoscopic microscope: (a) mounted needle; (b) fine brush; (c) tweezers; (d) small spatula.

plane surface which itself is placed on the glass stage. A glass slide is suitable for holding fine-grained mineral particles but a shallow, flat-bottomed Petri dish is a more useful container for coarse mineral grains (watch-glasses and small bowls are *not* suitable, because the instrument must be continually refocused when traversing the specimen). A mounted needle (an ordinary needle fitted into a wooden handle), a fine brush, a pair of tweezers and a small spatula are useful accessories for manipulating mineral particles under the microscope (Fig. A.1.11).

Conditions of observation

Specimen grains should be viewed only when they are completely dry or when they are completely immersed in a liquid. Damp particles tend to aggregate into clusters, which are difficult to study. Complete immersion in a liquid often provides a clearer view of the specimen than immersion in air because very fine-grained ("slime") particles, which would otherwise adhere to the larger particles, are dispersed by the liquid. In addition, less light is scattered from a rough mineral surface when it is wet. Any stable, non-toxic liquid that does not react with the minerals can be used to "wet" the specimen. Water is often the most suitable liquid and a drop or two of a detergent or mild soap solution will further improve the dispersion of slime particles. Propane-1,2,3-triol (glycerol – a liquid of high viscosity) is sometimes used to prevent particles moving about too freely when the specimen holder is moved.

Experimental procedure

Use a microscope capable of about ×60 magnification. Familiarise yourself with its various parts, i.e. the lenses, the focusing arrangement, the stage, the eye-piece graticules and so forth. Learn how to adjust the two separate optical units so that you can obtain clear, unstrained stereoscopic vision; many models have "click-stops" to indicate the correct alignment of the rotatable turret that carries the objective lenses. See how the light source can be varied both in intensity and in its position relative to the specimen.

Calibrate the eyepiece graticule (a ruled grating that is inserted into one of the eyepiece lens mountings). It should be calibrated against a standard length – preferably a specially prepared stage micrometer, which is usually in the form of a subdivided millimetre scale. If there is nothing

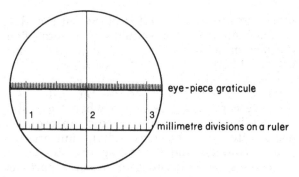

Fig. A.1.12 Representation of view down microscope fitted with an eyepiece graticule. What is the magnification of a microscope with these scales?

eye-piece graticule

millimetre divisions on a ruler

under the microscope at different magnifications. Note their characteristics – size, shape, colour, transparency and so on. Experiment with the type of illumination – vary the position and angle of the lamp, vary the light intensity, use coloured glass filters if they are available. Try using different backgrounds by placing sheets of different coloured paper under the specimen: use light transmitted through the specimen from a white sheet placed beneath the stage; try using only the light reflected from the mineral surfaces (i.e. use a matt black background). Tabulate the effects you obtain by using these different forms of lighting (Table A.1.4 shows a typical set of records).

Repeat your observations with the specimen grains immersed in water. Report any new effects that you see. Then add a drop or two of soap solution to the water and note the effect. In particular, note the effect of the soap solution on very fine-grained "slime" coatings which may occur on the larger mineral particles. Note the following features:

Opacity. Remember that most minerals become increasingly transparent as their thickness is reduced. Other minerals remain opaque even when they occur as very thin flakes (cf. Practical no 6, where specimens of *standard* thickness are used with the petrological microscope).

else available, the graticule may be calibrated against a simple millimetre ruler. If a millimetre on the ruler is equivalent to four major divisions on the graticule, then each of these divisions will be equivalent to 250 μm. If each major division is subdivided into ten equal parts, then each division on the graticule will be equivalent to 25 μm. (Remember, this applies only at the magnification provided by a particular set of lenses. What is the relative magnification of the image shown in Fig. A.1.12?)

Observe a selection of dry mineral particles

TABLE A.1.4
Typical tabulation of results

	Mineral A	Mineral B	Mineral C	Mineral D
Absorption of white light	Transparent	Opaque	Translucent	Opaque
Colour	Colourless	Black	Green	Blue grey
Streak	Colourless	Black	Green	Blue grey
Refractive index	Low	n/a	High	n/a
Reflection characteristics (lustre)	—	—	—	Metallic
Cleavage	—	—	—	3 at right angles
Fracture	Shell-like (conchoidal)	—	Irregular	Dominated by cleavages
Hardness	6	6	4	2 to 3
Tenacity	Brittle	Brittle	Soft	Soft
Crystal form	Prismatic	Octahedral?		Cubic
Particle shape	Irregular	Some cuboid shapes	Irregular	Cuboid
Particle-size distribution	100–200 μm	100–200 μm	50–100 μm	200–300 μm
Mineral name	Quartz	Magnetite	Malachite	Galena

Colour. No attempt should be made to describe the colour in detail. Colour can vary with grain size, grain orientation, composition and type of illumination while, in addition, surface coatings can alter the true colour of a mineral.

Streak. Crush a mineral fragment by firmly pressing on it with a metal spatula. Categorise the powder colours into "colourless", "dark" and "obviously coloured".

Reflection characteristics. These will depend on whether the surface that is observed is a cleavage face of the mineral or an irregular fracture through that mineral. Is the mineral highly reflective or earthy and dull?

Hardness. Carry out a Mohs hardness determination. Make sure which mineral is scratched, i.e. whether the unknown mineral is really scratching the standard mineral. Check by careful examination.

Tenacity is determined by pressing a small metal spatula on to the mineral and observing the result with the microscope. If the mineral is *brittle* it breaks easily and noisily and the broken fragments scatter; a *malleable* mineral, such as a native metal, will change shape but will not break under pressure; a *flexible* mineral, such as molybdenite, will bend instead of breaking; and an *elastic* mineral, such as mica, will bend but will recover its original shape when the pressure is removed.

Crystal form. Long, slender, needle-like crystals are called *acicular*, multiple-branched aggregates are *dendritic*, rounded masses of mineral grains are *mamillary*, and so on. These forms are seldom preserved in broken rock fragments.

Particle shape may mimic crystal form, or the particles can be *rounded*, *subrounded* or *platy* as a result of erosion and abrasion by natural agencies.

Particle size is determined by comparison with the calibrated eyepiece graticule. Remember that loose, unmounted particles will always assume their most stable positions and the average size obtained by microscope measurements of these particles will be larger than that obtained when the same particles are screened through a set of sieves (see Practical no 2).

Refractive index. Use the shadow test to estimate the refractive indices of unmounted mineral grains, and a refractometer to make a quantitative determination.

(1) The *shadow test* determines whether a transparent mineral grain has a higher or lower refractive index than that of the liquid in which it is immersed.

Take a lightly greased glass slide (drawing the hand over the slide is often enough to leave a film

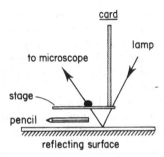

Fig. A.1.13 Determination of relative refractive index by the shadow method. The card is used to shield the specimen from any "top" light: the reflecting surface (which can be a sheet of white paper) reflects the light up through the specimen.

of grease). Place a single drop of liquid on the slide and immerse the mineral grain in the liquid. View the selected grain using transmitted light only – the unwanted "top" light from the microscope lamp must be cut off by using a card (or the hand) to shield the specimen (Fig. A.1.13). Note the appearance of the grain boundary – is this boundary very obvious, or is it difficult to see? (i.e. note whether or not the grain shows high relief).

Introduce a pencil obliquely beneath the specimen stage – i.e. push in a pencil from, say, the south-east corner of the stage (Fig. A.1.14). Shadows will then appear, both in the immersion fluid and in the mineral grain. If the shadow in the grain appears on the same side of the field of view as the shadow in the liquid, then the grain has a lower refractive index than that of the

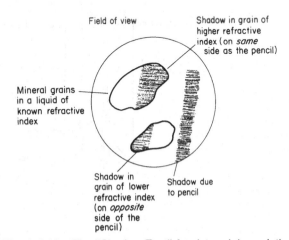

Fig. A.1.14 The "Shadow Test" for determining relative refractive indices. The diagram shows that the insertion of an opaque object (say a pencil) to one side of the stage will produce a shadow in the specimen. The position of this shadow shows whether the object is of lower or higher refractive index than the mounting medium.

liquid. Conversely, if the shadow appears in the grain on the opposite side from the shadow in the liquid, then the grain has the higher refractive index. The greater the difference between the refractive indices of the mineral and the immersion liquid, the clearer is the shadowing effect; if (either by chance or by design – see below) the grain and the liquid have the same refractive index, then the grain will be almost invisible in the liquid.

Remember that the refractive indices of amorphous materials and isometric minerals do not vary with the orientation of the specimen; the refractive indices of all other minerals may, however, vary with the direction in which they are viewed. This variation may be small, but it is the reason why birefringent minerals cannot completely disappear in any liquid.

Use the shadow test to determine the relative refractive indices of the following:

(1) quartz and water (r.i. = 1.00);
(2) quartz and α-monobromonaphthalene (r.i. = 1.62);
(3) zircon and α-monobromonaphthalene.

The mineral grain under study must be totally immersed in the liquid. If a mineral has a lower density than the liquid then the mineral will

TABLE A.1.5

Useful liquids for determining the refractive indices of minerals

Liquid	Refractive index
Water*	1.33
Propanone (acetone)	1.36
Methanol (formaldehyde)	1.41
Trichloromethane (chloroform)	1.46
Turpentine	1.47
Castor oil	1.48
Xylol	1.50
Cedarwood oil	1.52
Clove oil†	1.53
Methyl-2-nitrobenzene (nitroluene)	1.55
Tribromomethane (bromoform)	1.59
α-Monobromonaphthalene	1.66
Clerici solution	1.68
Di-iodomethane (methylene iodide)	1.74

* May dissolve some minerals.
† Varies from 1.53 to 1.54 – will mix with petroleum to give intermediate values from 1.45 to 1.54.

float; however, it can still be completely immersed by covering the liquid containing it with a thin glass cover-slip.

Table A.1.5 gives a list of the liquids that are commonly used to determine the refractive indices of mineral grains. It cannot be too often

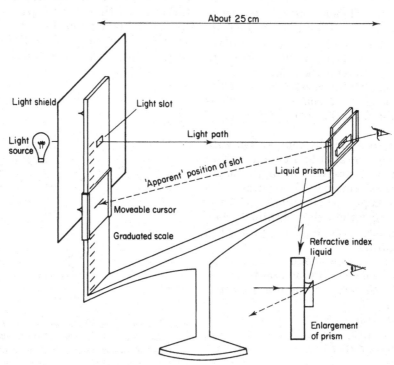

Fig. A.1.15 Leitz–Jelley refractometer for accurately determining the refractive indices of liquids (after M. H. Batley).

stressed that many of these organic liquids are potentially dangerous. Use them carefully, and dispose safely of any glass slides that are smeared with these liquids.

(2) The *Leitz–Jelley refractometer* is a simple instrument that can be used for accurately determining the refractive index of a liquid. If a liquid and a mineral have the same refractive index then the refraction of that mineral can be accurately established.

Immerse the mineral grain in a convenient liquid, and then determine whether the mineral or the liquid has the higher refractive index. Add, drop by drop, another miscible liquid of appropriately high (or low) refractive index until the mineral effectively "disappears" – i.e. until its outlines become very faint. At this stage the mineral and the liquid mixture have approximately the same refractive index.

Determine the refractive index of the liquid by using the Leitz–Jelley refractometer (Fig. A.1.15) as follows: take a drop of the liquid mixture (use a glass capillary tube to do this; do *not*, on any account, attempt to suck up the liquid with a pipette) and place the drop in the special hollow prism fitted to the instrument. The liquid-filled prism diffracts a narrow beam of white light and produces a spectrum of colours on a graduated scale. Read the refractive index of the liquid from the scale using the position of the yellow band in the spectrum (this is because reference works normally give refractive indices measured at the wavelength of yellow sodium light).

Practical no 4: Determination of mineral density

Special equipment and material needed: heavy liquids (tetrabromoethane, di-iodomethane); 25-ml pycnometer; soap solution; kerosene; 10–20 g of clean quartz grains.

The *density* of a mineral is its mass per unit volume; it is a property that does not vary with crystal orientation. The *specific gravity* of a mineral is the ratio if its mass to the mass of an equivalent volume of water.

Qualitative determination of mineral density by using heavy liquids

The approximate density of a small mineral grain only a few tens of micrometres in size can easily be determined by observing its behaviour in liquids of known densities. Use the stereoscopic microscope (see Practical no 3) as an aid when selecting a clean, non-porous mineral grain for the density test.

Place a *drop* of tetrabromoethane (density 2900 kg m^{-3}) on to a lightly greased glass slide; use a pair of tweezers to immerse a grain of quartz in the liquid. Use a stereoscopic microscope to make sure that the grain is properly "wetted" by the liquid or it may be supported by surface tension effects at the air–liquid interface, and make

sure that there are no air bubbles adhering to the mineral surface.

Observe whether the quartz grain floats or sinks in the liquid. It should float, because its density is less than that of the liquid; had it been greater, the grain would have sunk to the bottom. Grains of some minerals would appear to be suspended in the liquid and neither float nor sink: these minerals have the *same* density as the liquid.

Use the determinative scheme outlined in the flowsheet (Fig. A.1.16) to classify the following minerals into the appropriate density categories: (1) quartz, (2) apatite, (3) zircon, (4) rutile.

Note: Read and observe the safety precautions required when using heavy liquids which are listed in Practical no 5.

Accurate determination of mineral density using a pycnometer

The mean density of a population of mineral grains can be determined using a small bottle, called a pycnometer, which has a ground-in stopper pierced with a fine hole (see Fig. 4.2, p. 52).

The empty bottle is first weighed; the mineral grains are then introduced into the bottle and the bottle plus mineral are weighed; the bottle (plus mineral) is then filled with water to the top of the

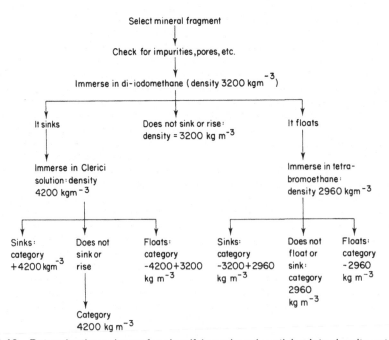

Fig. A.1.16 Determinative scheme for classifying mineral particles into density categories.

hole in the stopper and weighed again; finally, the mineral grains are removed, the bottle is filled with water and weighed. Thus:

Mass of empty bottle $= W$

Mass of bottle + mineral $= W_1$

Mass of bottle + mineral + water $= W_2$

Mass of bottle full of water $= W_3$

Mineral density $= \dfrac{\text{Mineral weight}}{\text{Mineral volume}} = \dfrac{W_m}{V_m}\ \text{kg m}^{-3}$

Mass of mineral $= (W_1 - W)$

Mass of water in full bottle $= (W_3 - W)$

Mass of mineral + water in full bottle $= (W_2 - W)$

Mass of water displaced by the mineral (this is equivalent to the *volume* of the mineral, V_m) $= (W_3 - W) - (W_2 - W_1)$

Mineral density $= \dfrac{W_m}{V_m}$

$= \dfrac{(W_1 - W) \times 1000}{(W_3 - W) - (W_2 - W)}\ \text{kg m}^{-3}$

Make sure that *all* the mineral grains are "wetted" by the water and that no air bubbles adhere to the grains or to the inner surface of the pycnometer bottle. A drop of soap solution helps to wet the grains, and air bubbles can be removed either by thoroughly shaking the bottle or by applying negative pressure with a small pump.

If the mineral to be measured is soluble in water then an organic liquid, such as kerosene, must be used instead of water (provided that the density of that liquid is known or can be established). The density calculation is then modified to:

Density of mineral =

$$\dfrac{(W_1 - W) \times \text{liquid density} \times 1000}{(W_3 - W) - (W_2 - W_1)}$$

Determine the density of a population of clean quartz grains using the pycnometer method (1) with water, (2) with kerosene. (*Note:* Select a bottle size so that the available mineral grains occupy about 30–40% of its volume.) How do the densities obtained by the two methods compare with one another?

Practical no 5: Fractionating minerals in heavy liquids

Aim: to illustrate some of the methods that are available for fractionating minerals in heavy liquids.

Special equipment and material needed: tetrabromoethane; filter funnels; pinch clips; filter papers; solvent (ethanol or propanone). The specimen material is a sample taken from a mixture of 90 g quartz and 10 g magnetite, similar to that used in earlier practical exercises. The quartz has a density of $2650\ kg\ m^{-3}$ and that of the magnetite is $5200\ kg\ m^{-3}$.

Separation of sized fractions

Screen the sample into three suitable size categories and separate the components of each category in tetrabromoethane. Set up the separation system as shown in Fig. 3.5. Introduce the feed material *slowly* into the liquid and make sure that all the particles are wetted (otherwise some of the magnetite may float because of surface tension effects).

When the separation is complete drain off first the "sinks", and then the "floats", into separate filter papers; these can be conveniently held in a single filter funnel. Let the heavy liquid drain through the papers back into its bottle (it is ready to be used again). Then carefully wash the mineral particles with either ethanol or propanone, and save the washings so that you may later recover the small amounts of heavy liquid that they contain.

Repeat the separation procedure for each size fraction, and record any problems encountered when separating the finest fraction.

Dry and weigh the mineral fractions, and then calculate the overall proportion of magnetite in the original specimen.

Separation of an unsized specimen

Recombine all the fractions obtained in the previous experiment and then separate the combined material in clean tetrabromoethane recovered from the earlier test. Drain, wash, dry and weigh the two products to establish the proportion of magnetite in the original material.

Comment on the overall magnetite values obtained by the two separation procedures, and compare these values with the known proportion of magnetite in the feed material that you prepared.

Safety precautions when using heavy liquids

As long as the heavy liquids and their solvents are handled carefully there is no health hazard involved. Careless or foolish behaviour can result in serious dangers, however. Observe the following rules at all times.

(1) *Avoid breathing the fumes* from these liquids: heavy liquids and solvents must *only* be used in well-ventilated fume cupboards. In many instances the toxic effects of these fumes are cumulative and symptoms may only develop slowly.

(2) *Avoid skin contact with the liquids* because many are carcinogenic or toxic by absorption through the skin. Always use disposable gloves (check that they have no leaks) and safety spectacles when handling these liquids. If any liquid does get on to your skin then wash it off immediately with the appropriate solvent followed by copious amounts of water.

(3) *Do not smoke in the vicinity of heavy liquids.* The solvents have low flash-points and are extremely flammable; for example, ethanol and propanone have flash-points below 30°C and can form explosive mixtures with air. (If the solvent does catch fire do *not* attempt to extinguish it with water: use only extinguishers suitable for Class B, liquid fires.) Moreover, the vapours from heavy liquids decompose at red flame temperatures (e.g. the flames produced by matches, cigarettes, cigarette lighters and so on). The decomposition products are often extremely toxic, and may include bromine gas, phosgene and carbon monoxide. The act of smoking draws these toxic products directly into the lungs.

(4) *Thallium salts* (e.g. Clerici solution) are extremely toxic and special care should be taken when using these solutions.

(5) *Use of "safe" liquids.* Tetrachloromethane (carbon tetrachloride) and benzene are, contrary to popular belief, very dangerous liquids and they should *not* be used during heavy liquid analyses. Trichloroethene can be used as a substitute for tetrachloromethane and either propanone or methylbenzene (toluene) can substitute for benzene – but with all the precautions mentioned above.

(6) *Use as little as possible* of the heavy liquid. On completing the test, store all liquids in a safe place.

Practical no 6: The transmitted-light polarising microscope

Aim: to gain familiarity in the setting up and use of the polarising microscope in the examination of mineralogical specimens.

Special equipment and materials needed: polarising microscope; thin sections of ores and/or mineral specimens.

Setting up the microscope

(1) Carry out a general check to ensure the completeness and the cleanliness of the optical parts – mirrors, lenses, polarisers. Remove any dust with a soft brush or with a special lens tissue so that the parts are not scratched.

(2) Check the mechanical parts for cleanliness and easy movement, i.e. check the stage, focusing arrangements, polariser slides and so on.

(3) Check that the analyser is "out", i.e. check that it is not included in the optical path. See that the substage diaphragm is open so that light can pass into the optical system. Move the Bertrand lens into the "out" position, and see that there are no accessory plates inserted in the microscope. (The Bertrand lens and the accessory plates are not described in this book but they are included in some microscopes and are of considerable value to petrologists. These "extras" are described in detail in many textbooks on petrography.)

(4) Switch on the microscope lamp to about half-power and direct the beam on to the flat side of the substage mirror.

(5) Select a low-power objective lens (about 10× magnification is suitable). Place a plain glass slide on the stage and adjust the position of the mirror to provide reasonably uniform illumination over the field of view; then adjust the stage diaphragm to provide sufficient light for comfortable viewing. Try not to adjust the light by adjusting the power to the lamp, because this will alter the colour of the light.

(6) Check the orientations of the analyser and the polariser in their rotatable mounts. First, rotate the polariser to its zero (east–west) position; this is generally indicated by a "click-stop" – a slight clicking sound as the polariser engages with a nick in the frame. With the glass slide still on the stage and the analyser also inserted, the view should become completely dark. If the field is not completely dark then slacken the analyser clamping screw, and gently rotate the mounting so as to produce the minimum amount of trans-

mitted light; re-clamp the screw. The analyser and polariser are then correctly aligned relative to each other.

(7) Use a high-power lens and centre the stage so that it rotates accurately about the axis of the microscope. The centring procedure should be carried out as follows:

(*a*) use any suitable specimen and place some recognisable small feature of that specimen at the intersection of the cross-hairs (see Fig. A.1.17*a*);

(*b*) rotate the stage and note the path followed by the selected feature (see Fig. A.1.17*b*);

(*c*) rotate the stage to the position where this feature is furthest from the centre of the cross-hairs and lock the stage-rotation locking screw (in a badly misaligned microscope this "furthest" position may be outside the field of view);

(*d*) use the two stage-centring screws to bring the selected feature *half*-way back to the intersection of the cross-hairs (Fig. A.1.17c).

(*e*) move the *specimen* so that the selected feature lies under the cross-hairs and repeat instructions (*a*), (*b*), (*c*) and (*d*). After two, or at most, three sequences of operations the selected feature will remain accurately centred at the cross-hair intersection throughout a complete rotation of the stage. If the centring operation is carried out with a high-power objective lens then the stage will, at the same time, be centred accurately for any lower-power lens.

(8) Defocus the substage condenser lens, by

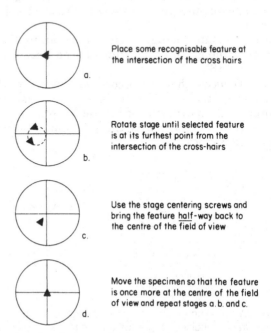

Fig. A.1.17   Stage-centring procedure – this ensures that the stage revolves around the optical axis of the microscope.

racking the lens to its top position and then racking down slightly with the focusing knob.

The microscope is now correctly aligned for normal use. With practice this setting-up procedure can be completed in a minute or two and it should be carried out before every series of observations.

Observations with ordinary transmitted light

Remove both the polariser and the analyser from the optical path of the microscope. It can now be used as a simple, ordinary microscope to examine thin sections of rocks, thin sections of particulate materials, and crushed unmounted mineral particles (cf. stereoscopic microscope).

Examine a typical thin section of, say, a granite. How many major mineral constituents can you distinguish? List the properties that enable you to distinguish these minerals. What can you learn about the sizes and the proportions of these minerals?

Refractive index. The refractive index of a mineral within a thin section can be estimated by using the *Becke line test.* The refractive index of the unknown mineral is compared either with the refractive index of its mounting medium or with that of a contiguous (touching) mineral grain. The procedure is as follows:

(1) Select a suitable mineral grain.

(2) Reduce the light intensity by closing the substage diaphragm, and use a high-power objective lens.

(3) A bright "fringe" will appear around the edges of the grain – this fringe is the Becke line (Fig. A.1.18).

(4) Defocus the microscope by lowering the stage slightly (or raise the objective lens) and observe the movement of the Becke line: it will appear to move *towards* the medium of higher refractive index.

(5) If the mineral and its mounting medium have closely similar refractive indices then the Becke line will be indistinct and it may split to form separate red and blue fringes which may move in opposite directions when the microscope is slightly defocused. Such an effect is obtained when the refractive indices of the mineral at different light wavelengths bracket the refractive index of the medium i.e. the refractive indices of mineral and mounting medium match for some intermediate wavelength.

(*Note:* the relative refractive index of a loose mineral grain can also be determined by the "shadow method" – see Practical no 3).

Cleavage traces show up as dark, straight lines which may be parallel to some prominent grain edge. Minerals can show more than one cleavage direction: the angles between sets of cleavages can be of diagnostic value (Chapter 4, section 5.7) and should be recorded whenever they are seen.

Observations with polarised light

First insert the polariser into the optical path. Rotate the specimen, and establish whether any of the minerals show pleochroism (a change of colour on rotation) or twinkling effects. Note the colours that occur at different orientations with pleochroic minerals.

Observations with crossed polars. Insert both the polariser and the analyser into the optical path. Determine whether the specimen contains any isotropic sections and also whether these sections represent isotropic minerals or are merely rare isotropic sections of anisotropic minerals).

Measure the extinction angles of as many different anisotropic minerals as you can distinguish, and note whether or not there are any twinned crystals in the specimen.

Tabulate the results of your observations as follows:

	Mineral A	Mineral B	Mineral C
Colour			
Refractive index			
Cleavage			
Habit or shape			
Twinning			
Alteration effects			
Any intergrowth phenomena?			
Size range			
Mineral proportions etc.			
⋮			

Becké line
Stage lowered (or distance between specimen and objective lens increased). This mineral has a higher refractive index than the medium.
Stage lowered – mineral has lower refractive index than the medium.

Fig. A.1.18 The Becke line method of determining relative refractive indices. Bright line around the perimeter of a grain appears to move into and out of the grain when the stage is moved up or down (on some microscopes, the objective lens is moved rather than the stage), at high magnifications.

Practical no 7: Preparation of thin sections

Aim: to illustrate the problems that are often encountered during the preparation of thin sections. (There is no guarantee that the student will be able to produce good thin sections after this single attempt, but constant practice will produce good results.)

Special equipment and material needed: thick glass plates; diamond rock-cutting blade; lapping machine; carborundum powders; epoxy resin; glass slides.

Solid specimens

Cut a parallel-sided slice some 3 to 5 mm thick from a granite block with a diamond saw. Make sure that the granite block is firmly held in a suitable vice before starting the cutting operation and observe all the safety instructions issued by the equipment manufacturer (i.e. wear safety goggles, close any safety doors fitted to the equipment, and so on).

Cut a roughly square piece about 1.5 × 1.5 cm in size from the granite slice and smooth one face of it by rubbing on fine carborundum powder (Grade 2F) on a thick glass plate.

Mount the smoothed face of the specimen on to a degreased and roughened standard glass slide, 7.5 × 2.5 cm in size. To do this, apply a little epoxy resin to the slide and to the smooth specimen, and press the glass and the specimen together. Move the two faces about a little to remove air bubbles. Place the section in a warm oven for about two hours for the resin to harden, according to the manufacturer's instructions (Fig. A.1.19).

Grind down the exposed face of the granite slice on a suitable lapping machine until the desired thickness of 30 μm is achieved. Attach a small rubber suction cup to the face of the glass slide to help you hold the specimen. The great difficulty, at this stage, is to retain as much as possible of the area of the original slice of granite. This problem can only be overcome by great care and by experience. Preliminary grinding should be done on a fast, coarse-grinding diamond lap, or on a glass plate armoured with coarse abrasive grains (e.g. carborundum). Check the thickness of the slice by observing the birefringence of a known mineral (usually quartz). The colour of quartz in polarised light changes rapidly from yellow to white when the correct thickness of 30 μm is achieved. In the later stages of the thinning procedure the specimen is transferred to a fine-grinding lap armoured with fine-grained car-

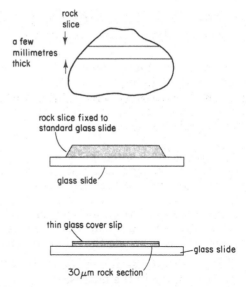

Fig. A.1.19 Stages in preparation of a thin section of a mineral or rock.

borundum or alumina abrasive. Compare the total area of your final thin section with the area of the original slice that you stuck on to the glass slide. What does the result show?

The grinding process is usually carried out using water as the lubricant. However, if there is any danger of some mineral within a specimen being soluble in water, then paraffin or a light oil should be used.

Friable specimens are first impregnated with an epoxy resin before being subjected to the diamond-sawing stage. The impregnation is carried out by immersing the specimen, under vacuum, in an epoxy-resin solution for several hours to allow the resin to soak into the pores of the specimen. The impregnated specimen is then baked for a few hours at 100°C to harden the resin. The specimen can then be treated like a coherent rock.

Thin sections are usually protected by a thin glass cover-slip. Temporary protection is possible by attaching the cover-slip with a small amount of thin oil. More permanent protection is afforded either by spraying the section with a clear, proprietory specimen-coating fluid, or by attaching the cover-slip with epoxy resin.

Particulate specimens

Mount the particles on to a thin cover-glass (about 1.5 × 1.5 cm in size) and carefully grind to produce a flat surface. Mount the particles (plus cover-slip) on to a standard glass slide, with the flat faces nearest the slide. Grind off the cover-glass and proceed as for a solid specimen.

Practical no 8: The reflected light polarising microscope

Aim: to familiarise the student with the use of vertically reflected light in the examination of minerals and to teach the student to measure and to interpret the mineralogical features that are displayed on polished sections.

Special equipment and material needed: reflected-light polarising microscope; polished mineral sections; specimen levelling device.

The procedure for setting up the microscope varies slightly with the particular model of microscope, but will include the following steps:

(1) if one is fitted, use the cross-bar to attach the microscope lamp to the microscope;

(2) focus the lamp filament as nearly as possible on to the polariser;

(3) place a levelled, isotropic polished section on the stage and adjust both the lamp and the mirror to give an image of the lamp iris diaphragm on the surface of the specimen;

(4) close the lamp iris diaphragm and adjust the mirror to bring its image into the centre of the field of view;

(5) open the lamp diaphragm sufficiently to illuminate the whole field of view and then defocus the lamp to give maximum uniform illumination across the field of view;

(6) set the analyser to "90°" and clamp it;

(7) rotate the polariser to the extinction position and clamp it.

The polariser scale should now read zero against the index arrow. If it does not, then the scale on the polariser should be moved to bring the zero mark into line: in this position, the vibration direction of the polariser is east–west.

Unpolarised, ordinary illumination

All sections must be viewed in such a way that the polished surface is perpendicular to the incident (and to the reflected) beam – i.e. the polished surface must be parallel to the surface of the stage. This is arranged by "levelling" the specimen with a levelling press (see Fig. A.1.20). Specimens tend to tarnish with age and with exposure to the laboratory atmosphere. Consequently, they should be buffed up on a polishing cloth before being viewed.

Most minerals look different in reflected light than in either transmitted light or oblique illumination, and initially are difficult to identify.

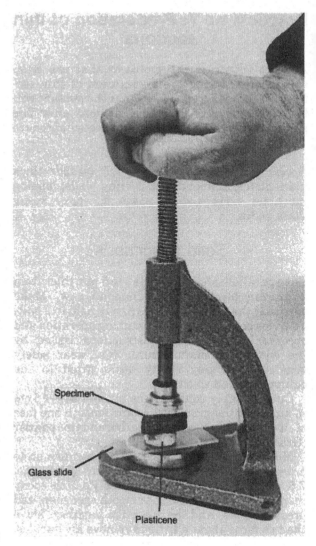

Fig. A.1.20 Levelling press.

Practice in the use of the reflecting microscope will overcome this problem.

Use the microscope to study (1) polished grains of ilmenite, pyrite, galena, chalcopyrite, sphalerite and quartz, and (2) a polished ore specimen containing some or all of these minerals. Observe the following features:

(*a*) grain and particle shapes;

(*b*) the effects of cleavage – these are especially obvious with galena grains, which tend to show many triangular etch pits and also right-angled corners to grain boundaries;

(*c*) mineral colour: this is a difficult feature to describe since its perception depends on the nature of the enclosing minerals; for instance, chalcopyrite on its own has a characteristic yellow colour but it can appear to be bright yellow when viewed against sphalerite, whilst the same mineral often looks a dull yellow-green when viewed against native gold. Most minerals show only shades of grey but a few show extremely pale tints (ilmenite, for example, when freshly polished, can be very faintly pink);

(d) comparative reflectivities of minerals – the brighter the mineral the higher its reflectivity;

(e) alteration and intergrowth features; for instance, galena (PbS) may be ringed by the alteration product, cerussite ($PbCO_3$);

(f) special polishing characteristics; pitting, etching and grain relief features – these features may not occur on perfectly polished specimens but few specimens are that good: for example, galena almost always shows triangular polishing pits (Fig. 7.7);

(g) grain sizes and grain proportions: these features can be fully quantified by various image-analysing procedures (see Practical no 10), but rough estimates can be obtained by careful scrutiny of the polished sections;

(h) relative hardness of the various minerals: the softer minerals will often be polished to a lower general level than the harder minerals.

Plane-polarised illumination

With the polariser inserted, but not analyser; this mode of operation can be used to observe all the features seen in ordinary light. In addition, it can show:

(1) bireflectance (i.e. the change of reflectivity that occurs with change of mineral orientation): this occurs when most sections through anisotropic minerals are rotated,

(2) reflection pleochroism (i.e. a change of colour with change of orientation) in a few anisotropic minerals.

With both the polariser and analyser inserted, the microscope can be used to observe:

(1) extinction effects (and other rotation properties) of minerals,

(2) polarisation colours – the colours shown by some anisotropic minerals in between extinction positions,

(3) internal reflections – these are caused by fractures or cleavages which exist at acute angles to the polished surface: these reflections appear as non-uniform, often coloured, illumination when the mineral is at its extinction position.

Note: it is clearly useful to have a set of mineral specimens of known colours and reflectances. Preliminary examinations of these specimens will help to provide standardised observing conditions and improve your qualitative judgement.

Examine a number of polished mineral specimens and tabulate the results of your observations in a similar style to that used in Practical no 6.

Practical no 9: Preparation of polished specimens

Aim: to allow the student to gain experience in the preparation of polished specimens for study in reflected light (these specimens will also be suitable for use in the electron probe X-ray micro-analyser).

Special equipment and material needed: diamond-armoured rock-cutter; specimen moulds; powdered graphite; epoxy resin; grinding and polishing laps; diamond pastes.

There is no well-documented, standard procedure for preparing polished specimens – every experimenter has his or her own favourite method. Specimens that contain different combinations of minerals require different polishing techniques, and it is usually a matter of intelligent trial and error to find the best method for any particular material.

One of the primary objects of the preparation procedure is to ensure that specimen features are not destroyed (or added to!) during the polishing operation. The chemical reactions that take place during a polishing process are not clearly understood: it is known, however, that the quality of the final polish depends on the lubricant used and on the type of abrasive (not only the particle size of the abrasive, but also the *kind* of abrasive). Artefacts (features introduced into the specimen by the polishing procedure) show up clearly in polished specimens. For example, galena almost invariably shows large numbers of triangular etch pits, harshly polished pyrite may become anisotropic (rather than isotropic), the reflectance of a mineral polished in a water-based medium is often lower than when it is polished in an oil-based medium, and so on. The method described below will nevertheless produce a satisfactory degree of polish on a wide range of the materials commonly encountered in the mineral industry.

Every polishing method comprises the following sequence of operations:

(1) specimen selection and preparation,
(2) specimen mounting,
(3) grinding a flat surface, and
(4) polishing that flat surface.

The success achieved at each step in this sequence depends on the successful completion of the previous step. Great care is therefore essential at every stage of the polishing procedure – the final result will otherwise be, at best, disappointing; at worst, the specimen will be quite useless.

Specimen selection and preparation

Solid specimens of rocks or ores are first cut to a suitable size with a diamond-armoured saw and one face is then carefully flattened. Friable specimens must be strengthened by impregnation with a suitable resin under vacuum. Use the largest convenient specimens, so that the spatial relationships between the minerals under study are retained as far as possible. It is sometimes possible to drill out cylindrical specimens of rock of suitable diameter with a diamond-armoured coring bit.

When particulate material is to be polished then, once again, the coarsest possible material should be selected so that spatial relationships are preserved. However, this selection procedure must not bias any earlier sampling procedures and it must produce specimens that fit the requirements of the experiment.

Mounting the specimen

Specimens are usually mounted in standard-size blocks of epoxy resin. This is done for ease of handling and to standardise the finished specimen size so that all specimens fit the specimen holders used in the polishing machines and in analytical instruments such as the electron probe microanalyser.

The most commonly used mounting medium is "Araldite": this gives good adhesion to the specimen material. Other epoxy resins can also be used, as can polyester resins, bakelite and acrylics. Unless the specimen mounting is specifically required in clear, transparent form, the Araldite is mixed with about 15–20% of colloidal graphite: this increases the electrical and thermal conductivities of the mountant and improves its performance when used in the electron probe X-ray microanalyser. The resin is mixed with the graphite; di-*n*-butyl phthalate is often added to reduce the surface tension of the Araldite, thus improving the dispersion of mineral grains and reducing the number of air bubbles trapped in the resin: any remaining air bubbles are removed by a small suction pump. (A stock of de-gassed graphite–resin mixture can be kept available for use.) When mounting a specimen, a suitable quantity of the resin is mixed with about 10% of hardener. This mixture has a long setting time at room temperature. It is usable for an hour or two and allows plenty of time for the final de-gassing of fine-grained particulate specimens. Furthermore, the slow setting ensures that the specimen does not become overheated and "delicate" min-

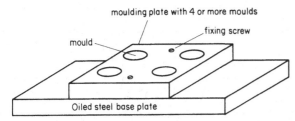

Fig. A.1.21 Moulds for preparing resin mounts of mineral or rock fragments. The fragments are immersed in an epoxy resin which is allowed to set in the moulds; the base plate is then unscrewed and the resin blocks are pushed out.

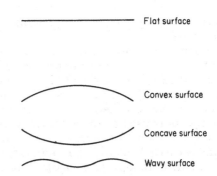

Fig. A.1.22 Exaggerated views of the type of undesirable surfaces that can be produced by careless grinding and polishing. The modern automatic measuring systems demand a flat surface right across a specimen.

erals are not damaged by local hot spots caused by the exothermic hardening process.

Particles are mounted in the epoxy resin in greased metal, glass or polythene moulds (Fig. A.1.21). An acceptable degree of randomisation of particle positioning and particle orientation can be achieved by first lining the base of the mould with sticky tape (sticky side uppermost): the particles are then sprinkled on to this tape where they stick in "random" positions and in "random" orientations to form a single-particle layer. The mould is then inverted, and any excess particles will fall out. The well-mixed epoxy-plus-graphite mountant is poured gently on to the mineral particles.

When lump specimens are being prepared for polishing a little of the resin–graphite mixture is poured on to the bottom of the mould. The roughly flattened specimen is then placed, face down, into this resin. When the specimen is positioned satisfactorily it is covered with more resin and is then treated in the same way as a particulate specimen.

If a specimen is known *not* to contain thermally delicate minerals then the resin-hardening process can be speeded up by gently warming the mould.

Grinding a flat surface

After hardening, the specimens are removed from the moulds and baked at about 60°C for an hour or two. The appropriate face is then ground on glass plates, or on paper laps, armoured with carborundum. Water is used as a lubricant. Every effort must be made to produce a surface that is *flat*, as well as polished (Fig. A.1.22 – the flat surface reduces the need for constant re-focusing of microscopes and microanalysers during subsequent examinations).

The grinding is repeated a number of times on progressively finer abrasive-coated paper laps.

The successive lapping procedures are used to remove the scoring and other surface damage produced by the previous, coarser size abrasive. Great care is, therefore, needed to ensure that all the abrasive powder and all the debris from one stage of the polishing procedure are *completely* removed before moving on to the next stage. This is done by careful washing with ethanol or by holding the specimen under running water, or by cleaning it in an ultrasonic bath. The grinding process is continued until the specimen surface has a flat, matt appearance.

Note: if any specimen is suspected to contain water-sensitive minerals then oil-based lubricants must be used and the specimen *must* be washed in ethanol rather than water.

Polishing the prepared surface

There are many variants of the polishing procedure. Polishing abrasives range from expensive diamond-loaded paste, to comparatively cheap alumina powder. Polishing laps are made of metal, wood or plastic: these laps may be covered with materials such as paper, cloth, synthetic fibres or aluminium foil, or they may be left uncovered. The lap speed can be varied, and the amount of pressure applied to the specimen and the amount of the lubricant are also variable.

Experienced operators select the most appropriate combinations of these variables. A student, or other first-time operator, is recommended to polish the specimen on a synthetic fibre lap using, initially, 8-μm diamond paste and ethanol as lubricant and with the lap rotating at 125 revolutions per minute. This first polishing step can take up to 10 minutes but the specimen should be periodically examined to ensure that all the deep scratches caused by the grinding operation have been removed.

The specimen is then moved to a second synthetic fibre lap where 6-μm diamond paste is used with emulsified oil as lubricant. This stage usually takes about 15 minutes. The third polishing stage uses *another* synthetic fibre lap where 3-μm diamond paste is used as the abrasive; this stage takes up a further 10 minutes or so. The final polishing is carried out with 1-μm diamond paste on a synthetic fibre lap using a spirit lubricant. If an even better surface is needed then an additional stage using sub-micron alumina paste can be employed. Remember that absolute cleanliness is essential at each stage so as to avoid contamination of a lap by coarse abrasive particles.

(*Note:* in order to avoid confusion the commercially available diamond pastes are dyed different colours: 8 μm = red; 6 μm = yellow; 3 μm = green; and 1 μm = blue.)

A booklet by Lister describes, in detail, the highly satisfactory procedures used by the Institute of Geological Sciences, London, where hundreds of specimens are prepared on a routine basis every week.

Specimen identification

When large numbers of specimens are being prepared in a laboratory it is important that each should be easily identifiable. A piece of paper showing the specimen number can be incorporated within a transparent plastic mount. This system will not work with graphite-loaded epoxy resin because the material is opaque. However, a layer of clear plastic can be poured on to the mineral-bearing opaque resin layer (Fig. A.1.23),

Fig. A.1.23 Identification of mounted specimens. Small pieces of paper bearing the specimen numbers are immersed in the resin. If opaque resin is necessary (for sample preparation for electron probe X-ray microanalysis) then a little clear resin can be poured on to the opaque material, immersing the paper labels.

and the paper label bearing the specimen number can be incorporated. Alternatively, the specimen details must be scratched on to the base of the graphitised layer with a mounted needle.

Practical exercise

Take a mixture of mineral particles and prepare a polished section containing a representative sample of this mixture.

Practical no 10: Quantitative assessment of mineral proportions by point- and particle-counting procedures

Aim: to provide an opportunity for the student to quantify those mineralogical features needed by the mineral engineer.

Special equipment and materials needed: mechanical stage for a stereoscopic microscope; coloured beads (ballatini).

Almost all investigations of mineralogical materials require quantitative assessments of the *proportions* of the various minerals in ores and in mineral products. A simple and effective way of making such an assessment is to carry out point or particle counts. These assessments can be carried out either directly on the test material, as provided, or on various fractions of the original material after appropriate pre-treatment procedures such as sizing or classification (see Practical no 2).

The appropriate sampling procedure has to be used to produce a suitable sample (see Practical no 1). Remember that solid, lump materials are difficult to sample. Fine-grained particulate materials are comparatively easy to sample and tens of thousands of particles can be mounted on to a single small specimen.

Point- and particle-counting procedures

A counting grid is produced by moving the specimen stage by fixed increments in the x-direction followed by known increments in the y-direction until an adequate number of points has been counted. Care must be taken to avoid, as far as possible, the errors which are likely to occur during particle- and point-counting exercises. These include *operational errors* such as incorrect identification, miscounting and misrecording of results. *Statistical errors* occur because of the sampling effect of the counting procedure and these can be kept within stated limits by counting a sufficient number of particles or points (see Chapter 6, section 7).

Counting can be carried out in one of the following ways:

(1) count *all* the particles in the specimen. This method can only be applied to specimens that contain small numbers of particles.

(2) Count *all* the particles in selected fields of view chosen either in a truly random fashion (using random number tables to give the x- and y-coordinates of each field), or in a regular pattern at the intersections of a grid. If the specimen is suspected to show a regular structure or pattern then it is preferable to choose the fields at random.

(3) Count only those particles which coincide with the intersections of a regular grid (for instance, the grid pattern of the eyepiece graticule). This procedure analyses only a few points per field of view and requires that a large number of fields be examined. These fields should be selected as described in (2) above.

When solid specimens are being analysed each counting point will fall on a mineral (occasionally, a point will fall on to a pore: when this happens the pore should be counted as such). With particulate material many of the points fall on the mounting medium – these points can be ignored in subsequent calculations, except when it is necessary to establish the "particle density" of a specimen of this kind.

Interpretation of measured data

Counting procedures only provide *number* proportions and these values are often converted into *volume* and *mass* proportions by the mineral engineer.

Point proportions are numerically equivalent to volume proportions if the minerals are randomly positioned in the measured specimen and a sufficient number of points has been counted; i.e. if mineral A occupies 50% of the measured points then it will also occupy 50% of the specimen by volume.

Volume proportions may be converted into mass proportions if the densities of all the constituents are known:

$$W = \frac{P_1 \times D_1 \times 100}{D_m(100 - P_1) + (P_1 \times D_1)}$$

where W = mass percent of the measured mineral,

P_1 = volume percent of the measured mineral,

D_1 = density of the measured mineral, and

D_m = mean density of the other minerals in the specimen, i.e.

$$D_m = \frac{P_2 D_2 + P_3 D_3 + P_4 D_4 + \ldots}{P_2 + P_3 + P_4 + \ldots}$$

where P_2, D_2 = volume percent and density of mineral 2, and so forth.

Particle counts can be converted into volumetric proportions if the material is closely sized and all the particles are of the same shape, i.e.

particle proportion ≡ volumetric proportion

If there are large differences in shape or size (or both) then this simple conversion is not valid (see Chapter 6, section 5.4), and great care is needed when interpreting the results of such an analysis. Thus, particle-counting procedures are best used for analysing *sized and* liberated materials such as those found in sized beach sand and alluvial specimens. It is usually convenient to count the unmounted particles rather than spend time and effort in preparing mounted, polished specimens.

Practical exercise

(a) Take 0.9 g of white ballatini (small glass beads) and 0.1 g of green ballatini of the same size. Determine the proportions (by number) of each colour by:

(1) counting all the beads,
(2) counting all the beads in selected fields of view,
(3) counting the beads that occur at the intersections of the eyepiece graticule in selected fields of view.

In this instance the proportion of ballatini by number will also be the proportion by volume and by weight. Compare your calculated values of the weight proportions with the known composition of the measured specimen.

(If ballatini are unavailable, screened grains of quartz and magnetite (or hematite, or galena) can be used instead. The conversion of numerical proportions into volumetric or mass proportions is much less accurate than with the glass beads, however – see section (b) below.)

Calculation of errors
Use the following formula to establish the statistical errors involved in your work:

Standard deviation $(\sigma) = \sqrt{\dfrac{pq}{N}}$

where p = proportion of the selected phase,
$q = (1 - p)$, and
N = number of particles counted.
The acceptable error in most mineralogical analyses is at the 95 percent confidence limit (i.e. at $\pm 2\sigma$) and the formula for the acceptable absolute error (e) is:

$$e = 2\sqrt{\frac{pq}{N}}$$

Thus, the number of particles that must be counted to reduce e to acceptable limits is:

$$N = \frac{4pq}{e^2}$$

where e is the specified allowable absolute error, e.g. 0.1 at the 95% confidence level. The relative error E on p is a more useful indicator of precision; then

$$E = \frac{e}{p} \quad \text{and} \quad N = \frac{4q}{pE^2}$$

(*Note*: p and N must be obtained by iterative calculation – see Chapter 6, section 7.1).

Make full notes of your counting procedures and show how you assess your statistical errors.

(b) Take 0.9 g of closely sized quartz particles and 0.1 g of similarly sized galena. Determine the numerical proportions of the two minerals by counting those particles that occur at the intersections of the eyepiece graticule. You should aim for a statistical error less than 10% relative to the proportion of galena. Assume that the numerical proportion of galena equals its mass proportion. Explain why your results do not match the known value to 10% (by mass) of galena.

(c) Measure a "low-grade" specimen by taking 0.9 g of quartz and only 0.01 g of galena. Count the number of particles of galena (y) in N_F individual fields of view. If the *average* number of particles in each field of view is $\bar{x}$, then the proportion of galena, p, is:

$$p = \frac{y}{N_F \bar{x}}$$

The relative error (E_1) on p is given by:

$$E_1^2 = \frac{4q}{pN} = \frac{4(1 - p)}{pN_F \bar{x}} = \frac{4(1 - p)}{y}$$

If you use $y/N_F\bar{x}$ as an estimate of p then you must also consider the error on $\bar{x}$; $\bar{x}$ can be estimated by counting the total number of grains in each of N_c fields of view:

$$\bar{x} = \frac{1}{N_c}\sum_{i=1}^{N_c} x_i \quad \text{and} \quad \sigma^2 = \frac{1}{N_c - 1}\sum_{i=1}^{N_c}(x_i - \bar{x})^2$$

The relative error on $\bar{x}$ is given by E_2 where

$$E_2 = \frac{2\sigma}{\bar{x}\sqrt{N_c}} \quad \text{(with 95\% confidence)}$$

Now E_2 is independent of E_1, i.e. the number of particles in a field of view is not influenced by the mineral proportion, and therefore:

$$E_{total} = E = \sqrt{(E_1^2 + E_2^2)}$$

$$\therefore E^2 = \frac{4(1-p)}{y} + \frac{4\sigma^2}{\bar{x}^2 N_c}$$

$$\therefore y \left(E^2 - \frac{4\sigma^2}{\bar{x}^2 N_c} \right) = 4(1-p)$$

Hence,

$$y = \frac{4(1-p)}{E^2 - (4\sigma^2/\bar{x}^2 N_c)}$$

How many fields of view must you count to determine p to a relative error of 10% at the 95% confidence limit?

Practical no 11: Mineralogical analysis of a beach sand

Aim: to provide an opportunity to prepare a mineralogical balance showing the mineralogical compositions and the distribution of all the minerals in the various size fractions of a specimen.

Special equipment and materials needed: mineralised beach sand; stereoscopic microscope; hand magnet; isodynamic magnetic separator; heavy liquids; laboratory screens.

Take about 1 kg of beach sand known to contain heavy minerals such as ilmenite, magnetite, zircon, etc. (if no mineralised beach sand is available then it is possible to use any sand with added

heavy minerals to carry out the experiment). Screen the sand into the following size fractions:

>400 μm; 400–300 μm; 300–200 μm; 200–100 μm; <100 μm.

Weigh the fractions. Identify the various mineral species, then determine the mineralogical contents of each size fraction, using any method that appears appropriate – by density separation, for instance, or by magnetic fractionation or particle counting, or by some other method – and present the results in the form of a mineralogical balance (see Table A.1.6).

Use the information presented in your mineral balance to devise an outline flowsheet for recovering the potentially valuable components of the beach sand.

TABLE A.1.6
Mineralogical analysis

Size fraction (μm) (a)	Mass (g) (b)	Mass (%) (c)	Proportions of minerals:				Total	Distribution of minerals:			
			A (d)	B (e)	C (f)	D (g)		A	B	C	D
>400							100				
400–300							100				
300–200							100				
200–100							100				
<100							100				
Total		100					100	100	100	100	100

Notes:
(1) The measured values occur in columns *b*, *d*, *e*, *f* and *g*; all the other values must be calculated.
(2) The values in columns *d*, *e*, *f* and *g* must add to 100 for each size fraction.
(3) The values in the last four columns must add to 100 for each mineral.

Practical no 12: Photomicrography

Aim: to show some of the problems involved in producing photographic records of mineralogical materials.

Special equipment needed: microscope; 35 mm camera with adapter lens.

Photographs and photomicrographs of polished sections (and of thin sections) are often required as permanent records of the experimental work carried out and to illustrate mineralogical reports. Such photographs are usually taken with a microscope with a 35 mm camera fitted into the optical system. As with all other forms of photography the quality and the value of the final photographic image depend upon the type of illumination used, the accuracy of focus, the use of the correct exposure time and the suitability of the area selected to represent the specimen.

Most microscopes are fitted with good-quality, high-intensity lamps and, by using suitable filters, these can provide satisfactory illumination for most purposes. Good filters can improve phase discrimination; they can also be used to correct for imbalances in the film emulsion. For example, a blue filter is often used with "normal" colour film to prevent loss of detail in grey-brown tones.

The focusing of the image is carried out through the microscope eyepiece since the focus of the ocular lens is usually the focus of the photographic film (but always check that this is true). Exposure times are automatically determined with modern instruments but manual settings may be preferred for especially "difficult" specimens that contain unusually bright or unusually dark areas.

Using any photographic equipment that you have available, take black and white photographs of a typical polished section (try using the ones you prepared in Practical no 9) under different conditions of light, and with different filters and exposures. Check the developed negatives for good focusing and contrast, and see whether you have correctly chosen the most appropriate field of view. Record carefully all your procedures and camera settings. Mount the photographs in your notebook and comment on their quality.

Appendix 2

Determinative scheme: identification of mineral grains

1. Summary of determinative scheme

One of the more constant physical properties of a mineral is its specific gravity. In this scheme the relative specific gravity of the mineral is determined by immersing the mineral grain in a succession of three heavy liquids. As a result of this test it is possible to place the mineral in one of the four specific gravity groups that form the major subdivisions of the scheme. The fact that the specific gravities of some minerals are variable does not detract from the value of this initial subdivision, for any mineral that has a wide range of density may be included in more than one group in the tables. Similarly, a mineral having the same density as one of the liquids used will be included in two specific gravity groups – the one above and the one below the liquid density.

If the mineral grain is transparent, it is possible to determine its refractive index relative to that of the heavy liquids during the specific gravity tests. This information, although valuable, is not used for further subdivision because a large proportion of the minerals listed in the tables are opaque and do not have a refractive index value. However, the refractive index of a transparent mineral may be useful as a final, confirmation test.

After a grain has been removed from the heavy liquids its hardness can be compared with two standard materials. This usually allows a mineral specimen to be placed in one of three hardness categories. If the mineral has a hardness equal to that of a standard it will be included in two of the hardness groups, one above and one below the standard value.

The colour of a mineral powder is usually of greater diagnostic value than the colour of the mineral itself and forms the basis of the next subdivision of the determinative scheme.

The magnetism of the powder is determined and, although this property is not used for classification, it is often useful in the final identification.

Chemical tests are then made on the mineral powder. These tests, together with the physical properties already determined, are usually sufficient to identify the mineral if it is one of those included in the tables.

However, confirmatory tests should always be carried out. For each mineral, the tables give properties which are not used in the selection procedure and these may be augmented by reference to an up to date standard mineralogical textbook. So far as the available equipment permits, all critical properties must be checked.

Outline of determinative scheme

Procedure 1
Carefully select the mineral specimen.

Procedure 2.
Note general information about the mineral, including colour, shape and cleavage.

Procedure 3
Determine specific gravity group:

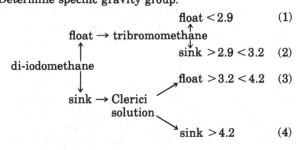

	float <2.9	(1)
float → tribromomethane		
di-iodomethane	sink >2.9 <3.2	(2)
	float >3.2 <4.2	(3)
sink → Clerici solution		
	sink >4.2	(4)

Procedure 4
Determine refractive index group:

>di-iodomethane .... =   >1.74   (d)

<di-iodomethane >Clerici
solution .... =   <1.74 >1.68   (c)

<Clerici solution >tribromo-
methane .... =   <1.68 >1.60   (b)

<tribromomethane .... =   <1.60   (a)

Procedure 5
Determine hardness group:

soft (will not scratch calcite) =   <3   (s)

medium .... =   >3   <7   (m)

hard (will scratch quartz) ... =   >7   (h)

Procedure 6
Determine the colour of the mineral powder:

white or colourless .... =   white   (W)

distinctly coloured, e.g., red,
yellow, blue or green .... =   coloured   (C)

black, grey or brown .... =   dark   (Dk)

Procedure 7
Determine magnetic permeability:

highly magnetic, moderately magnetic, weakly magnetic or non-magnetic.

Procedure 8
Determine solubility in acids:

concentrated HCl → effervesces   (E)

                soluble   (S)

                decomposed   (D)

      not attacked   (I)

concentrated HNO_3 → soluble   (S)

                decomposed   (D)

      not attacked   (I)

concentrated H_2SO_4 → soluble   (S)

                decomposed   (D)

             not attacked   (I)

Procedure 9
Spot tests on the acid solution using solutions of ammonium oxalate, phosphate and molybdate, barium chloride or hydrochloric acid to determine the major components of the minerals.

Procedure 10
Additional tests including fusibility, flame test, etc.

Procedure 11
Confirmatory tests by all available means.

2. List of equipment and reagents

The following equipment is needed to carry out the tests required for the determinative scheme:

(1) Microscope; preferably stereoscopic with total magnification ×20 or greater.

(2) Hand magnet; the Eclipse "Quick-Release" is very convenient and is obtainable from leading British laboratory supply houses.

(3) Streak plate (a piece of unglazed porcelain).

(4) Waxed or greased glass microscope slides.

(5) Mounted needles; it is useful to have one with a "chisel" edge.

(6) Metal forceps (*not* platinum; lead, arsenic and antimony minerals, especially, are easily reduced to the metal and in a reducing flame will combine with platinum to give low-melting-point alloys).

(7) Metal spatula.

(8) Bunsen or other gas burner; alternatively an alcohol lamp may be used.

(9) Standard quartz and calcite samples: to be used as hardness standards.

(10) Small boiling tubes or small lengths of 0.5 cm glass tubing closed at one end.

(11) Sample tubes.

(12) Small camel-hair brush.

(13) Glass capillary tubing, drawn from ordinary glass tubing.

(14) Reagents:

     di-iodomethane (methylene iodide),
     tribromomethane (bromoform),
     Clerici solution,
     methylated spirit,
     concentrated hydrochloric, nitric and sulphuric acids,
     concentrated ammonia solution,
     ammonium oxalate, ammonium phosphate, ammonium molybdate, barium chloride and silver nitrate solutions,
     powdered zinc metal,
     metallic tin granules,
     dimethylglyoxime,
     filter paper soaked in turmeric solution.

Note: some of these reagents are toxic, others are corrosive. Care and common sense are necessary to protect both the operator and the microscope.

3. Procedure

3.1. Selection of mineral specimen

Careful selection of the mineral grains is a vital preliminary to any identification procedure. The chosen grains must be homogeneous and free from such features as cavities, intergrowths, inclusions or surface films; otherwise, the determinative scheme will not be applicable. It is often difficult to ensure that a mineral grain consists of only a single phase and the best available methods must be applied to confirm the purity of the fragments before they are used. For most purposes it is sufficient to examine a mineral under a simple stereoscopic microscope.

Occasionally it is found that grains which appear to be homogeneous when examined under low magnification are, in fact, composed of two or more phases when seen under high magnification. Ideally, therefore, the minerals should be examined with the best possible magnification, even with a petrological microscope if one is available.

The grains should be observed either when completely dry or when fully immersed in a liquid and they must be turned over so that all surfaces can be checked for purity. Transparent minerals should be checked by examination in transmitted light, whilst opaque minerals are best examined under reflected light.

The selection of suitable, clean grains is often simplified by preparing a concentrate containing a reasonably high proportion of the desired mineral. Heavy-liquid separation, panning, selective solution, magnetic separation or any other appropriate mineralogical technique may be used for this purpose. However, the final selection of the grains used during the determinative procedure must be carried out by careful hand sorting using a mounted needle or a pair of fine tweezers.

It cannot be too strongly emphasised that the greatest possible care must be used at this stage or the identification scheme will be invalidated.

The tests that follow have been designed for use on single mineral fragments weighing about 0.01 g, i.e., approximately 25 B.S. mesh in size. It is prudent, where possible, to select a number of clean grains before commencing the identification procedure.

3.2. General characteristics of the mineral

The following information should be noted before commencing the determinative scheme: date, specimen number, description of specimen, locality, nature of the mineral occurrence, the nature of the minerals associated with the unknown mineral, the treatment (if any) undergone by the specimen, and the colour, shape and cleavage of the unknown mineral.

The *nature of the mineral occurrence* states whether the mineral is alluvial, pegmatitic, from the oxidised zone of an ore body and so forth, and can be a useful guide to the identification of the mineral. The "nature of the associated minerals" is also frequently useful in suggesting the identity of the unknown mineral; e.g. most tellurides are found associated with ores that contain native gold, while many of the platinum-group minerals may be found together in a single ore.

The *colour* of a mineral is often a most obvious property but it may vary greatly and be of little diagnostic value. Thus, tourmaline may be black, blue, green, red or colourless. When the colour of a mineral is of diagnostic value it has been included in the "Remarks" column of the tables, e.g. cinnabar, cochineal-red colour.

The *shape* of mineral grains found in eluvial or alluvial deposits is often a useful diagnostic characteristic and, where appropriate, this information has been included in the tables, e.g. tourmaline forms grains of triangular cross-section. Well-formed minerals may be chemically altered by natural agencies into some other compound which occasionally retains the form of the original mineral. These pseudomorphic grains can often be recognised by a lack of sharpness in the angles of a grain and by the dull, earthy lustre of the mineral faces.

The *cleavage* controls the tendency of a mineral to split along definite planes to give distinctively shaped fragments. Some minerals have one perfectly developed cleavage so that the grains always occur as flakes, e.g. mica or graphite. Other minerals have three well-developed cleavages at right angles to one another and these minerals commonly occur in cubical fragments or as fragments that show numerous 90° angles, e.g. galena. When the cleavage is of diagnostic value it has been noted in the tables.

3.3. Determination of specific gravity group

A mineral grain is first immersed in a drop of di-iodomethane (specific gravity/s.g. 3.2) on a waxed glass slide. (If the grain is large it is better to crush it into fragments before placing it in the heavy liquid. In this way the purity of the fragments can be checked by examination under a microscope; also, a single drop of the liquid will be sufficient for the test.) Care must be taken to ensure that the mineral fragment is "wetted" by the liquid and is not buoyed on the surface by surface tension effects. If the mineral floats on di-iodomethane a fragment is immersed in a drop of tribromomethane (s.g. 2.9) in which it will either sink or float. Minerals that sink in the di-iodomethane are then immersed in Clerici solution (s.g. 4.2) in which they will either sink or float. In this way, and using only two liquids for any one mineral grain, the mineral can be placed in one of the following groups which form the first subdivision of the determinative tables:

Clerici solution: "sinks" . . . . . s.g. >4.2

Clerici solution: "floats";
   di-iodomethane: "sinks" . . . s.g. <4.2 >3.2

di-iodomethane: "floats";
   tribromomethane: "sinks" . s.g. <3.2 >2.9

tribromomethane: "floats" . . . s.g. <2.9

Remember that these heavy liquids are toxic and should be handled with care.

3.4. Determination of refractive index group

The refractive index of a transparent mineral may be determined while the grain is immersed in the heavy liquid by the "shadow" method (see Practical no 3).

The refractive indices of the heavy liquids are as follows: di-iodomethane 1.74, Clerici solution 1.68, tribromomethane 1.60.

3.5. Determination of hardness group

The hardness of a mineral in the context of these determinative tables is its resistance to abrasion. It is important that the hardness determined is that of the true mineral and not that of an impurity or a surface coating. In addition, hardness must not be confused with brittleness; a granular material may appear to be scratched when the grains have only been torn apart. A definite scratch must be produced in the standard materials by the sample and this must be confirmed by examining the scratch under a microscope. The hardness of a small grain is best determined by placing it on the standard surface against which it is to be tested, pressing downward on the grain with a flat wooden implement like the end of a mounted-needle holder, and drawing the grain across the surface. In the tables the minerals have

been divided into three groups: the "soft" minerals will not scratch calcite and the "hard" minerals will scratch quartz; the "medium-hard" minerals will scratch calcite but will not scratch quartz.

If a mineral has a wide range of hardness values or if its hardness happens to coincide with that of a standard (quartz or calcite), the mineral will appear in more than one group in the tables.

3.6. Determination of the streak

The streak is the colour of the mineral powder and it is of much greater diagnostic value than the colour of the mineral itself. It is convenient to determine the colour of the powder of a single grain by crushing the grain between a streak-plate and a metal spatula held in the hand. The mineral should be crushed to a very fine powder and the colour of the streak classified into one of the following groups:

(a) white or colourless;
(b) has a definite colour, e.g. red, blue, green or yellow;
(c) dark and indefinite colours, e.g., black, grey or brown.

3.7. Determination of magnetism

All minerals are affected by powerful magnetic fields but few are sufficiently magnetic to be attracted by a hand-magnet. An Eclipse "Quick-Release" hand-magnet (see Fig. 3.20 , page 40) is useful for making a rapid assessment of the magnetism of a mineral powder.

The values quoted in the tables have been determined as follows. A polythene bag is drawn tightly around the jaws of an Eclipse magnet and, with the lever in the "off" position, the magnet is drawn through the mineral powder. Magnetite and metallic iron will adhere to the outside of the polythene bag and will only be removed when the bag is pulled away from the magnet. The bag is then replaced and the lever is moved half-way to the "on" position; moderately magnetic minerals such as ilmenite and franklinite will adhere to the bag until the lever is returned to the "off" position. Weakly magnetic minerals such as monazite will be attracted to the magnet only when it is in the fully "on" position. Minerals that are not attracted to the magnet even when it is fully "on" have been called non-magnetic. The settings on the hand-magnet are summarised below:

(a) Highly magnetic minerals are attracted when the magnet is in the "off" position.

(b) Moderately magnetic minerals are attracted when the magnet is in the "half-on" position.

(c) Weakly magnetic minerals are attracted only when the magnet is in the fully "on" position.

(d) Non-magnetic minerals are not attracted even when the magnet is in the fully "on" position.

When the magnetism of a mineral is of diagnostic importance, that value has been included in the tables. Care is necessary, however, when interpreting the results of the magnetic tests, as small amounts of iron oxide inclusions can greatly alter the magnetic properties of a mineral; e.g. zircon, which is normally a non-magnetic mineral, can become moderately magnetic due to micrometre-sized inclusions of magnetite or γ-hematite.

3.8. Determination of solubility in acids

The solubility determinations are carried out on the powdered mineral used for the streak and the

magnetic tests. It is important that the mineral is very finely powdered or the rate of reaction with the acids may be too slow for the purposes of the tests. A few milligrams of the powder is carefully brushed into a dry 10-cm test-tube (or into a small pyrex dish) and about 2 ml of acid are added (i.e., about half the test-tube full). The acid is heated gently over a small flame or on a water bath until about one-third of the original volume remains and is then cooled and brought back to the original volume by adding distilled water. (Great care must be taken with sulphuric acid which must always be added to the water and not vice versa.)

Observations of the solution should reveal that:

(a) the mineral powder is unchanged, or

(b) the mineral dissolved with effervescence and the liquid remains clear on dilution, or

(c) the mineral dissolved slowly and the liquid remains clear on dilution, or

(d) the mineral decomposed and a precipitate was formed which does not dissolve on diluting the solution.

In the tables, the results of the solubility tests are

TABLE A.2.1
Summary of results of acid attack

Acid	Mineral	Effect	Colour: Precipitate	Solution after dilution
Conc. HCl	Carbonate	Effervescence		
	Oxide of Mn, Cr or V	Pungent smell of chlorine gas		
	Sulphide minerals	Smell of H_2S		
	Co-rich minerals			Pink*
	Iron(III) minerals			Yellow or yellow-brown
	Ni, Cu or Cu–Fe minerals			Green
	Cu minerals			Green or blue
	Ag, Pb, Hg, Ti, W, Mo, V or SiO_2 minerals		White or pale-yellow	
Conc. HNO_3	Sulphide, oxide or native element	NO_2 given off (dark brown fumes)		
	Sulphide minerals		Yellow (floating)	
	Co minerals			Pink*
	Iron(III) minerals			Yellow, brown or red-brown
	Ni, Cu or Cu–Fe minerals			Green
	Cu minerals			Blue
	W minerals		Yellow	
	Sn, Sb, As, Pb, Ag or SiO_2 minerals		White	
Conc. H_2SO_4	Ca, Ba, Sr, Hg, Ag or Pb minerals		White	

* The pink colour due to cobalt will be concealed if any iron is present in the sample.

shown thus:

(a) unchanged I
(b) soluble with effervescence .... E
(c) slowly soluble .... S
(d) decomposed .... D

The solubility of the mineral in concentrated hydrochloric acid is determined first. If the mineral dissolves or decomposes, the solution is used for spot tests as described in the following section. If the mineral is *not* attacked, the acid is decanted away (or removed with capillary glass tubing) and is replaced with concentrated nitric acid. If the mineral is also insoluble in nitric acid, the acid is again decanted away and replaced with concentrated sulphuric acid. If sufficient clean mineral is available, it is, of course, much easier to use different portions for each acid test.

The results of acid attack on various minerals are summarised in Table A.2.1.

The water-soluble minerals have been excluded from the main tables because they are only infrequently encountered by the mineral technologist – and even then only under special circumstances. These minerals occur in arid regions, as deposits near hot springs, and in trona or similar pans, and are shown separately in Table A.2.2.

3.9. Spot tests

The acid solution from the previous procedure should be diluted with about an equal volume of distilled water. The spot tests must be carried out on the *solution* only and every effort must be made to exclude any precipitate or unattacked fragments of the mineral. With the aid of a piece of glass tubing fitted with a rubber bulb, six separate drops of the diluted acid solution are placed on a glass slide set on the stage of a stereoscopic microscope. Drops of six different reagents are then placed alongside and, when the reagent is run into an adjacent drop of mineral solution, the reaction is observed through the microscope. The observation of light-coloured precipitates can be improved by using a dark background beneath the stage.

The acid reagents should be kept in glass-stoppered dropping bottles; the other solutions should be stored in rubber-stoppered bottles and a hole can be drilled in each stopper for inserting a glass rod to be used as a dropper.

The following is a summary of the more useful and distinctive effects seen during the spot tests.

TABLE A.2.2
Water-soluble minerals

Mineral	Theoretical composition	$\bar{Z}$	Crystal system*	Specific gravity	Hardness	Colour of powder	Refractive index	Remarks
Alum	$4[KAl(SO_4)_2.12H_2O]$	9.9	Is	1.76	2	white	1.43–1.46	"Potash alum"
Borax	$4[Na_2B_2O_7.10H_2O]$		Mo	1.7	2–2.5	white	1.45–1.47	
Carnallite	$12[KMgCl_3.6H_2O]$	13.0	Or	1.6	2.5	white	1.46–1.49	Deliquescent
Chalcanthite	$2[CuSO_4.5H_2O]$	14.1	Tr	2.1–2.3	2.5	white	1.51–1.55	"Blue vitriol"; sky-blue colour
Copiapite	$R''Fe_4(SO_4)_6(OH)_2.nH_2O$ (where R'' includes Fe, Mg, Al, Cu or Na)		Or	2.1	2.5	yellow	1.50–1.60	Sulphur-yellow colour
Epsomite	$4[MgSO_4.7H_2O]$	9.0	Or	1.75	2–2.5	white	1.43–1.46	"Epsom salt"
Glauberite	$Na_2Ca(SO_4)_2$	12.1	Mo	2.8	2.5–3	white	1.51–1.54	Dissolves only in large excess of water
Halite	$4[NaCl]$	14.6	Is	2.2	2–2.5	white	1.54	Cubic cleavage
Mirabilite	$NaSO.10HO$	8.8	Mo	1.48	1.5–2	white	1.43–1.44	"Glauber salt"
Natron	$Na_2CO_3.10H_2O$	7.9	Mo	1.42	1–1.5	white		Effervesces with acid; usually only found in solution
Nitratine	$2[NaNO_3]$	8.6	He	2.29	1.5–2	white	1.34–1.59	"Chile saltpetre"; "soda nitre"; very deliquescent
Nitre	$4[KNO_3]$	12.1	Or	1.9–2.1	2	white	1.33–1.50	Saltpetre
Sal-ammoniac	$4[NH_4Cl]$	10.8	Is	1.52	1.5–2	white	1.64	
Sassolite	$4[B(OH)_3]$		Tr	1.48	1	white	1.34–1.46	"Boric acid"; very easily fusible
Sylvine	$4[KCl]$	18.0	Is	2.0	2	white	1.49	Cubic cleavage
Trona	$Na_3H(CO_3)_2.2H_2O$		Mo	2.1	2.5–3	white	1.41–1.54	Effervesces with acid

* For key to these abbreviations see page 190.

TABLE A.2.3
Reaction of various elements to excess ammonia
solution

Element	Reaction
Iron(II)	dirty-green precipitate
Iron(III)	red-brown precipitate
Uranium	yellow precipitate
Chromium	grey-green precipitate
Mercury	black precipitate
Bi, Ti, Zr, Th, Al, Be, Sn, Pb, As, (? Mo, V)	white precipitate which is readily coloured by small amounts of iron minerals
Copper	deep-blue solution
Nickel	blue-green solution
Cobalt	yellow-brown solution
Manganese	dirty-white or pale brown precipitate

For a more detailed treatment of chemical spot tests the reader is referred to Feigl (1958).

During these spot tests the operator should be concerned only with definite, positive results as the object is to determine the major components of the minerals.

The addition of concentrated ammonia solution
The presence of excess ammonia precipitates a number of metallic hydroxides and gives coloured solutions or precipitates. The results are summarized in Table A.2.3.

The addition of 10% ammonium oxalate solution and excess ammonia
This reagent throws down the insoluble oxalates of calcium, barium and strontium as white precipitates.

The addition of 10% ammonium phosphate solution
This reagent throws down the insoluble phosphates of magnesium and manganese as white precipitates (the manganese phosphate is occasionally slightly pink in colour). Magnesium phosphate may be rather difficult to precipitate.

The addition of ammonium molybdate-nitric acid solution
This reagent is made up by adding a few drops of concentrated nitric acid to a 10% ammonium molybdate solution. The presence of phosphates or arsenates is shown by the formation of a yellow precipitate in the test solution. The reaction takes

place slowly at room temperature but can be speeded by gently warming the glass slide.

The addition of 10% barium chloride solution
In the presence of a sulphate this reagent forms a white precipitate. The test, of course, must not be used on the sulphuric acid solution.

The addition of concentrated hydrochloric acid to a nitric acid solution of a mineral
This throws down the insoluble chlorides of mercury, silver, and lead as white precipitates. The silver precipitate is dissolved with some difficulty on the addition of ammonium solution and the lead chloride is soluble in hot water.

3.10. Additional tests

The results obtained from the above procedures should reduce the possible minerals to one, two or possibly three species. The final column of Table A.2.5 (below) lists characteristic properties and further simple diagnostic tests which, where appropriate, should be carried out. The tests are given below.

Closed-tube test
This test is only used in the determinative scheme to differentiate between hydrated and non-hydrated minerals of similar composition, e.g. gypsum and anhydrite. A quantity of the powdered mineral is placed in a narrow glass tube that is closed at one end (a small test-tube will do). The closed end of the tube is then gently heated and any water driven off from a hydrated mineral will condense and collect as small drops on the walls near the mouth of the tube.

Burning test
When a fragment of native sulphur is held in a match flame it burns readily and gives off fumes of sulphur dioxide.

Fusibility test
For this test a fragment of mineral is held in metal forceps or tweezers (*not* platinum-tipped) and heated in a low-temperature flame, e.g. match or candle. The following minerals fuse readily under these circumstances: stibnite, mimetite, pyromorphite, anglesite, jamesonite, descloizite and bismuthinite.

Flame test
A fragment of the mineral is heated in a hot flame, e.g. Bunsen burner or alcohol lamp, and the colour of the flame is noted. This is a useful

confirmatory test for strontium, which colours the flame crimson.

Malleability test

All the native metals listed in the tables are malleable with the exception of bismuth. Naumannite – a silver–lead selenide – is also malleable. The malleability of a small mineral grain can be checked by rolling a mounted needle over the particle while maintaining a steady downward pressure. If the particle is malleable it will readily deform without any sign of cracking.

Fluorescence and radioactivity tests

These tests require specialised equipment but, when such equipment is available, they can provide useful confirmatory information about several minerals. When the fluorescence or radioactivity of a mineral is of diagnostic value this is noted in the determinative table. The fluorescence effects are given for radiation in short-wave ultraviolet light (250 nm).

Additional chemical tests

A number of simple chemical tests mentioned in the "Remarks" columns of the tables are described below.

(a) Carbonate

Many carbonates are soluble with effervescence in cold 1:1 hydrochloric acid. All carbonates are soluble with effervescence in hot 1:1 hydrochloric acid.

(b) Chloride

A white precipitate is formed if silver nitrate solution is added to an acid solution containing a chloride. This test obviously cannot be used with a solution of a mineral in hydrochloric acid.

(c) Cassiterite

The "tinning" test is specific for this mineral. The test is carried out by intimately mixing the mineral with a little powdered zinc metal and adding dilute hydrochloric acid. Cassiterite is the *only* mineral that becomes coated with a dull-grey skin of metallic tin. It is sometimes difficult to obtain the above reaction when treating alluvial cassiterite that has been coated with iron oxides. The iron oxides may, however, be removed by boiling the cassiterite with concentrated hydrochloric acid for a few minutes.

(d) Nickel

An ammoniacal solution of nickel plus dimethylglyoxime gives a scarlet precipitate. Iron, cobalt and copper interfere with this test.

(e) Zinc

When dilute ammonia solution is added to a hydrochloric acid solution containing zinc, a white precipitate is formed which is soluble in excess ammonia.

(f) Silver

The silver chloride precipitated from an acid silver-bearing solution by the addition of hydrochloric acid is slowly soluble in ammonia.

(g) Titanium, tungsten and niobium

When granules of metallic tin are added to a boiling solution containing these elements, the following reactions occur:

> titanium gives a violet solution,
> tungsten gives a blue solution and this colour remains on dilution,
> niobium gives a blue solution but the colour disappears on dilution.

(h) Borate

When turmeric paper is moistened with a dilute hydrochloric acid solution containing boron and dried at about 100°C, the paper becomes reddish-brown. If ammonia is added to the paper the colour changes to black.

(i) Helvine

When the mineral is boiled for 1–2 minutes with 1:5 H_2SO_4 and As_2O_3, and then washed with water, helvine gives a brilliant canary-yellow deposit of As_2S_3.

(j) Manganese

Most of the manganese oxides decompose hydrogen peroxide. When a fragment of mineral is placed in a drop of 3% hydrogen peroxide it produces effervescence but the mineral is not noticeably attacked. This test can be used to confirm the presence of the manganese oxides hausmannite, manganite, psilomelane and pyrolusite.

(i) Differentiation of beryl and quartz

These two minerals are often difficult to differentiate. If beryl is suspected in the mineral sample the following test will usually distinguish it. A fragment of known quartz is placed along with the unknown mineral in a drop of tribromomethane. Both minerals will float on the surface of the liquid. Methylated spirit is then added drop-by-drop to the tribromomethane until the quartz just begins to sink; if the unknown mineral has sunk before the quartz then it is likely to be beryl. Unfortunately, the specific gravity of beryl is very occasionally almost the same as that of quartz and the test is then no longer applicable. (A similar test can be devised to differentiate between many pairs of minerals, e.g. feldspar and petalite.)

A preferential stain for beryl has been used by Jedwab (1957), which involves etching the particles in hot sodium hydroxide solution and then staining with a boiling aqueous solution of quinalizarin. This gives an intense blue colour on the surface of beryl.

3.11. Confirmatory tests

The tables include most of the minerals which commonly occur in ore deposits, but the list is by no means exhaustive, nor is the determinative scheme infallible (because of poor experimental skills or the very variable nature of some minerals). Confirmatory tests, therefore, are essential and these should be as detailed and thorough as time and facilities permit. Once a tentative identification has been made, reference to Table A.2.4 will show additional properties which should be checked as far as possible. Although the system given in Table A.2.5 has been devised for use with inexpensive equipment and a minimum of experience, the information in Table A.2.4 will permit more sophisticated tests to be carried out if the necessary equipment is available. Chemical and X-ray analysis or tests for fluorescence and radioactivity may help to confirm an identification. Precise determination of the index of refraction or reflectivity, which is shown in Table A.2.4, may be useful with transparent minerals (Winchell, 1942). The crystal system to which each mineral belongs is given and, with the aid of a polarising microscope, it is easy to distinguish the isotropic (isometric and amorphous) from the anisotropic minerals. If conoscopic, polarised illumination can be used then the biaxial minerals (orthorhombic, monoclinic and triclinic) can be differentiated from the uniaxial minerals (tetragonal and hexagonal) (Dana, 1949b). The reflectivity of all the opaque minerals is also listed in Table A.2.4 and can be a valuable diagnostic property if vertical illumination of polished specimens can be employed (Bowie and Taylor, 1958). Reference to standard mineralogical texts will augment the information given in the tables.

4. Recording test results

Only definite results are significant in the determination of a mineral from the tables. Therefore,

only clear-cut, decided reactions should be taken into account. Weak or uncertain results may be due to poor manipulation or the presence of small amounts of impurities in the specimen.

The results of each test must be recorded as soon as the test is carried out.

5. Explanation of the determinative tables

In Table A.2.4 the minerals are listed alphabetically and many of their properties are given. The square brackets used in the mineral formulae indicate that the unit cell has been established by X-ray methods and is either the formula within the brackets or a multiple indicated by the figure outside the brackets. Minerals shown in round brackets in column 2 of Table A.2.4 are not included in the determinative scheme.

In Table A.2.5 the minerals are divided into four major groups based on their specific gravities (groups 1, 2, 3 and 4). Within each of these groups, a division based on hardness is indicated by a solid line extending across the page. Further subdivision into groups of different coloured streaks is indicated by broken lines extending across the page.

Minerals shown in brackets in Table A.2.5 do not *normally* occur in the groups in which they are listed.

The common names of some minerals are given in inverted commas in Table A.2.4, e.g. sphalerite, "zinc blende".

The following abbreviations have been used in the tables:

Crystal system

Is	= isometric
Te	= tetragonal
Or	= orthorhombic
He	= hexagonal
Mo	= monoclinic
Tr	= triclinic

Solubility

E	= soluble with marked effervescence
S	= soluble
D	= decomposed, leaving an insoluble residue
I	= insoluble
HCl	= concentrated hydrochloric acid

H_2SO_4 = concentrated sulphuric acid
HNO_3 = concentrated nitric acid

Refractive index
r.i. = refractive index
a = r.i. <1.60
b = r.i. $>1.60 <1.68$
c = r.i. $>1.68 <1.74$
d = r.i. >1.74

Reflectivity
The values for reflectivity given in Table A.2.4 are always underlined

Hardness
H = hardness
s = soft H <3
m = medium H $>3 <7$
h = hard H >7

Streak
Bl = blue
Br = brown
C = coloured, red, blue, green or yellow
Dk = dark, indefinite colour
G = green
R = red
W = white or colourless
Y = yellow
G-bl = green-blue
G-br = green-brown
O-r = orange-red
O-y = orange-yellow
R-br = red-brown
Y-br = yellow-brown
Y-r = yellow-red

Acid solution plus
When a mineral has been attacked by an acid

the resultant acid solution is used plus the listed reagents, and the reactions are noted.

NH_4OH = concentrated ammonia solution

Ppt = colour of precipitate obtained on adding NH_4OH to the acid solution of the mineral: W = white, Br = brown, Bk = black, Y = yellow

Sol = colour of solution obtained on adding NH_4OH to the acid solution of the mineral: Bl = blue, G = green, Y = yellow, G-bl = green-blue

Ox = white precipitate formed on adding 10% ammonium oxalate to the acid solution of the mineral

Mo = yellow precipitate formed on adding 10% ammonium molybdate plus nitric acid solution to the acid solution of the mineral

Ba = white precipitate formed on adding 10% barium chloride solution to the acid solution of the mineral

HCl = white precipitate formed on adding HCl to the acid solution of the mineral

General
X = positive reaction
x = positive reaction but unusual with this particular mineral
s.g. = specific gravity
C.T. = closed tube test
$<$ = less than . . .
$>$ = more than . . .
$\geqslant$ = much greater than . . .
H_2S = hydrogen sulphide gas
cf. = compare with . . .
$\simeq$ = approximately equal to . . .

TABLE A.2.4

Alphabetical list of minerals

Mineral	Theoretical composition and % by mass of saleable component	$\bar{Z}$	Crystal system	Specific gravity	Hardness (Mohs scale)	Colour of powder	Solubility: HCl	HNO₃	H₂SO₄	Refractive index or reflectivity	General remarks and distinguishing features
Alabandine	4[MnS]	Mn = 63 / 21.7	Is	4.0	4	green	S			23̲	Black colour; cubic cleavage; H₂S given off in HCl.
Allanite	(Ca,Fe^{II})₂(R,Al,Fe^{III})₃Si₁₂O₁₂OH (R = rare earth elements)	28.8–30.0	Mo	3.0–4.2	5.5–6	pale brown	D			1.64–1.80	Slightly radioactive.
Altaite	4[PbTe]	Pb = 62 Te = 38 / 70.6	Is	8.2	3	white	I	S		65̲	Tin-white colour; cubic cleavage; may show slight effervescence with HNO₃.
Amblygonite	2[(Li,Na)AlPO₄(F,OH)]	Li₂O ≏ 10 / 10.3̲	Tr	3.0–3.1	6	white	S			1.57–1.60	"Hebronite"; slowly soluble; easily fusible.
Amphibole	complex Mg,Fe silicate, RSiO₃, with R usually Ca, Mg or Fe	11.4 ± 0.8	Or, Mo or Tr	2.8–3.4	6	white or green	I	–	–	1.64–1.73	Commonly green colour; 2 well-marked cleavages at 60°; fibrous grains more common than with pyroxenes; often moderately to weakly magnetic.
Andalusite	4[Al₂SiO₅]	10.7	Or	3.2	7.5	white	I			1.64	Grains frequently show a square cross-section.
Anglesite	4[PbSO₄]	Pb = 68 / 59.4	Or	6.3–6.4	3	white	I	D	–	1.89	High s.g.; difficult to dissolve; very easily fusible.
Anhydrite	4[CaSO₄]	13.4	Or	2.9–3.0	3–3.5	white	S			1.57–1.61	Pseudo-cubic cleavage; C.T. gives no water, cf. gypsum.
Ankerite	[Ca(Mg,Fe)(CO₃)₂]	Mg:Fe < 5 / 12.5–12.9	He	3.0–3.1	3.5	white	E			1.53–1.73	Frequently associated with iron ores.
Apatite	2[Ca₅(PO₄)₃F]	P₂O₅ = 42 / 14.1	He	3.2	5	white	S			1.63–1.65	Hexagonal crystals common; wide range of composition found.
Argentite	2[Ag₂S]	Ag = 87 / 43.0	Is	7.2–7.4	2–2.5	lead-grey	I	D		29̲	"Silver glance"; soft; sectile; often pseudomorphed by monoclinic low-temperature form called acanthite.
Argyrodite	32[Ag₈GeS₆]	Ag = 74 Ge = 6 / 40.7	Is	6.1–6.3	2.5	grey-black	I	D		27	Difficult to decompose in HNO₃.
Arsenopyrite	8[FeAsS]	As = 46 / 27.3	Or	5.9–6.2	5.5–6	grey-black	I	D		52–56̲	"Mispickel"; tin-white colour; grey-black streak.
Atacamite	4[Cu₂Cl(OH)₃]	Cu = 60 / 21.9	Or	3.8	3–3.5	apple-green	S	D		1.8 +	Green colour; C.T. gives water; secondary after cuprite and malachite.
Autunite	Ca(UO₂)₂(PO₄)₂ . 10H₂O	U₃O₈ ≏ 61 / 50.9	Te	3.1	2–2.5	yellow	S			1.55–1.58	"Lime uranite"; yellow colour; radioactive; fluorescent.

Mineral	Formula	Composition		System	S.G.	H	Colour/Streak				R.I.	Remarks
Azurite	$2[Cu_3(CO_3)_2(OH)_2]$	Cu = 55	19.4	Mo	3.8–3.9	3.5–4	blue	E	–	–	1.73–1.84	Distinctive blue colour.
Barite	$4[BaSO_4]$	BaO = 66	37.3	Or	4.3–4.6	2.5–3.5	white	–	–	–	1.64–1.65	"Heavy spar".
Beryl	$2[Be_3Al_2Si_6O_{18}]$	BeO = 14	10.2	He	2.6–2.8	7.5–8	white	–	–	D	1.57–1.60	Frequently greenish-white; s.g. usually 2.69–2.70; emerald and aquamarine are gem varieties.
Beryllonite	$12[NaBePO_4]$	BeO = 20	10.0	Or	2.8	5.5–6	white	S	–	–	1.56	White or yellow colour.
Biotite	complex K,Mg,Fe aluminosilicate		11.4–15.8	Mo	2.7–3.1	2.5–3	white	–	–	D	1.58–1.68	"Black mica"; elastic flakes; moderately magnetic; demagnetised by HCl.
Bismuth	$2[Bi]$	Bi = 100	83.0	He	9.8	2–2.5	silver-white	–	S	–	$\underline{68}$	Not malleable, cf. other native metallic elements; easily fusible.
Bismuthinite	$4[Bi_2S_3]$	Bi = 81	70.5	Or	6.4–6.5	2	lead-grey	–	S	–	$\underline{42}$–$\underline{49}$	"Bismuth glance"; fuses in a match flame.
Bismutite	$2[(BiO)_2CO_3]$	Bi = 82	69.4	–	7.0	4	green-grey	E	–	D	2.1 +	Non-crystalline; easily fusible; secondary after bismuthinite.
Bornite	$8[Cu_5FeS_4]$	Cu = 63	25.3	Is	4.9–5.4	3	grey-black	–	D	–	$\underline{22}$	"Peacock ore"; acicular crystals or fibrous masses; shows iridescent tarnish; becomes magnetic on heating in a reducing flame.
(Boulangerite)	$Pb_5Sb_4S_{11}$	Pb = 55, Sb = 26	61.5	Or	5.7–6.3	2.5–3	red-brown	–	–	–	$\underline{36}$–$\underline{38}$	
Bournonite	$4[CuPbSbS_3]$	Pb = 42, Cu = 13, Sb = 25	54.4	Or	5.7–5.9	2.5–3	grey-black	–	D	D		"Wheel ore".
Brannerite	$(U,Y,Ca,Fe,Th)_3Ti_5O_{16}$?	$U_3O_8 < 40$	52.2–60.4	Or ?	4.5–5.4	4.5	green-brown	–	–	–	2.3	Weakly magnetic; strongly radioactive.
Braunite	$8[Mn^{II}_3Mn^{III}_2Si_3O_{12}]$	Mn = 50	17.4	Te	4.8	6–6.5	dark brown	D	–	D	$\underline{18}$–$\underline{20}$	Chlorine given off in HCl; commonly of secondary origin.
(Bravoite)	$(Fe,Ni)S_2$	Fe = 15, Ni ≈ 20	20.8 ± 0.1	–				–	–	–		
Breithauptite	$2[NiSb]$	Ni = 33	43.5	He	7.5	5.5	red-brown	–	S	–	$\underline{45}$–$\underline{55}$	Light copper-red colour.
Brochantite	$4[Cu_4SO_4(OH)_6]$	Cu = 56	20.3	Or	3.9	3.5–4	pale-green	S	–	–	1.73–1.80	Green colour; secondary after other copper minerals.
Calamine	$2[Zn_4Si_2O_7(OH)_2 \cdot H_2O]$	Zn = 54	20.6	Or	3.4–3.5	4.5–5	white	D	D	–	1.61–1.64	"Hemimorphite"; C.T. gives water.
Calaverite	$2[AuTe_2]$	Au = 44	63.8	Mo	9.0	2.5–3	yellow-grey	–	S	–	$\underline{63}$	Deep red solution in H_2SO_4; contains only traces of Ag, cf. sylvanite; readily fusible to a gold button.
Calcite	$2[CaCO_3]$		12.6	He	2.7	3	white	E	–	–	1.49–1.66	Effervesces in cold, dilute HCl; rhombohedral cleavages.

continued over

TABLE A.2.4
Alphabetical list of minerals (continued)

Mineral	Theoretical composition and % by mass of saleable component		$\bar{Z}$	Crystal system	Specific gravity	Hardness (Mohs scale)	Colour of powder	Solubility: HCl	HNO₃	H₂SO₄	Refractive index or reflectivity	General remarks and distinguishing features
Calomel	2[Hg₂Cl₂]	Hg = 85	70.5	Te	6.5	1–2	white	–	–	–	2.0 +	"Horn quicksilver"; very soft; dissolves in aqua regia.
Carnotite	KUO₂VO₄·⅓H₂O	U₃O₈ = 65 V = 12	57.3	Or	4.5	about 1.5	yellow	S	–	–	1.75–2.1	Yellow colour; strongly radioactive.
Cassiterite	2[SnO₂]	Sn = 79	41.1	Te	6.8–7.1	6–7	white	–	–	–	12	"Tinstone"; colour very variable; usually non-magnetic but may show wide range of magnetism; "tinning" test.
Celestine	4[SrSO₄]	Sr = 48	23.7	Or	4.0	3–3.5	white	–	–	–	1.62–1.63	S.g. lower than barite; red flame test.
Cerussite	4[PbCO₃]	Pb = 78	65.3	Or	6.5	3–3.5	white	E	(E)	–	1.8 +	"White lead"; very high s.g. for a carbonate; effervesces only in hot, concentrated HCl; readily effervesces in HNO₃.
Chalcocite	96[Cu₂S]	Cu = 80	26.4	Or	5.5–5.8	2.5–3	lead-grey	–	S	–	32	"Copper glance"; more brittle than argentite.2[Ag₂S]; dissolves in HNO₃ with evolution of NO₂.
Chalcophyllite	[Cu₁₈Al₂(AsO₄)₃(SO₄)₃(OH)₂₇·36H₂O]	Cu = 38 As = 7	18.0	He	2.4–2.7	2	pale green	–	S	–	1.57–1.63	"Copper mica"; green colour; soluble in NH₄OH.
Chalcopyrite	4[CuFeS₂]	Cu = 35	23.5	Te	4.1–4.3	3.5–4	dark green	–	D	–	42–46	"Copper pyrites"; brass-yellow colour (deeper yellow than pyrite); a most important ore of copper.
Chlorargyrite	4[AgCl]	Ag = 75	39.6	Is	5.6	1–1.5	white	–	–	–	2.1	"Horn silver"; streak turns brown in sunlight; soluble in NH₄OH.
Chromite	8[FeCr₂O₄]	Cr₂O₃ = 68	19.9	Is	4.1–4.9	5.5	brown	–	–	–	1.56	A chromium spinel; magnetic but less so than the iron spinel magnetite.
Chrysoberyl	4[BeAl₂O₄]	BeO = 20	9.8	Or	3.5–3.8	8.5	white	–	–	–	1.75	Usually greenish colour; very hard; much denser than beryl.
Chrysocolla	CuSiO₃·2H₂O (approx.)	Cu < 36 usually ≈ 20	16.4	—	2.0–2.2	2	white or pale green or pale blue	D	–	–	1.46–1.57	Usually green or blue colour; mineral gel of very variable composition.
Cinnabar	3[HgS]	Hg = 86	71.2	He	8.0–8.2	2–2.5	scarlet	–	S	–	26	Cochineal-red colour; the only common mineral of Hg.

Mineral	Formula	Composition	Comp. no.	Composition (elements)	Crystal system	S.G.	H	Colour of streak	(a)	(b)	(c)	R.I.	Remarks
Clausthalite	4[PbSe]	Pb=72 Se=28	68.8		Is	7.6–8.8	2.5–3	black		S	I	50	Cubic cleavage; resembles galena (PbS).
Cobaltite	4[CoAsS]	Co=36 As=45	27.6		Is	6.0–6.3	5.5	grey-black	D	D	S	53	Colour often silver-white but streak is grey-black; cubic cleavages; forms isomorphous series with gersdorffite (NiAsS).
Colemanite	2[Ca₂B₆O₁₁.5H₂O]	$B_2O_3=51$	9.7		Mo	2.4	4–4.5	white		S	S	1.59–1.61	Separation of boric acid on cooling the HCl solution.
Columbite	4[(Fe,Mn)(Nb,Ta)₂O₆] with Nb > Ta Nb₂O₆ < 79		29.0–55.8		Or	5.3–6.3	6	dark brown	D	I	—	16–18	Moderately magnetic; forms isomorphous series with tantalite.
Copper	4[Cu]	Cu=100	29.0		Is	8.8	2.5–3	reddish		S	S	81	Copper-red colour; malleable; dissolves readily in HNO_3 with evolution of nitrous fumes.
Corundum	2[Al₂O₃]	Al=53	10.6		He	4.0–4.1	9	white		I	—	1.76	Very hard; some varieties may be ruby- or sapphire-colour.
Cosalite	2[CuPb₇Bi₈S₂₂]	Pb=37 Bi=43 Cu=2	69.6		Or	6.4–6.7	2.5–3	black	D	D	D	43	Lead- or steel-grey colour; Cu may not be an essential constituent.
Covelline	6[CuS]	Cu=67	24.6		He	4.6	1.5–2	purple-black		D	D	1.45	Purple or indigo-blue colour; thin flexible plates; opaque except in very thin splinters.
Crocoite	4[PbCrO₄]	Pb=64 Cr=16	58.0		Mo	5.9–6.1	2.5–3	orange-yellow		—	D	2.3+	Bright red colour.
Cryolite	2[Na₃AlF₆]	F=54 Na=33	10.2		Mo	3.0	2.5	white		—	S	1.34	R.i. same as that of water; readily fusible.
Cuprite	2[Cu₂O]	Cu=89	26.7		Is	5.8–6.2	3.5–4	brown-red	S	S	—	27	"Red copper ore"; commonly secondary after Cu sulphides.
Danburite	4[CaB₂Si₂O₈]	$B_2O_3=28$	11.1		Or	3.0	7	white		—	—	1.63	Very slowly attacked by HCl and may give reaction for B with turmeric paper.
Descloizite	4[PbZnVO₄OH]	V₂O₅=23 Pb=53 Zn=16	51.3		Or	5.9–6.2	3.5	orange-brown to red		S	—	2.1+	Cherry-red to brown-red colour; easily fusible.
Diamond	8[C]		6.0		Is	3.5	10	white		—	—	2.4	Very hard; colour very variable.
Diaspore	4[AlO.OH]		10.1		Or	3.3–3.5	6.5–7	white		—	—	1.70–1.75	Alteration product of corundum; C.T. gives water.
Dioptase	6[CuSiO₂(OH)₂]	Cu=41	17.4		He	3.3	5	green	D	—	D	1.64–1.71	Emerald-green, cf. chrysocolla.
Dolomite	[CaMg(CO₃)₂]		10.9		He	2.8–2.9	3.5–4	white	E	—	—	1.50–1.68	Does not effervesce as readily as calcite and gives Mg reaction.
Enargite	2[Cu₃AsS₄]	Cu=48 As=19	25.5		Or	4.4	3	grey-black		S	I	25–28	Often contains some Sb.

continued over

TABLE A.2.4
Alphabetical list of minerals (continued)

Mineral	Theoretical composition and % by mass of saleable component	$\bar{Z}$	Crystal system	Specific gravity	Hardness (Mohs scale)	Colour of powder	Solubility: HCl	HNO₃	H₂SO₄	Refractive index or reflectivity	General remarks and distinguishing features
Epidote	$2[Ca_2(Al,Fe)_3Si_3O_{12}OH]$	12.1–16.3	Mo	3.2–3.5	6–7	white	D			1.73–1.77	Pistachio-green (yellow-green) colour.
Erythrite	$2[Co_3(AsO_4)_2 \cdot 8H_2O]$ Co = 30, As = 25	19.7	Mo	3.0	1.5–2.5	pale red	S			1.63–1.70	"Cobalt bloom"; pale red colour.
Euclase	$4[BeAlSiO_4OH]$ BeO = 20	9.8	Mo	3.1	7.5	white	I			1.65–1.67	Sometimes pale green or pale blue.
Euxenite	$(Y,Er,Ce,La,U)(Nb,Ti,Ta)_2(O,OH)_6$ $U_3O_8 = 1{-}20$	54.9–59.3	Or	4.7–5.2	6.5	yellow-red	D	I	I	2.0+	Strongly radioactive.
Feldspar	Na,K,Ca, aluminosilicates	≈10.0	Mo or Tr	2.5–2.9	6	white	I	I	I	1.52–1.59	Large group of rock-forming minerals; monoclinic feldspars include orthoclase and celsian; triclinic feldspars include microcline, anorthoclase and the plagioclase varieties.
Fergusonite	$8[(Y,Er)(Nb,Ta)O_4]$ U_3O_8 up to 8 $(Nb,Ta)_2O_5$ up to 65	31.7–60.9	Te	4.3–5.8	5.5–6	pale brown	I	I	D	2.2	S.g. usually about 5.8; moderately magnetic; often weakly radioactive.
Fluorite	$4[CaF_2]$ F = 49, Ca = 51	14.6	Is	3.0–3.2	4	white	I	I	S	1.43	"Fluorspar"; cubic or octahedral cleavages; commonly purple colour.
Franklinite	$8[(Zn,Mn,Fe^{II})(Fe^{III},Mn^{III})_2O_4]$ Zn up to 27	20.3–22.3	Is	5.1–5.2	5.5–6.5	red-brown or black	S	I		19	Chlorine evolved from HCl solution; weakly magnetic.
Gadolinite	$2[Be_2FeY_2Si_2O_{10}]$ BeO = 10	22.5	Mo	4.0–4.5	6.5–7	grey-green	D	I		1.77+	Associated with other rare-earth minerals, e.g. allanite.
Gahnite	$8[ZnAl_2O_4]$	17.3	Is	4.0–4.6	7.5–8	grey	I	I	S	1.82	"Zinc spinel"; green colour; very hard; difficultly soluble.
Galena	$4[PbS]$ Pb = 87	73.2	Is	7.4–7.6	2.5	lead-grey	D	I		43	"Lead glance"; H₂S given off in HCl; cubic cleavages.
Garnet	complex silicate	13.0–15.6	Is	3.4–4.3	6.5–7.5	white	I	I	I	1.7–1.9	All varieties are isometric; often moderately magnetic but shows wide range of magnetic properties.
Garnierite	$(Ni,Mg)_3Si_2O_5(OH)_4$ Ni up to 46	10.1–18.1	—	2.3–2.8	soft	white or pale green	D			1.6	Green colour.
Gersdorffite	$4[NiAsS]$ Ni = 36	27.9	Is	5.6–6.2	5.5	grey-black	I	D		47	Isomorphous series with cobaltite, $4[CoAsS]$; silver-white colour, grey-black streak.
Goethite	$4[FeO \cdot OH]$ Fe = 63	19.2	Or	4.3	5–5.5	yellow-brown	S			16–18	Yellow-red colour; slowly soluble; C.T. gives water.

Gold	4[Au]	Au = 100	[illegible]	Is	[illegible]	[illegible]	gold yellow	[illegible]	[illegible]	[illegible]	[illegible]	Distinctive yel[illegible] colour; high density; malleable; insoluble.
Graphite	4[C]		6.0	He	2.1–2.2	1–2	black	I	I	I	12	"Black lead"; flexible, inelastic flakes; low density; streak is darker than that of molybdenite.
Greenockite	2[CdS]	Cd = 78	40.9	He	4.9–5.0	3–3.5	orange to red	S			19	Yellow colour; H_2S given off in HCl.
Gypsum	8[$CaSO_4 . 2H_2O$]		12.1	Mo	2.3	1.5–2	white	S			1.52–1.53	C.T. gives water, cf. anhydrite.
Hausmannite	8[Mn_3O_4]	Mn = 72	20.2	Te	4.9	5–5.5	brown	S			16–19	Primary Mn mineral; chlorine given off in HCl.
Helvine	[$(Mn,Fe,Zn)_8Be_6Si_6O_{24}S_2$]	BeO ≃ 8	16.1	Is	3.2–3.4	6–6.5	white	D			1.74	H_2S given off in HCl; dilute HCl solution plus As_2O_3 gives canary-yellow precipitate of As_2S_3.
Hematite	2[Fe_2O_3]	Fe = 70	20.6	He	4.9–5.3	5.5–6.5	red-brown	S			25–30	Commonly specular; sometimes moderately magnetic due to admixed magnetite; slowly soluble.
Hessite	Ag_2Te	Ag = 63	48.9	Is	8.3–8.5	2.5–3	lead-grey	I	D		38	Easily fusible.
Hydrozincite	2[$Zn_5(CO_3)_2(OH)_6$]	Zn = 60	20.9	Mo	3.6–3.8	2–2.5	white	E			1.63–1.75	Usually secondary after sphalerite; lower density and softer than smithsonite, 2[$ZnCO_3$].
Idocrase	Ca,Mg,Fe,Al,OH silicate		13.2	Te	3.3–3.5	6.5	white	D			1.70–1.74	"Vesuvianite"; commonly brown or green.
Ilmenite	2[$FeTiO_3$]	TiO_2 = 53	19.0	He	4.5–5.0	5–6	black	S			18–21	Frequently platy; moderately magnetic.
Ilvaite	4[$CaFe^{II}_2Fe^{III}Si_2O_8OH$]		17.4	Or	4.0	5.5–6	black	D			1.9	Non-magnetic but frequently occurs with magnetite; may contain MnO and occasionally Sn.
Iron	2[Fe]	Fe = 100	26.0	Is	7.3–7.8	4–5	grey	S			62	Malleable; highly magnetic; readily soluble in HCl.
Jamesonite	2[$Pb_4FeSb_6S_{14}$]	Sb = 35 Pb = 40	55.2	Mo	5.5–6.0	2–3	grey-black	S			36–40	"Brittle feather ore"; H_2S given off with HCl and $PbCl_2$ precipitated on cooling the HCl solution: easily fusible.
Kermesite	8[Sb_2S_2O]	Sb = 75	41.9	Mo	4.5–4.6	1–1.5	brown-red	S			27	"Pyrostibite"; cherry-red colour; alteration product of stibnite; very soft; easily fusible.
Kyanite	4[Al_2SiO_5]	Al_2O_3 = 63	10.7	Tr	3.6	5–7	white	I	I	I	1.72	Frequently blue colour.
Lepidolite	4[$K(Li,Al)_3(Si,Al)_4O_{10}(F,OH)_2$]	Li_2O ≃ 4	9.3–10.9	Mo	2.8–3.3	2.5–4	white	D			1.53–1.56	"Lithia mica"; elastic flakes; commonly violet colour.
Limonite	hydrated mixed oxides of Fe^{III}	Fe = var.	≃19	—	3.6–4.0	5–5.5	yellow-brown	S			16–25	"Yellow ochre"; often earthy; C.T. gives water; slowly soluble; a mineral colloid.

continued over

TABLE A.2.4
Alphabetical list of minerals (continued)

Mineral	Theoretical composition and % by mass of saleable component		$\bar{Z}$	Crystal system	Specific gravity	Hardness (Mohs scale)	Colour of powder	Solubility: HCl	HNO$_3$	H$_2$SO$_4$	Refractive index or reflectivity	General remarks and distinguishing features
Linnaeite	8[Co$_3$S$_4$]	Co = 58	22.4	Is	4.8–5.0	5.5	black-grey	I	I	D	47	Tin-white colour; tarnishes to copper-red.
Livingstonite	4[HgSb$_4$S$_7$]	Hg = 22	48.8	Or	4.8	2	red	I	D		32	Red-grey colour; red streak.
Löllingite	2[FeAs$_2$]		31.1	Or	7.0–7.4	5–5.5	grey-black	I	D		53–55	Silver-white colour; grey-black streak.
Magnesite	2[MgCO$_3$]	Mg = 29	8.9	He	3.0–3.1	3.5–4.5	white	E			1.52–1.72	Resembles dolomite but gives Mg reaction with only traces of Ca.
Magnetite	8[Fe$_3$O$_4$]	Fe = 72	21.0	Is	5.2	5.5–6.5	black	S			21	Very slowly soluble in HCl; highly magnetic.
Malachite	4[Cu$_2$CO$_3$(OH)$_2$]	Cu = 58	19.9	Mo	3.9–4.0	3.5–4	pale green	E			1.65–1.91	Green colour; copper reaction.
Manganite	8[MnO.OH]	Mn = 62	18.5	Or	4.2–4.4	4	red-brown	S			14–20	Chlorine given off in HCl; frequently alters to pyrolusite.
Marcasite	2[FeS$_2$]	S = 53	20.7	Or	4.9	6–6.5	grey-black	I	D		49–55	"White iron pyrites"; pale yellow colour, cf. pyrite, 4[FeS$_2$]; unstable and readily altered.
Miargyrite	8[AgSbS$_2$]	Sb = 41 Ag = 37	41.9	Mo	5.1–5.3	2–2.5	cherry-red	I	D		32–36	Iron-black to steel-grey colour.
Microlite	8[(Ca,Na)$_2$(Ta,Nb)$_2$(O,OH,F)$_7$] Ta$_2$O$_5$ up to 80 U$_3$O$_8$ up to 15		48.4–51.2	Is	5.5	5.5	pale brown	I	I	I	1.9	Frequently contains U and is radioactive; Ta > Nb, Ca ≫ Na; cf. pyrochlore.
Millerite	3[NiS]	Ni = 65	23.8	He	5.3–5.7	3–3.5	green-black	I	D		54–60	"Capillary pyrites"; usually forms capillary crystals.
Mimetite	2[Pb$_5$(AsO$_4$)$_3$Cl]	Pb = 70	63.5	He	7.0–7.2	3.5	white	I	D		2.1	Easily fusible; chloride.
Molybdenite	2[MoS$_2$]	Mo = 60	31.6	He	4.7–4.8	1–1.5	blue-grey	I	D		15–37	Laminae flexible but not elastic; denser than graphite.
Monazite	4[(La,Ce)PO$_4$] with thorium	ThO$_2$ up to 9	36.5–38.1	Mo	4.9–5.3	5–5.5	white	S			1.79–1.84	Moderately radioactive; slowly soluble; moderately magnetic.
Muscovite	4[KAl$_3$Si$_3$O$_{10}$(OH)$_2$]		11.3	Mo	2.8–3.0	2–2.5	white	I			1.56–1.59	"Common mica"; non-magnetic; elastic flakes.
Naumannite	Ag$_2$Se	Ag = 71	43.5	Is	6.5–8.0	2.5	iron-black	I	D		31–34	Cubic cleavages; malleable; may contain appreciable Pb.
Niccolite	2[NiAs]	Ni = 44	30.8	He	7.3–7.7	5–5.5	brown-black	I	I		52–58	"Copper nickel"; pale copper-red colour; soluble in aqua regia; may grade towards breithauptite, 2[NiSb].

Mineral	Formula	% element	System	S.G.	H	Colour				R.I.	Remarks
Olivenite	$4[Cu_2(AsO_4)_2 \cdot 5H_2O]$	Cu = 26	Or	4.1–4.4	3	olive-green to brown	I	—	S	1.79	Commonly green.
Olivine	$4[(Mg,Fe)_2SiO_4]$ with Mg > Fe	10.5–18.7	Or	3.3–3.4	6.5–7	white	D	D	—	1.63–1.69	"Chrysolite"; greenish colour; commonly weakly magnetic.
Orpiment	$4[As_2S_3]$	As = 61	Mo	3.4–3.5	1.5–2	yellow	I	S	D	20–25	"Yellow arsenic"; flexible plates.
Pentlandite	$4[(Fe,Ni)_9S_8]$	usually Ni $\rightleftharpoons$ 22	Is	4.6–5.0	3.5–4	bronze-brown	I	—	—	52	Unlike pyrrhotite this mineral is non-magnetic.
Periclase	$4[MgO]$	Mg = 60	Is	3.7–3.9	6	white	S	S	—	1.74	Cubic cleavages.
Perovskite	$8[CaTiO_3]$	TiO_2 = 59	Is	4.0	5.5	white	—	—	D	17	Cubic cleavages; non-magnetic.
Petalite	$4[LiAlSi_4O_{10}]$	Li_2O = 5	Mo	2.4	6–6.5	white	—	—	I	1.50–1.52	Low density; blue phosphorescence when gently heated.
Phenakite	$6[Be_2SiO_4]$	BeO = 46	He	3.0	7.5–8	white	—	—	I	1.66	Very hard; usually found in pegmatites; can be mistaken for quartz.
Platinum	$4[Pt]$	Pt = 100	Is	14–21.5	4–4.5	pale green	—	—	I	70	Malleable; may be moderately magnetic due to included iron.
Polybasite	$16[Ag_{16}Sb_2S_{11}]$	Ag = 74	Mo	6.0–6.2	2–3	black	—	—	D	35	Thin splinters may show cherry-red colour.
Powellite	$8[CaMoO_4]$	Mo = 48	Te	4.3	3.5	pale green-yellow	D	—	—	2.0	Alteration product of molybdenite.
Proustite	$2[Ag_3AsS_3]$	Ag = 65	He	5.6	2–2.5	scarlet	—	—	D	25–28	"Ruby silver ore"; scarlet colour, cf. pyrargyrite, $2[Ag_3SbS_3]$.
Psilomelane	$2[(Ba,Mn^{II})Mn_4^{IV}O_8(OH)_2]$	Mn = var.	—	3.3–4.7	5–7	brown-black	S	—	S	24	H greater than pyrolusite, $2[MnO_2]$; chlorine given off in HCl; C.T. gives water.
Pyrargyrite	$2[Ag_3SbS_3]$	Ag = 60	He	5.8–5.9	2.5	purple-red	—	—	D	28–31	"Ruby silver ore"; dark red, cf. proustite, $2[Ag_3AsS_3]$.
Pyrite	$4[FeS_2]$	S = 53	Is	5.0–5.1	6–6.5	green-black	—	—	D	54	"Iron pyrites"; pale-brass yellow colour, cf. chalcopyrite and marcasite; non-magnetic, cf. pyrrhotine; grains frequently striated.
Pyrochlore	$8[(Ca,Na,Ce)(Nb,Ti,Ta)_2(O,OH,F)_7]$ $Nb_2O_6 \rightleftharpoons 70$	14.1–56.3	Is	4.2–4.4	5–5.5	light brown	—	—	1 or D	2.0	Commonly octahedral form; Ta always small, cf. microlite; occasionally contains Th and is radioactive; usually insoluble in acids.
Pyrolusite	$2[MnO_2]$	Mn = 63	Or	4.7–4.9	2–2.5	black	S	S	—	30–41	Soft; commonly pseudomorphous after manganite; chlorine given off in HCl.
Pyromorphite	$2[Pb_5(PO_4)_3Cl]$	Pb = 76	He	5.9–7.1	3.5–4	white	—	—	S	2.0	"Green lead ore"; easily fusible; phosphate.

continued over

TABLE A.2.4

Alphabetical list of minerals (*continued*)

Mineral	Theoretical composition and % by mass of saleable component	$\bar{Z}$	Crystal system	Specific gravity	Hardness (Mohs scale)	Colour of powder	Solubility: HCl	HNO₃	H₂SO₄	Refractive index or reflectivity	General remarks and distinguishing features
Pyroxene	complex metasilicate, $RSiO_3$, with R = Ca, Mg, Fe, Al, etc.	12.3–13.9	Or, Mo or Tr	3.1–3.6	5–7	white to grey	I	I	I	1.65–1.69	Commonly green colour; often weakly magnetic; two well-marked cleavages at almost 90°.
Pyrrhotine	$2[FeS]$, Fe = 64	22.4	He	4.6	3.5–4.5	grey-black	S	I		38–45	"Magnetic pyrites": bronze-yellow to copper-red colour; highly magnetic; H_2S given off in HCl.
Quartz	$3[SiO_2]$	10.8	He	2.6–2.7	7	white	I	I	I	1.54–1.55	Commonly colourless or white but may be almost any colour.
Rammelsbergite	$2[NiAs_2]$, Ni = 28	31.6	Or	6.9–7.2	5.5–6	grey-black	I		I	58–60	Tin-white colour; grey-black streak.
Realgar	$16[AsS]$, As = 70	27.9	Mo	3.6	1.5–2	red	I	D		18	Red colour; very soft.
Rhodochrosite	$2[MnCO_3]$, Mn = 48	15.9	He	3.5–3.6	3.5–4.5	white	E			1.60–1.82	"Manganese spar"; pink colour; changes superficially to black MnO_2 on roasting in an oxidising atmosphere.
Rhodonite	$10[MnSiO_3]$, Mn = 42	16.4	Tr	3.4–3.7	5.5–6.5	white	D			1.73–1.74	Frequently pink colour; harder than rhodochrosite; unaltered by oxidising roast, cf. rhodochrosite.
Rutile	$2[TiO_2]$, TiO_2 = 100	16.4	Te	4.2	6–6.5	white or pale brown	I	I		2.6	Non-magnetic: anatase and brookite have same composition but different structure; very variable colour.
Scheelite	$8[CaWO_4]$, W = 64	51.8	Te	5.9–6.1	4.5–5	white	D			1.92–1.93	Fluoresces in short-wave ultraviolet light; W reaction.
Scorodite	$8[FeAsO_4 \cdot 2H_2O]$, As = 5	21.3	Or	3.1–3.3	3.5–4	white	S			1.74–1.92	Readily fusible; C.T. gives water.
Senarmontite	$16[Sb_2O_3]$, Sb = 84	43.9	Is	5.3	2	white	S			2.1	Secondary after stibnite.
Serpentine	hydrated Mg silicate	var.	Mo	2.5–2.7	2.5–5.5	white	D			1.49–1.57	Group name, includes chrysotile and antigorite; usually green; often fibrous.
Siderite	$2[FeCO_3]$, Fe = 48	16.5	He	3.9	3.5–4	white	E			1.63–1.87	"Spathic iron"; "chalybite"; weakly magnetic.
Silver	$4[Ag]$, Ag = 100	47.0	Is	10.1–11.1	2.5–3	silver-white	I	S		95	Malleable.
Smaltite	$8[CoAs_2]$, Co = 28	31.3	Is	5.7–6.8	5.5–6	grey-black	I	S		58	Isomorphous series with chloanthite ($NiAs_2$); tin-white colour; grey-black streak.
Smithsonite	$2[ZnCO_3]$, Zn = 52	19.3	He	4.3–4.5	5.5	white	E			1.62–1.85	"Dry-bone ore"; high density for a carbonate.

continued over

Mineral	Formula	Composition	Crystal system	S.G.	Hardness	Colour					No.	Remarks
Sphalerite	$4[ZnS]$	Zn = 67	Is	3.9–4.1	3.5–4	brown to white	S	I	—	—	2.4	"Zinc blende"; ferriferous variety called marmatite; H_2S given off in HCl.
Sphene	$4[CaTiSiO_5]$	$TiO_2 = 41$	Mo	3.4–3.6	5–5.5	white	D	—	—	—	1.9 +	"Titanite"; Ti reaction.
Spinel	$8[MgAl_2O_4]$		Is	3.5–4.1	8	white	—	—	S	—	1.72–2.0	Mg and Al can be replaced by Zn, Mn, Fe, etc. to give many varieties of spinel; difficultly soluble.
Spodumene	$4[LiAlSi_2O_6]$	$Li_2O = 8$	Mo	3.1–3.2	6.5–7	white	—	—	—	—	1.65–1.68	Li pyroxene; some varieties fluoresce in ultraviolet light.
Stannite	$2[Cu_2FeSnS_4]$	Sn = 28, Cu = 30	Te	4.3–4.5	3.5	black	—	D	—	—	28	"Tin pyrites"; frequently found in cassiterite-bearing veins; SnO_2 precipitated from HNO_3 solution.
Staurolite	$[(Fe,Mg)_4Al_{18}Si_8O_{46}(OH)_2]$		Or	3.7–3.8	7–7.5	white to grey	—	—	D	—	1.74	Weakly magnetic; only partly decomposed in H_2SO_4; frequently shows cruciform twinning in alluvial deposits.
Stephanite	$4[Ag_5SbS_4]$	Ag = 69, Sb = 15	Or	6.2–6.3	2–2.5	black	—	D	—	—	30	"Brittle silver ore"; Ag test.
Stibnite	$4[Sb_2S_3]$	Sb = 72	Or	4.5–4.6	2	lead-grey	S	D	S	—	30–40	"Antimonite"; fuses in a match flame; most common antimony mineral.
Stolzite	$8[PbWO_4]$	W = 40	Te	7.9–8.1	3	white	—	D	—	—	2.2	Orange-yellow colour; readily fusible.
Strontianite	$4[SrCO_3]$	Sr = 59	Or	3.7	3.5–4	white	E	—	—	—	1.52–1.67	Colours a flame crimson.
Sulphur	$16[S_8]$	S = 100	Or	2.1	1.5–2.5	white	—	—	—	—	1.9 +	Yellow; melts at 108°C; burns readily; soluble in CS_2.
Sylvanite	$2[AgAuTe_4]$	Au = 24, Ag = 13	Mo	7.9–8.3	1.5–2	white or grey	—	D	—	—	54	"Graphic tellurium"; contains appreciable Ag, cf. calaverite; decomposed in HNO_3 and leaves a residue of rusty gold.
Talc	$4[Mg_3Si_4O_{10}(OH)_2]$		Or or Mo	2.7–2.8	1–1.5	white	—	—	—	—	1.54–1.59	Very soft.
Tantalite	$4[(Fe,Mn)(Ta,Nb)_2O_6]$	Ta > Nb, Ta_2O_6 up to 86	Or	6.3–7.3	6	dark brown	—	—	D	—	16–18	Isomorphous series with columbite; moderately magnetic.
Tennantite	$8[Cu_3AsS_3]$	Cu = 53	Is	4.4–4.5	3–4	black	—	D	—	—	29	Isomorphous series with tetrahedrite, $8[Cu_3SbS_3]$.
Tenorite	$4[CuO]$	Cu = 80	Tr	6.5	3–4	black	S	S	S	—	20–27	Oxidation product of other copper minerals.
Tetradymite	$[Bi_2Te_2S]$	Bi = 59, Te = 36	He	7.2–7.6	1.5–2	steel-grey	—	S	I	—	57	Flexible laminae; very soft.
Tetrahedrite	$8[Cu_3SbS_3]$	Cu = 47, Sb = 30	Is	4.4–5.1	3–4	black to cherry-red	—	D	—	—	31	"Grey copper ore"; isomorphous series with tennantite, $8[Cu_3AsS_3]$; tetrahedral habit.

TABLE A.2.4
Alphabetical list of minerals (continued)

Mineral	Theoretical composition and % by mass of saleable component		$\bar{Z}$	Crystal system	Specific gravity	Hardness (Mohs scale)	Colour of powder	Solubility: HCl	HNO₃	H₂SO₄	Refractive index or reflectivity	General remarks and distinguishing features
Thorianite	$4[ThO_2]$	$Th = 88$	80.1	Is	9.3	6.5	brown	I	S		14–15	Cubic fragments; strongly radioactive and may contain U.
Thorite	$ThSiO_4$	$Th = 71$	67.2	Te	4.5–5.4	4.5–5	white	D			1.68–1.8	Usually non-magnetic but grains may be moderately magnetic due to included iron oxides; radioactive; orange-coloured variety called orangite; metamict.
Topaz	$4[Al_2SiO_4(F,OH)_2]$		10.5	Or	3.4–3.6	8	white	I		D	1.63–1.64	Usually colourless; only slightly soluble.
Torbernite	$Cu(UO_2)_2(PO_4)_2.8H_2O$	$U_3O_8 = 60$	52.4	Te	3.2	2–2.5	pale green	S	I		1.58–1.59	Green colour; radioactive.
Tourmaline	complex B,Fe,Al silicate of variable composition		17.8	He	3.0–3.2	7–7.5	white	I	I		1.62–1.64	Frequently greenish-brown; triangular cross-section.
Triphylite	$4[LiFePO_4]$	$Li_2O = 10$	15.5	Or	3.4–3.6	4.5–5	white	S			1.68–1.69	Isomorphous series with lithiophyllite ($LiMnPO_4$).
Triplite	$8[(Mn,Fe)_2PO_4F]$		17.4–18.0	Mo	3.4–3.8	4.5–5	yellow-grey to brown	S			1.65–1.68	Easily fused to a magnetic globule; phosphate test.
Ullmannite	$4[NiSbS]$	$Ni = 28$	39.4	Is	6.2–6.7	5–5.5	grey-black	I	D		47	Silver-white colour; grey-black streak.
Uraninite	UO_2	$U_3O_8 = 104$	82.0	Is	9.0–9.7	5.5	brown to olive-green	I	S		17	"Pitchblende"; black colour; non-magnetic; strongly radioactive; often impure.
Uranophane	$Ca(UO_2)_2Si_2O_7.6H_2O$	$U_3O_8 = 66$	55.6	Or	3.8–3.9	2–3	yellow	D			1.64–1.67	Yellow colour; strongly radioactive.
Vanadinite	$2[Pb_5(VO_4)_3Cl]$	$Pb = 73$ $V_2O_6 = 13$	64.0	He	6.7–7.1	3	white	D			2.2	Commonly ruby-red colour; readily fusible.
Willemite	$6[Zn_2SiO_4]$	$Zn = 59$	21.7	He	3.9–4.2	5.5	white	D			1.69–1.72	Some varieties fluoresce strongly in ultraviolet light.
Witherite	$4[BaCO_3]$	$BaO = 78$	41.3	Or	4.3	3–4	white	E			1.53–1.68	Cf. barite, $4[BaSO_4]$, which is insoluble in acids.
Wolframite	$2[(Fe,Mn)WO_4]$	max. $W = 60$	51.1–51.3	Mo	7–7.5	5–5.5	black	D			16–18	Includes hubnerite ($MnWO_4$) and ferberite ($FeWO_4$); slowly attacked by HCl; moderately to weakly magnetic.
Wulfenite	$8[PbMoO_4]$	$Pb = 56$ $Mo = 26$	58.6	Te	6.7–7.0	3	white	D			2.3+	"Yellow lead ore", but colour rather variable; easily fusible.

Xenotime	8[YPO_4]	Y_2O_3=61	24.2	Te	4.4–5.1	4–5	pale brown	—	—	—	1.72–1.81	Rare-earth phosphate; pseudo-cubic cleavage; moderately magnetic; may be very slowly attacked by sulphuric acid.
Zaratite	[$Ni_3CO_3(OH)_4.4H_2O$]	Ni=47	17.1	—	2.6	3	pale green	E	—	—	1.56–1.61	"Emerald nickel".
Zeolite	hydrous aluminosilicate, usually of Ca and Na		13.9	Or, Mo or Tr	2.0–2.5	3.5–5.5	white	D (or S)	—	—	1.47–1.53	C.T. readily gives water; all varieties "boil" on heating.
Zincite	2[ZnO]	Zn=80	25.7	He	5.4–5.7	4–4.5	orange-yellow	S	—	—	11	"Red oxide of zinc".
Zircon	4[$ZrSiO_4$]	ZrO_2=67	24.8	Te	4.7	7.5	white	—	—	D	1.9+	Common constituent of alluvial deposits; usually non-magnetic but rarely is moderately magnetic.

Note: The "saleable component" of a mineral is the element or oxide that is quoted when a concentrate of that mineral is sold.

TABLE A.2.5
Determinative Table

Group 1. Specific gravity <2.9

Mineral	Hardness			Streak			Solubility			Acid solution plus:						Refractive index	Diagnostic tests and general remarks
	s <3	m	h >7	W	C	Dk	HCl	HNO₃	H₂SO₄	NH₄OH Ppt	NH₄OH Sol	Ox	Ph	Mo	Ba		
(Calcite)	x	X		X			E					X				a,b	Carbonate; no Mg.
Gypsum	X			X			S					X			X	a	C.T. gives water, cf. anhydrite.
(Anhydrite)	x	X		X			S					X			X	a,b	C.T. gives no water, cf. gypsum; s.g. 2.9–3.0; H 3–3.5.
Chrysocolla	X			X	Bl		D				Bl					a	S.g. 2.0–2.2; green or blue colour.
Garnierite	X			X	G		D				G	X				a,b	Streak pale green or white.
(Serpentine)	x	X		X			D						X			a	H 2.5–5.5; usually green colour.
(Lepidolite)	x	X		X			D			W						a	H 2.5–4; elastic flakes; usually violet colour.
Biotite	X	x		X			I	I	D	Br			X			a,b,c	S.g. usually >2.9; elastic flakes.
Sulphur	X			X			I	I	I							a	Burns in match flame; soluble in CS₂.
Talc	X			X			I	I	I							a	Very soft; soapy feel.
Muscovite	X			X			I	I	I							a	Elastic flakes.
(Zaratite)	x	X			G		E				G					a,b	Emerald colour; Ni test.
Chrysocolla	X			X	Bl		D				Bl					a	Streak usually blue; s.g. 2.0–2.2.
Garnierite	X			X	G		D				G	X				a,b	Streak pale green or white.
Chalcophyllite	X				G		I	S			Bl				X	a,b	Pale green colour; C.T. gives much water.
Graphite	X					X	I	I	I							—	Flexible, inelastic flakes.
Calcite	x	X		X			E					X				a,b	Carbonate; no Mg, cf. dolomite.
Dolomite		X		X			E					X	X			a,b,c	Carbonate.
Zeolite		X		X			S (or D)					X				a	C.T. readily gives water.
Colemanite		X		X			S					X				a,b	Turmeric test for borate.
Beryllonite		X		X			S			W			X			a	Phosphate.
Anhydrite	x	X		X			S					X			X	a,b	S.g. 2.9–3.0; C.T. gives no water.
Zeolite		X		X			D (or S)					X				a	C.T. readily gives water.
Serpentine	x	X		X			D						X			a	Usually green colour.
Lepidolite	x	X		X			D			W						a	Elastic flakes; usually violet colour.
(Biotite)	X	x		X			I	I	D	Br			X			a,b,c	S.g. usually <2.9; elastic flakes.
Petalite		X		X			I	I	I							a	S.g. 2.4; blue phosphorescence.
Feldspar		X		X			I	I	I							a	S.g. 2.5–2.9; common rock-forming mineral.
(Quartz)		x	X	X			I	I	I							a	H 7; usually colourless; common rock-forming mineral.

Group 1. (*continued*)

Mineral	Hardness			Streak			Solubility			Acid solution plus: NH$_4$OH		Ox	Ph	Mo	Ba	Refractive index	Diagnostic tests and general remarks
	s <3	m	h >7	W	C	Dk	HCl	HNO$_3$	H$_2$SO$_4$	Ppt	Sol						
(amphibole)		X		X	G		I	I	I							b,c,d	S.g. usually >2.9; commonly green colour.
(zaratite)	x	X			G		E				G					a,b	Emerald colour; Ni test.
(quartz)		x	X	X			I	I	I							a	S.g. 2.6–2.7; usually colourless; common rock-forming mineral.
(beryl)			X	X			I	I	I							a,b	S.g. 2.6–2.9; commonly pale green colour; quinalizarin test will distinguish beryl from quartz.

Group 2. Specific gravity >2.9 <3.2

Mineral	Hardness			Streak			Solubility			Acid solution plus: NH$_4$OH		Ox	Ph	Mo	Ba	Refractive index	Diagnostic tests and general remarks
	s <3	m	h >7	W	C	Dk	HCl	HNO$_3$	H$_2$SO$_4$	Ppt	Sol						
(anhydrite)	x	X		X			S					X			X	a,b	H 3–3.5; C.T. gives no water.
(lepidolite)	x	X		X			D			W						a	H 2.5–4; elastic flakes; usually violet colour.
(cryolite)	X			X			I	I	S	W						a	R.i. same as water; easily fusible.
(biotite)	X	x		X			I	I	D	Br				X		a,b,c	Elastic flakes; moderately magnetic.
(muscovite)	X			X			I	I	I							a	Elastic flakes.
(erythrite)	X					R	S			Y				X		b,c	Pale red colour.
(autunite)	X				Y		S			Y		X		X		a	Yellow colour; radioactive; fluorescent.
(torbernite)	X					G	S			Y	Bl			X		a	S.g. 3.2; green colour; radioactive.
(dolomite)		X		X			E					X	X			a,b,c	S.g. 2.8–2.9; carbonate; strong Ca reaction.
(ankerite)		X		X			E			Br		X	X			a,b,c,d	Carbonate; frequently associated with iron ores.
(magnesite)		X		X			E						X			a,b,c	Carbonate; strong Mn reaction, only traces of Ca, cf. dolomite.
(anhydrite)	x	X		X			S					X			X	a,b	C.T. gives no water.
(amblygonite)		X		X			S			W				X		a,b	Slowly soluble; easily fusible.
(scorodite)		X		X			S			Br				X		d	Arsenate; easily fusible; C.T. gives water.
(Apatite)		X		X			S					X		X		b	S.g. 3.2; hexagonal, prismatic crystals common.
(lepidolite)	x	X		X			D			W						a	H 2.5–4; elastic flakes; usually violet colour.
(Helvine)		X		X			D			Br			X			c	S.g. 3.2–3.4.
(Epidote)		X	x	X			D			Br		X				d	S.g. 3.2–3.5; H 6–7; green colour.
(Fluorite)		X		X			I	I	S			X				a	Commonly purple colour; cubic cleavage.
(Biotite)	X	x		X			I	I	D	Br				X		a,b,c	Elastic flakes; moderately magnetic.

continued over

TABLE A.2.5 (*continued*)

Group 2. (*continued*)

Mineral	Hardness			Streak			Solubility			Acid solution plus:						Refractive index	Diagnostic tests and general remarks
	s <3	m	h >7	W	C	Dk	HCl	HNO$_3$	H$_2$SO$_4$	NH$_4$OH		Ox	Ph	Mo	Ba		
										Ppt	Sol						
(Feldspar)		X		X			I	I	I							a	S.g. usually <2.9.
Amphibole		X		X	G		I	I	I							b,c,d	Often green colour; 60° cleavages; moderately to weakly magnetic.
Danburite	X		x	X			I	I	I							b	H 7; very slowly attacked by HCl and may give B reaction with turmeric paper.
(Tourmaline)	x	X		X			I	I	I							b	H 7–7.5; frequently greenish-brown colour.
Spodumene	X		x	X			I	I	I							b,c	Li pyroxene.
Pyroxene	X		x	X		x	I	I	I							b,c	Often green colour; 90° cleavages; often weakly magnetic.
(Amphibole)		X		X	G		I	I	I							b,c,d	Streak usually white; often green colour; 60° cleavages; moderately to weakly magnetic.
Allanite		X				X	D			Br		X					b,c,d	S.g. usually >3.2; slightly radioactive.
Pyroxene	X		x	X		x	I	I	I							b,c	Streak commonly white; often green colour; 90° cleavages; often weakly magnetic.
(Epidote)	X		x	X			D			Br		X					d	S.g. 3.2–3.5: H 6–7; green colour.
(Danburite)	X		x	X			I	I	I							b	H 7; very slowly attacked by HCl and may give B reaction with turmeric paper.
Phenakite			X	X			I	I	I							b	S.g. 3.0; can be mistaken for quartz.
Tourmaline	x		X	X			I	I	I							b	H 7–7.5; frequently greenish-brown colour; triangular cross-section.
Euclase			X	X			I	I	I							b,c	May be pale green or pale blue colour.
(Spodumene)	X		x	X			I	I	I							b,c	H 6.5–7; Li pyroxene.
(Pyroxene)	X		x	X		x	I	I	I							b,c	H 5–7; often green colour; 90° cleavages; often weakly magnetic.
Andalusite			X	X			I	I	I							b	S.g. 3.2; square cross-section.
(Pyroxene)	X		x	X		x	I	I	I							b,c	H 5–7; streak usually white; often green colour; 90° cleavages; often weakly magnetic.

roup 3. Specific gravity >3.2 <4.2

eral	Hardness			Streak			Solubility			Acid solution plus:						Refractive index	Diagnostic tests and general remarks
	s <3	m	h >7	W	C	Dk	HCl	HNO_3	H_2SO_4	NH_4OH Ppt	Sol	Ox	Ph	Mo	Ba		
drozincite	X			X			E									b,c,d	Carbonate; secondary after sphalerite.
pidolite)	x	X		X			D			W						a	S.g. usually 2.8–2.9; elastic flakes.
elestine)	x	X		X			I	I	I							b	H 3–3.5; flame test crimson.
rbernite	X				G		S			Y	Bl			X		a	Green colour; radioactive.
acamite)	x	X			G		S				Bl					d	Green colour; chloride.
anophane	X	x		Y			D			Y		X				b	Yellow colour; radioactive.
ivenite)	x	X			G		I	S			Bl			X		d	H 3; commonly green; arsenate.
oiment	X			Y			I	D								—	Yellow colour; flexible flakes; As_2S_3.
algar	X				R		I	D								—	Red colour; AsS, cf. orpiment.
ivenite)	x	X				X	I	S			Bl			X		d	H 3; streak usually green.
odochrosite		X		X			E						X			a,b,c,d	Carbonate; pink colour.
ontianite		X		X			E					X				a,b	Carbonate; flame test crimson.
derite		X		X			E			Br						b,c,d	Carbonate; weakly magnetic.
orodite		X		X			S			Br				X		d	C.T. gives water; easily fusible; arsenate.
atite		X		X			S					X		X		b	S.g. 3.2; frequently hexagonal form; phosphate.
phylite		X		X			S			Br			X	X		b,c	Phosphate.
riclase		X		X			S					X				d	Cubic cleavage.
phalerite)		X		(X)		X	S									d	Streak usually dark; H_2S given off in HCl.
pidolite)	x	X		X			D			W						a	S.g. usually 2.8–2.9; elastic flakes.
lvine		X		X			D			Br		X				c	Usually honey-yellow colour; H_2S given off in HCl; dilute HCl solution plus As_2O_3 gives canary-yellow As_2S_3.
idote		X	x	X			D			Br		X				d	Green colour.
vine		X	x	X			D						X			b,c	Green colour; commonly weakly magnetic.
crase		X		X			D			W		X				c,d	Commonly brown or green colour.
lamine		X		X			D									b	White colour; C.T. gives water; Zn test.
hene		X		X			D					X				d	Ti test.
odonite		X		X			D						X			d	Pink colour.
llemite		X		X			D									c	Fluorescent; Zn test.
uorite)		X		X			I	I	S			X				a	S.g. 3.0–3.2; cubic cleavage; commonly purple colour.

continued over

TABLE A.2.5 (*continued*)

Group 3. (*continued*)

Mineral	Hardness			Streak			Solubility			Acid solution plus:						Refractive index	Diagnostic tests and general remarks
	s <3	m	h >7	W	C	Dk	HCl	HNO$_3$	H$_2$SO$_4$	NH$_4$OH		Ox	Ph	Mo	Ba		
										Ppt	Sol						
(Staurolite)		x	X	X			I	I	D	Br						c,d	H 7–7.5; r.i. same as di-iodomethane
Perovskite		X		X			I	I	D			X				—	Cubic cleavage; Ti test.
Amphibole		X		X	G		I	I	I							b,c,d	Commonly green colour; 60° cleavages; moderately to weakly magnetic.
(Tourmaline)		x	X	X			I	I	I							b	S.g. 3.2 or less; H 7–7.5; usually black or brown green colour; triangular cross-section.
(Spodumene)		X	x	X			I	I	I							b,c	S.g. 3.1–3.2; Li pyroxene.
Pyroxene		X	x	X			I	I	I							b,c	Often green colour; 90° cleavages; often weakly magnetic.
Garnet		X	X	X			I	I	I							c,d	Often moderately magnetic; isometric.
Diaspore		X	x	X			I	I	I							c,d	C.T. gives water; alteration product of corundum.
Kyanite		X	x	X			I	I	I							c	Commonly blue.
Celestine		X		X			I	I	I							b	H 3–3.5; flame test crimson.
Rutile		X		X			I	I	I							d	S.g. 4.2.
Azurite		X			Bl		E				Bl					d	Carbonate; blue colour.
Malachite		X			G		E				Bl					b,c,d	Carbonate; green colour.
Limonite		X			Y-br		S			Br						—	Yellow colour; C.T. gives water.
Atacamite		X			G		S				Bl					d	Green colour; chloride.
Brochantite		X			G		S								X	d	Green colour; sulphate.
Alabandine		X			G		S						X			—	Black colour; streak green; cubic cleavage.
Triplite		X			Y	X	S			Br			X	X		b,c	Easily fused to a magnetic globule; phosphate test.
(Manganite)		X			R-br		S						X			—	S.g. 4.2–4.4.
Dioptase		X			G		D				Bl					b,c	Emerald-green colour.
(Uranophane)	X	x			Y		D			Y		X				b	H 2–3; yellow colour; radioactive.
(Gadolinite)		X	x		G		D			Br						d	S.g. usually >4.2.
Olivenite	x	X			G	X	I	S			Bl			X		d	Commonly green colour; arsenate.
Chalcopyrite		X			G		I	D		Br	Bl					—	Brass-yellow colour; s.g. 4.1–4.3.
Amphibole		X		X	G		I	I	I							b,c,d	Commonly green colour; 60° cleavages; moderately to weakly magnetic.
Psilomelane		X	x			X	S						X			—	Chlorine given off in HCl; C.T. gives water.
Triplite		X				X	S			Br			X	X		b,c	Easily fused to magnetic globule; phosphate test.
Sphalerite		X				X	S									d	H$_2$S given off in HCl.

Group 3. (continued)

Mineral	Hardness			Streak			Solubility			Acid solution plus:						Refractive index	Diagnostic tests and general remarks
	s <3	m	h >7	W	C	Dk	HCl	HNO_3	H_2SO_4	NH_4OH Ppt	Sol	Ox	Ph	Mo	Ba		
Allanite		X				X	D			Br		X				b,c,d	Slightly radioactive.
Yvaite		X				X	D			Br		X				d	Non-magnetic; frequently occurs with magnetite.
Olivenite	x	X				X	I	S								d	H 3; commonly green colour; arsenate.
Pyroxene		X	x			X	I	I	I							b,c	Often green colour; 90° cleavages; often weakly magnetic.
(Chromite)		X				X	I	I	I							—	S.g. usually >4.2; moderately magnetic.
Pyrochlore		X				X	I	I	I							d	S.g. 4.2–4.4; isometric; Nb test.
(Epidote)		X	x	X			D			Br		X				d	H 6–7; green colour.
(Olivine)		X	x	X			D			Br			X			b,c	H 6.5–7; commonly green colour and weakly magnetic.
Spinel		X		X			I	I	S	Br			X			c,d	Octahedral form common.
Topaz		X		X			I	I	D	W						b	Usually colourless; slightly soluble.
Staurolite		X		X			I	I	D	Br						d	R.i. same as di-iodomethane.
(Tourmaline)	x	X		X			I	I	I							b	S.g. 3.2 or less; usually greenish-brown colour; triangular cross-section.
(Spodumene)		X	x	X			I	I	I							b,c	H 6.5–7; Li pyroxene.
(Pyroxene)		X	x	X			I	I	I							b,c	H 5–7; often green colour; 90° cleavages; often weakly magnetic.
Andalusite		X		X			I	I	I							b	Square cross-section is common.
(Diaspore)		X	x	X			I	I	I							c,d	H 6.5–7; C.T. gives water.
Garnet		X	X	X			I	I	I							c,d	Often moderately magnetic; isometric.
(Kyanite)		X	x	X			I	I	I							c	H 5–7; frequently blue.
Diamond		X		X			I	I	I							d	H 10; colour may be very variable.
Chrysoberyl		X		X			I	I	I							d	H 8.5; usually green; much denser than beryl.
Corundum		X		X			I	I	I							d	Occasionally ruby or sapphire colour.
(Gadolinite)		X	x		G		D			Br						d	H 6.5–7; s.g. usually >4.2.
(Psilomelane)		X	x			X	S						X			—	H 5–7; chlorine given off in HCl; C.T. gives water.
(Gahnite)		X				X	I	I	S	W						d	S.g. usually >4.2; green colour.
(Pyroxene)		X	x	X		X	I	I	I							b,c	Often green colour; 90° cleavages: often weakly magnetic.

TABLE A.2.5 (*continued*)

Group 4. Specific gravity >4.2

Mineral	Hardness			Streak			Solubility			Acid solution plus:						Refractive index	Diagnostic tests and general rema
	s <3	m	h >7	W	C	Dk	HCl	HNO₃	H₂SO₄	NH₄OH Ppt	NH₄OH Sol	Ox	Ph	Mo	HCl		
(Witherite)	x	X		X			E					X					Carbonate; HCl solution plus H_2SO_4 gives white precipitate of $BaSO_4$.
(Cerussite)	x	X		X			E			W							H 3–3.5; carbonate; $PbCl_2$ formed in HCl.
Senarmontite	X			X			S										Secondary after stibnite.
Wulfenite	X			X			D			W							H 3; easily fusible; molybdate and tests.
Vanadinite	X			X			D			W							H 3; commonly ruby-red colour; easily fusible.
Altaite	X			X			I	S		W					X		Tin-white colour; cubic cleavage; Pb test.
Bismuth	X			X			I	S									Easily fusible; not malleable.
Silver	X			X			I	S							X		Malleable; Ag test.
Anglesite	X			X			I	D		W					X		H 3.
Stolzite	X			X			I	D		W					X		Commonly orange-yellow colour; easily fusible; Pb test.
(Sylvanite)	X			X		X	I	D							X		Streak usually grey; Ag test.
Barite	X	X		X			I	I	I								H 2.5–3.5.
Chlorargyrite	X			X			I	I	I								Soluble in NH₄OH; streak turns brown in sunlight.
Calomel	X			X			I	I	I								Soluble in aqua regia.
Carnotite	X				Y		S			Y							Yellow colour; radioactive.
Kermesite	X				R-br		S			W							Cherry-red colour; very soft.
(Greenockite)	x	X			O-r		S			W							Yellow colour.
(Olivenite)	x	X			G		I	S			Bl		X				Commonly green colour and dark streak.
Cinnabar	X				R		I	S		Bk					X		Cochineal-red colour.
Copper	X	x			R		I	S			Bl						Copper colour; malleable.
Calaverite	X	x			Y	X	I	S									Yellow-grey streak; easily fusible to give a button of gold.
Tetrahedrite	x	X			R	X	I	D		W	Bl						H 3–4; "grey copper ore".
Livingstonite	X				R		I	D							X		Red-grey colour; Hg test.
Miargyrite	X				R		I	D							X		Iron-black to steel-grey colour; cherry-red streak; Ag test.
Proustite	X				R		I	D							X		Scarlet colour and streak; Ag test.
Pyrargyrite	X				R		I	D							X		Dark red colour; Ag test.
Crocoite	X	x			O-y		I	I	D						X		Red colour; Pb test.
Gold	X	x			Y		I	I	I								Malleable.
Stibnite	X					X	S			W							Fuses in a match flame.
Pyrolusite	X					X	S							X			Chlorine given off in HCl; very soft.

Group 4. (continued)

Mineral	Hardness			Streak			Solubility			Acid solution plus:						Refractive index	Diagnostic tests and general remarks
	s <3	m	h >7	W	C	Dk	HCl	HNO$_3$	H$_2$SO$_4$	NH$_4$OH Ppt	NH$_4$OH Sol	Ox	Ph	Mo	HCl		
mesonite	X	x				X	S			W							Easily fusible; PbCl$_2$ precipitated on cooling the HCl solution.
enorite)	x	X				X	S				Bl						H 3–4; Cu test.
osalite	X	x				X	D			W							Pb test.
alena	X					X	D			W							Cubic cleavages; Pb test.
Olivenite)	x	X			G	X	I	S			Bl			X			H 3; commonly green colour.
nargite)	x	X				X	I	S			Bl						H 3; Cu test.
nalcocite	X	x				X	I	S			Bl						Cu test.
ismuthinite	X					X	I	S		W							Fuses in a match flame.
etradymite	X					X	I	S		W							Steel-grey; flexible flakes; very soft.
lausthalite	X	x				X	I	S		W					X		H 2.5–3; cubic cleavage; resembles galena.
alaverite	X	x			Y	X	I	S									Easily fusible to give a button of gold.
Tennantite)	x	X				X	I	D		W	Bl						H 3–4; Cu$_3$AsS$_3$; Cu test.
Tetrahedrite)	x	X			R	X	I	D		W	Bl						H 3–4; Cu$_3$SbS$_3$, may grade into tennantite.
ovelline	X					X	I	D			Bl						Purple colour; flexible plates.
olybdenite	X					X	I	D									Grey colour; flexible flakes.
ornite	X	X				X	I	D		Br	Bl						"Peacock ore"; iridescent tarnish.
Millerite)	x	X				X	I	D			G-bl						H 3–3.5; Ni test.
ournonite	X	x				X	I	D		W	Bl				X		"Wheel ore"; Pb test.
olybasite	X	x				X	I	D							X		May be cherry-red colour; Ag test.
rgyrodite	X					X	I	D							X		Ag test.
tephanite	X					X	I	D		W					X		Brittle; black streak; Ag test.
aumannite	X					X	I	D		W					X		Cubic cleavage; malleable; Ag and Pb tests.
rgentite	X					X	I	D							X		Lead-grey streak; Ag test.
ylvanite	X			X		X	I	D							X		HNO$_3$ solution has residue of "rusty" gold; Ag test.
essite	X	x				X	I	D							X		Ag test.
itherite	x	X		X			E										Carbonate; HCl solution plus H$_2$SO$_4$ gives white precipitate of BaSO$_4$.
mithsonite		X		X			E										Carbonate; Zn test.
erussite	x	X		X			E								X		Carbonate; PbCl$_2$ precipitated in HCl; effervesces only slowly in HCl.
onazite		X		X				S		W							Slowly soluble; phosphate; frequently radioactive.

continued over

TABLE A.2.5 (continued)

Group 4. (continued)

Mineral	Hardness			Streak			Solubility			Acid solution plus:						Refractive index	Diagnostic tests and general remarks
	s <3	m	h >7	W	C	Dk	HCl	HNO_3	H_2SO_4	NH_4OH		Ox	Ph	Mo	HCl		
										Ppt	Sol						
(Willemite)		X		X			D										S.g. 3.9–4.2; fluoresces; Zn test.
Thorite		X		X			D			W							Radioactive; sometimes moderately magnetic but demagnetised by acid
Scheelite		X		X			D				X						Fluoresces in short-wave ultraviolet light; W reaction.
Wulfenite	x	X		X			D			W			X		X		Easily fusible; molybdate; Pb test.
Vanadinite	x	X		X			D			W							Commonly ruby-red colour; easily fusible; chloride.
Cerussite	x	X		X			E	E							X		Effervesces readily in HNO_3 but only slowly in HCl.
Pyromorphite		X		X			I	S		W				X	X		Green colour; easily fusible; phosphate Pb test.
Altaite	x	X		X			I	S		W					X		Cubic cleavage; tin-white colour; Pb test.
Silver	x	X		X			I	S							X		Malleable; Ag test.
Anglesite	x	X		X			I	D		W					X		Easily fusible; sulphate.
Mimetite		X		X			I	D		W				X	X		Easily fusible; arsenate.
Stolzite	x	X		X			I	D		W					X		Commonly orange-yellow colour; easily fusible; tungstate.
Garnet		X	X	X			I	I	I								Isometric; often moderately magnetic
Rutile		X		X		X	I	I	I							2.6	S.g. 4.2; non-magnetic.
Barite	X	X	X				I	I	I							1.64	H 2.5–3.5
Cassiterite		X	X	X			I	I	I								"Tinning" test; occasionally magnetic
Bismutite		X			G	X	E										Easily fusible; secondary after bismuthinite; streak green-grey.
Manganite		X			R-br		S						X				Mn reaction; C.T. gives water; alters to pyrolusite.
Hematite		X			R-br		S			Br							Red colour; C.T. does not give water cf. goethite; slowly soluble.
Cuprite		X			R-br		S				Bl						Red colour; Cu test.
Franklinite		X			R-br		S			Br			X				Weakly magnetic.
Goethite		X			Y-br		S			Br							Yellow-red colour; C.T. gives water cf. hematite.
Greenockite	x	X			O-r		S			W							Yellow colour.
Zincite		X			O-y		S										Red colour; Zn test.
Gadolinite		X	X		G		D			Br							Usually associated with other rare-earth minerals.
Powellite		X			G		D						X				Alteration product of molybdenite.
Euxenite		X			Y-r		D			Y			X				Radioactive; Ti and Nb tests.
Olivenite	x	X			G		I	S			Bl			X			Commonly green colour; Cu test.

Group 4. (continued)

Mineral	Hardness s<3	m	h>7	Streak W	C	Dk	Solubility HCl	HNO₃	H₂SO₄	Acid solution plus: NH₄OH Ppt	Sol	Ox	Ph	Mo	HCl	Refractive index	Diagnostic tests and general remarks
escloizite		X			O-y		I	S		W	Bl				X		Commonly red colour; easily fusible; Pb test.
reithauptite		X			R-br		I	S			G-bl						Light copper-red colour; Ni test.
(Copper)	X	x			R		I	S			Bl						H 2.5–3; copper colour; malleable; Cu test.
(Calaverite)	X	x			Y	X	I	S									H 2.5–3; yellowish-grey streak; readily fusible to give a gold button.
raninite		X			G	X	I	S		Y							Black colour; strongly radioactive.
(Tetrahedrite)	x	X			R	(X)	I	D		W	Bl						"Grey copper ore"; Cu test.
(Crocoite)	X	x			O-y		I	I	D								H 2.5–3; red colour; Pb test.
rannerite		X			G-br		I	I	D								Radioactive; weakly magnetic.
(Gold)	X	x			Y		I	I	I								Malleable.
ismutite		X			G	X	E										Carbonate; easily fusible; secondary after bismuthinite.
silomelane		X	x			X	S						X				Chlorine given off in HCl; C.T. gives water.
menite		X				X	S			Br							Moderately magnetic; Ti test.
Pyrrhotine		X				X	S			Br							Moderately magnetic.
Hausmannite		X				X	S						X				Brown streak; chlorine given off in HCl; C.T. does not give water, cf. psilomelane.
ranklinite		X			R-br	X	S			Br			X				Weakly magnetic.
Magnetite		X				X	S			Br							Very slowly soluble; highly magnetic.
(Jamesonite)	X	x				X	S			W					X		PbCl₂ precipitated on cooling the HCl solution; easily fusible; Pb test.
enorite		X				X	S				Bl						Cu test.
ron		X				X	S			Br							Malleable; highly magnetic.
(Allanite)		X				X	D			Br		X					S.g. usually <4.2; slightly radioactive.
Braunite		X				X	D						X				Chlorine given off in HCl; commonly secondary after Mn minerals.
(Cosalite)	X	x				X	D			W					X		Pb test.
Wolframite		X				X	D			Br			X				Yellow precipitate in HCl solution; moderately to weakly magnetic; W test.
(Olivenite)	X	x				X	I	S			Bl			X			H 3; commonly green colour; Cu test.
(Enargite)	X	x				X	I	S			Bl						Cu test.
Pentlandite		X				X	I	S		Br	G-bl						Ni test.
(Chalcocite)	X	x				X	I	S			Bl						Cu test.
Smaltite		X				X	I	S			Y						Pink solution in HCl; Co test.

continued over

TABLE A.2.5 (*continued*)

Group 4. (*continued*)

Mineral	Hardness			Streak			Solubility			Acid solution plus:						Refractive index	Diagnostic tests and general remarks
	s <3	m	h >7	W	C	Dk	HCl	HNO₃	H₂SO₄	NH₄OH Ppt	Sol	Ox	Ph	Mo	HCl		
(Clausthalite)	X	x				X	I	S		W					X		H 2.5–3; cubic cleavage, resembles galena.
Uraninite		X			G	X	I	S		Y							Black colour; strongly radioactive.
Thorianite		X				X	I	S		W							Isometric; radioactive.
Chalcopyrite		X				X	I	D		Br	Bl						S.g. 4.1–4.3; brass-yellow colour.
Stannite		X				X	I	D		Br	Bl						SnO₂ precipitated from HNO₃ solution.
Tennantite		X				X	I	D		W	Bl						Cu₃AsS₃; Cu test.
Tetrahedrite	x	X			R	X	I	D		W	Bl						"Grey copper ore"; Cu₃SbS₃, may grade into tennantite (Cu₃AsS₃).
Marcasite		X				X	I	D		Br							Pale yellow, cf. pyrite.
Bornite	X	X				X	I	D		Br	Bl						"Peacock ore"; iridescent tarnish.
Pyrite		X				X	I	D		Br							Cubic; often striated; pale brass-yellow.
Millerite		X				X	I	D			G-Bl						Ni test.
Gersdorffite		X				X	I	D			G						NiAsS; solid solution series with cobaltite, CoAsS; silver-white colour.
(Bournonite)	X	x				X	I	D		W	Bl			X			"Wheel ore"; Pb test.
Arsenopyrite		X				X	I	D		Br							Tin-white colour; grey-black streak.
(Polybasite)	X	x				X	I	D						X			H 2–3; may be cherry-red colour.
Cobaltite		X				X	I	D			Y						Pink solution in HCl; Co test.
Ullmannite		X				X	I	D			G-Bl						Silver-white colour; Ni test.
Löllingite		X				X	I	D		Br							Silver-white colour; grey-black streak, cf. arsenopyrite.
(Hessite)	X	x				X	I	D						X			H 2.5–3; easily fusible.
(Calaverite)	X	x				X	I	D									H 2.5–3; easily fusible to give a gold button.
Pyrochlore		X				X	I	I	D or I								Usually insoluble; isometric; light-brown streak; may be radioactive; Nb test.
Fergusonite		X				X	I	I	D								Moderately magnetic; sometimes weakly radioactive; Nb test.
Brannerite		X			G-br	X	I	I	D								Streak usually green-brown; weakly magnetic; radioactive.
Linnaeite		X				X	I	I	D		Y						Tin-white colour; Co test.
Columbite		X				X	I	I	D	Br			X				Moderately magnetic; Nb test; solid solution series with tantalite.
Tantalite		X				X	I	I	D	Br			X				Moderately magnetic; Ta test; solid solution series with columbite.
Chromite		X				X	I	I	I								S.g. 4.1–4.9; moderately magnetic.
(Rutile)		X				X	I	I	I								S.g. 4.2; streak may be white; non-magnetic.

Group 4. (continued)

Mineral	Hardness			Streak			Solubility			Acid solution plus:						Refractive index	Diagnostic tests and general remarks
	s <3	m	h >7	W	C	Dk	HCl	HNO₃	H₂SO₄	NH₄OH Ppt	Sol	Ox	Ph	Mo	HCl		
yrochlore		X				X	I	I	I or D								Isometric; light-brown streak; may be radioactive; Nb test.
enotime		X				X	I	I	I								Light-brown streak; pseudo-cubic cleavages; moderately magnetic; phosphate.
icrolite		X				X	I	I									Commonly radioactive; Nb test.
ammelsbergite		X				X	I	I	I								Tin-white colour; black streak.
iccolite		X				X	I	I	I								Pale copper-red colour.
perrylite		X	x			X	I	I	I								Tin-white colour; black streak.
latinum		X				X	I	I	I								Malleable; often moderately magnetic.
ircon			X	X			I	I	D	W							On rare occasions may be magnetic; often found in alluvials.
arnet			X	X			I	I	I								S.g. 3.4–4.3; isometric; frequently moderately magnetic.
assiterite	x	X		X			I	I	I								H 6–7; may be magnetic; "tinning" test.
(Gadolinite)		X	x		G		D			Br							H 6.5–7; s.g. 4.0–4.5; usually associated with other rare-earth minerals.
(silomelane)		X	x			X	S										H 5–7; s.g. 3.3–4.7; chlorine given off in HCl; C.T. gives water.
ahnite			X			X	I	I	S	W							S.g. 4.0–4.6; usually green colour; Zn test.
(perrylite)		X	x			X	I	I	I								H 6–7; tin-white colour; black streak.

Appendix 3

Elements in alphabetical order, with their atomic numbers and relative atomic masses

Name	Symbol	Atomic number	Relative atomic mass	Name	Symbol	Atomic number	Relative atomic mass
Actinium	Ac	89	227	Neodymium	Nd	60	144.24
Aluminium	Al	13	26.98	Neon	Ne	10	20.18
Antimony	Sb	51	121.75	Nickel	Ni	28	58.71
Argon	Ar	18	39.92	Niobium	Nb	41	92.91
Arsenic	As	33	74.92	Nitrogen	N	7	14.01
Barium	Ba	56	137.34	Osmium	Os	76	190.2
Beryllium	Be	4	9.01	Oxygen	O	8	16.00
Bismuth	Bi	83	209				
Boron	B	5	10.8	Palladium	Pd	46	106.40
Bromine	Br	35	79.91	Phosphorus	P	15	30.97
				Platinum	Pt	78	195.09
Cadmium	Cd	48	112.40	Polonium	Po	84	210
Caesium	Cs	55	132.90	Potassium	K	19	39.10
Calcium	Ca	20	40.08	Praesodymium	Pr	59	140.91
Carbon	C	6	12.01	Promethium	Pm	61	[145]
Cerium	Ce	58	140.12	Protactinium	Pa	91	231.00
Chlorine	Cl	17	35.45				
Chromium	Cr	24	52.00	Radium	Ra	88	226.05
Cobalt	Co	27	58.93	Radon	Rn	86	222
Copper	Cu	29	63.54	Rhenium	Re	75	186.2
				Rhodium	Rh	45	102.90
Dysprosium	Dy	66	162.5	Rubidium	Rb	37	85.47
				Ruthenium	Ru	44	101.76
Erbium	Er	68	167.26				
Europium	Eu	63	152.0	Samarium	Sm	62	150.35
				Scandium	Sc	21	44.96
Fluorine	F	9	19.00	Selenium	Se	34	78.96
Francium	Fr	87	(223)	Silicon	Si	14	28.09
				Silver	Ag	47	107.87
Gadolinium	Gd	64	157.2	Sodium	Na	11	22.99
Gallium	Ga	31	69.72	Strontium	Sr	38	87.62
Germanium	Ge	32	72.59	Sulphur	S	16	32.06
Gold	Au	79	197				
				Tantalum	Ta	73	180.95
Hafnium	Hf	72	178.49	Technetium	Tc	43	98.91
Helium	He	2	4.00	Tellurium	Te	52	127.60
Holmium	Ho	67	164.93	Terbium	Tb	65	158.72
Hydrogen	H	1	1.01	Thallium	Tl	81	204.37
				Thorium	Th	90	232.04
Indium	In	49	114.82	Thulium	Tm	69	169.4
Iodine	I	53	126.90	Tin	Sn	50	118.69
Iridium	Ir	77	192.2	Titanium	Ti	22	47.90
Iron	Fe	26	55.85	Tungsten	W	74	183.85
Krypton	Kr	36	83.8	Uranium	U	92	238.03
Lanthanum	La	57	138.91	Vanadium	V	23	50.94
Lead	Pb	82	207.19				
Lithium	Li	3	6.94	Xenon	Xe	54	131.3
Lutetium	Lu	71	174.97				
				Ytterbium	Yb	70	173.04
Magnesium	Mg	12	24.3	Yttrium	Y	39	88.90
Manganese	Mn	25	54.94				
Mercury	Hg	80	200.59	Zinc	Zn	30	65.37
Molybdenum	Mo	42	95.94	Zirconium	Zr	40	91.22

Appendix 4

Elements in order of increasing atomic number

Atomic number	Relative atomic mass	Symbol	Name	Atomic number	Relative atomic mass	Symbol	Name
1	1.008	H	Hydrogen	47	107.87	Ag	Silver
2	4.003	He	Helium	48	112.40	Cd	Cadmium
3	6.940	Li	Lithium	49	114.82	In	Indium
4	9.012	Be	Beryllium	50	118.69	Sn	Tin
5	10.8	B	Boron	51	121.75	Sb	Antimony
6	12.01	C	Carbon	52	127.60	Te	Tellurium
7	14.007	N	Nitrogen	53	126.90	I	Iodine
8	16.000	O	Oxygen	54	131.3	Xe	Xenon
9	19.00	F	Fluorine	55	132.90	Cs	Caesium
10	20.183	Ne	Neon	56	137.34	Ba	Barium
11	22.989	Na	Sodium	57	138.91	La	Lanthanum
12	24.3	Mg	Magnesium	58	140.12	Ce	Cerium
13	26.98	Al	Aluminium	59	140.91	Pr	Praesodymium
14	28.09	Si	Silicon	60	144.24	Nd	Neodymium
15	30.974	P	Phosphorus	61	(145)	Pm	Promethium
16	32.064	S	Sulphur	62	150.35	Sm	Samarium
17	35.453	Cl	Chlorine	63	152.0	Eu	Europium
18	39.948	Ar	Argon	64	157.2	Gd	Gadolinium
19	39.102	K	Potassium	65	158.72	Tb	Terbium
20	40.08	Ca	Calcium	66	162.5	Dy	Dysprosium
21	44.96	Sc	Scandium	67	164.93	Ho	Holmium
22	47.90	Ti	Titanium	68	167.26	Er	Erbium
23	50.94	V	Vanadium	69	169.4	Tm	Thulium
24	52.00	Cr	Chromium	70	173.04	Yb	Ytterbium
25	54.94	Mn	Manganese	71	174.97	Lu	Lutetium
26	55.85	Fe	Iron	72	178.49	Hf	Hafnium
27	58.93	Co	Cobalt	73	180.95	Ta	Tantalum
28	58.71	Ni	Nickel	74	183.85	W	Tungsten
29	63.54	Cu	Copper	75	186.2	Re	Rhenium
30	65.37	Zn	Zinc	76	190.2	Os	Osmium
31	69.72	Ga	Gallium	77	192.2	Ir	Iridium
32	72.59	Ge	Germanium	78	195.09	Pt	Platinum
33	74.92	As	Arsenic	79	197	Au	Gold
34	78.96	Se	Selenium	80	200.59	Hg	Mercury
35	79.91	Br	Bromine	81	204.37	Tl	Thallium
36	83.80	Kr	Krypton	82	207.19	Pb	Lead
37	85.47	Rb	Rubidium	83	209	Bi	Bismuth
38	87.62	Sr	Strontium	84	210	Po	Polonium
39	88.90	Y	Yttrium	85	(210)	At	Astatine
40	91.22	Zr	Zirconium	86	222	Rn	Radon
41	92.91	Nb	Niobium	87	(223)	Fr	Francium
42	95.94	Mb	Molybdenum	88	226.05	Ra	Radium
43	98.91	Te	Technetium	89	227	Ac	Actinium
44	101.76	Ru	Ruthenium	90	232.04	Th	Thorium
45	102.90	Rh	Rhodium	91	231	Pa	Protactinium
46	106.40	Pd	Palladium	92	238.03	U	Uranium

Values in parentheses are approximate because the element is a very short-lived product of radioactive decay.

Appendix 5

Selected minerals in ascending order of mean atomic number

Note: Calculations of the mean atomic numbers have been based on the theoretical compositions of the minerals.

Mineral	$\bar{Z}$ value	Mineral	$\bar{Z}$ value	Mineral	$\bar{Z}$ value	Mineral	$\bar{Z}$ value
Coal macerals	3 to 5	Calcite	12.6	Hydrozincite	20.9	Naumannite	43.5
Napalite	5.5	Ankerite	12.7 ± 0.2	Scorodite	21.3	Senarmontite	43.9
Amber	5.7	Hypersthene	12.9 ± 0.7	Willemite	21.7	Fergusonite	44.1 ± 0.9
Diamond	6.0	Carnallite	13.0	Alabandine	21.7	Silver	47
Graphite	6.0	Pyroxene	13.1	Atacamite	21.9	Amalgam (Ag)	48.6
Natron	7.9	Idocrase	13.2	Linnaeite	22.4	Livingstonite	48.8
Natratine	8.6	Anhydrite	13.4	Gadolinite	22.5	Hessite	48.9
Phenakite	8.9	Biotite	13.6 ± 2.2	Pentlandite	23.4 ± 0.1	Tin	50.0
Mirabilite	8.8	Wollastonite	13.6	Chalcopyrite	23.5	Microlite	50.3 ± 1.9
Magnesite	8.9	Zeolite	13.9	Celestine	23.7	Autunite	50.9
Boracite	9.0	Chalcanthite	14.1	Millerite	23.8	Antimony	51.0
Epsomite	9.0	Apatite	14.1	Xenotime	24.2	Wolframite	51.2
Brucite	9.4	Epidote	14.2 ± 2.1	Covelline	24.6	Descloizite	51.3
Lepidolite	9.8 ± 0.3	Garnet	14.3 ± 1.3	Tenorite	24.8	Scheelite	51.8
Gibbsite	9.5	Olivine	14.6 ± 4.1	Zircon	24.8	Torbernite	52.4
Colemanite	9.7	Fluorite	14.6	Bornite	25.3	Pyrochlore	52.9 ± 1.7
Euclase	9.8	Halite	14.6	Sphalerite	25.4	Zincenite	54.3
Bauxite	9.8	Garnierite	14.7 ± 0.8	Enargite	25.5	Bournonite	54.4
Chrysoberyl	9.8	Sphene	14.7	Zincite	25.7	Tantalite	54.8 ± 0.9
Alum	9.9	Triphylite	15.5	Iron	26.0	Jamesonite	55.2
Beryllonite	10.0	Jarosite	15.8	Tennantite	26.4	Uranophane	55.6
Diaspore	10.1	Rhodochrosite	15.9	Orpiment	26.4	Brannerite	56.3 ± 4.1
Beryl	10.2	Sulphur	16.0	Chalcocite	26.4	Euxenite	57.1 ± 2.2
Kaolinite	10.2	Helvine	16.1	Cuprite	26.7	Carnotite	57.3
Cryolite	10.2	Rutile	16.4	Arsenopyrite	27.3	Sylvanite	57.9
Spodumene	10.3	Rhodonite	16.4	Cobaltite	27.6	Crocoite	58.0
Natrolite	10.3	Perovskite	16.5	Realgar	27.9	Sperrylite	58.5
Amblygonite	10.3	Siderite	16.5	Gersdorffite	27.9	Wulfenite	58.6
Montmorillonite	10.4 ± 0.1	Zaratite	17.1	Allanite	28.9 ± 0.1	Anglesite	59.4
Analcime	10.4	Galinite	17.3	Copper	29.0	Boulangerite	61.5
Periclase	10.4	Dioptase	17.4	Malayite	29.1	Mimetite	63.5
Topaz	10.5	Ilvaite	17.4	Stannite	30.5	Calaverite	63.8
Petalite	10.5	Braunite	17.4	Niccolite	30.8	Vanadinite	64.0
Talc	10.5	Triplite	17.7 ± 0.1	Löllingite	31.1	Pyromorphite	65.2
Pyrophyllite	10.6	Tourmaline	17.8	Smaltite	31.3	Cerussite	65.3
Serpentine	10.6 ± 0.4	Chalcophyllite	18.0	Rammelsbergite	31.6	Thorite	67.2
Spinel	10.6	Sylvine	18.0	Molybdenite	31.6	Stolzite	68.4
Forsterite	10.6	Manganite	18.5	Tetrahedrite	32.5	Tetradymite	68.7
Corundum	10.6	Limonite	18.7	Arsenic	33.0	Clausthalite	68.8
Enstatite	10.6	Fayalite	18.7	Columbite	34.7 ± 1.7	Coloradoite	69.1
Kyanite	10.7	Pyrolusite	18.7	Monazite	37.3	Bismutite	69.4
Andalusite	10.7	Ilmenite	19.0	Borite	37.3	Cosalite	69.6
Albite	10.7	Goethite	19.2	Proustite	38.9	Bismuthinite	70.5
Sal-ammoniac	10.8	Psilomelane	19.2 ± 0.5	Ullmanite	39.4	Calomel	70.5
Quartz	10.8	Smithsonite	19.3	Chlorargyrite	39.6	Altaite	70.6
Dolomite	10.9	Azurite	19.4	Cerargyrite	39.6	Cinnabar	71.2
Danburite	11.1	Erythrite	19.7	Argyrodite	40.7	Amalgam (Hg)	71.3
Sodalite	11.1	Malachite	19.9	Greenockite	40.9	Electrum	>72.6
Alunite	11.2	Chromite	19.9	Stibnite	41.1	Galena	73.2
Muscovite	11.3	Hausmannite	20.2	Cassiterite	41.1	Platinum	78.0
Amphibole	11.4 ± 0.8	Brochantite	20.3	Witherite	41.3	Gold	79.0
Orthoclase	11.8	Calamine	20.6	Kermesite	41.9	Mercury	80.0
Anorthite	12.0	Hematite	20.6	Miargyrite	41.9	Thorianite	80.1
Nitre	12.1	Pyrite	20.7	Pyrargyrite	42.4	Lead	82.0
Gypsum	12.1	Marcasite	20.7	Stephanite	42.6	Uraninite	82.0
Glauberite	12.1	Franklinite	20.8 ± 0.1	Polybasite	42.7	Bismuth	83.0
Datolite	12.3	Olivenite	20.8	Argentite	43.0		
Staurolite	12.5 ± 0.1	Bravoite	20.8 ± 0.1	Breithauptite	43.5		

Appendix 6

Minerals arranged according to elemental composition

Note: All the minerals within a single group contain a common element and are listed by contents of that element, in decreasing order. (RE = rare-earth element.)

Aluminium-bearing minerals

Name	Structural formula and chemical composition	Percentage aluminium	$\bar{Z}$ value
Corundum	$2[Al_2O_3]$	53	10.6
Diaspore	$4[AlO.OH]$	45	10.1
Chrysoberyl	$4[BeAl_2O_4]$	43	9.8
Bauxite	$Al_2O_3.2H_2O$	39	9.8
Spinel	$8[MgAl_2O_4]$	38	10.6
Gibbsite	$Al_2O_3.3H_2O$	35	9.5
Andalusite	$Al_2SiO_5 \quad 4[Al_2SiO_5]$	33	10.7
Kyanite	$Al_2SiO_5 \quad 4[Al_2SiO_5]$	33	10.7
Sillimanite	Al_2SiO_5	33	10.7
Topaz	$4[Al_2SiO_4(F,OH)_2]$	30	10.5
Gahnite	$8[ZnAl_2O_4]$	29	17.3
Staurolite	$[(Fe,Mg)_4Al_{18}Si_8O_{46}(OH_2)]$	29	12.5 ± 0.1
Kaolinite	$Al_4(Si_4O_{10})(OH)_3$	21	10.2
Muscovite	$4[KAl_3Si_3O_{10}(OH_2)]$	20	11.3
Euclase	$4[BeAlSiO_4.OH]$	19	9.8
Alunite	$K_2Al_6(OH)_2(SO_4)_4$	19	11.2
Amblygonite	$2[(Li,Na)AlPO_4(F,OH)]$	18	10.3
Epidote	$2[Ca_2(Al,Fe)_3Si_3O_{12}OH]$	up to 18	14.2
Garnet	$R_2^{II}R_2^{III}Si_3O_{12}$	up to 18	14.3 ± 1.3
Sodalite	$Na_8(AlSiO_4)_6Cl_2$	17	11.1
Pyrophyllite	$Al_2Si_4O_{10}(OH)_2$	15	10.6
Spodumene	$4[LiAlSi_2O_6]$	15	10.3
Lepidolite	$4[K(Li,Al)_3(Si,Al)_4O_{10}(F,OH)_2]$	13	9.8 ± 0.3
Natrolite	$Na_2Al_2Si_3O_{10}.2H_2O$	14	10.3
Cryolite	$2[Na_3AlF_6]$	13	10.2
Analcime	$Na_2Al_2Si_4O_{12}.2H_2O$	12	10.4
Beryl	$2[Be_3Al_2Si_6O_{18}]$	10	10.2
Albite (feldspar)	$NaAlSi_3O_8$	10	10.7
Orthoclase (feldspar)	$KAlSi_3O_8$	10	11.8
Anorthite (feldspar)	$CaAlSi_3O_8$	10	12.0
Idocrase	Ca,Mg,Fe,Al,OH silicate	10	13.2
Montmorillonite	$Al_2(Si_4O_{10})(OH)_2.nH_2O$	10	10.4
Petalite	$4[LiAlSi_4O_{10}]$	9	10.5
Tourmaline	Complex B,Fe,Al silicate	8	17.8

continued over

Aluminium-bearing minerals (continued)

Name	Structural formula and chemical composition	Percentage aluminium	$\bar{Z}$ value
Alum	$4[KAl(SO_4)_2.12H_2O]$	6	9.9
Biotite	Complex K,Mg,Fe aluminosilicate	6	13.6 ± 2.2
Chalcophyllite	$[Cu_{18}Al_2(AsO_4)_3(SO_4)_3(OH)_{27}.36H_2O]$	2	18.0

Antimony-bearing minerals

Name	Structural formula and chemical composition	Percentage antimony	$\bar{Z}$ value
Antimony (native)	Sb	100	51.0
Senarmontite	$16[Sb_2O_3]$	84	43.9
Kermesite	$8[Sb_2S_2O]$	75	41.9
Stibnite	$4[Sb_2S_3]$	72	41.1
Breithauptite	$2[NiSb]$	67	43.5
Ullmannite	$4[NiSbS]$	57	39.4
Livingstonite	$4[HgSb_4S_7]$	53	48.8
Zincenite	$PbSb_2S_4$	42	54.3
Miargyrite	$8[AgSbS_2]$	41	41.9
Jamesonite	$2[Pb_4FeSb_6S_{14}]$	35	55.2
Tetrahedrite	$8[Cu_3SbS_3]$	30	32.5
Dyscrasite	Ag_3Sb	27	48.1
Boulangerite	$Pb_5Sb_4S_{11}$	26	61.5
Bournonite	$4[CuPbSbS_3]$	25	54.4
Pyrargyrite	$2[Ag_3SbS_3]$	22	42.4
Stephanite	$4[Ag_5SbS_4]$	15	42.6
Polybasite	$16[Ag_{16}Sb_2S_{11}]$	11	42.7

Arsenic-bearing minerals

Name	Structural formula and chemical composition	Percentage arsenic	$\bar{Z}$ value
Arsenic (native)	As	100	33.0
Löllingite	$2[FeAs_2]$	73	31.1
Smaltite	$8[CoAs_2]$	72	31.3
Rammelsbergite	$2[NiAs_2]$	72	31.6
Realgar	$16[AsS]$	70	27.9
Orpiment	$4[As_2S_3]$	61	26.4
Niccolite	$2[NiAs]$	56	30.8
Arsenopyrite	$8[FeAsS]$	46	27.3
Cobaltite	$4[CoAsS]$	45	27.6
Gersdorffite	$4[NiAsS]$	45	27.9
Sperrylite	$4[PtAs_2]$	43	58.5
Scorodite	$8[FeAsO_4.2H_2O]$	35	21.3
Olivenite	$4[Cu_2(AsO_4)_2.5H_2O]$	30	20.8
Erythrite	$2[Co_3(AsO_4)_2.8H_2O]$	25	19.7
Tennantite	$8[Cu_2AsS_3]$	21	26.4
Enargite	$2[Cu_3AsS_4]$	19	25.5
Mimetite	$2[Pb_5(AsO_4)_3Cl]$	15	63.5
Proustite	$2[Ag_3AsS_3]$	15	38.9
Chalcophyllite	$[Cu_{18}Al_2(AsO_4)_3(SO_4)_3(OH)_{27}.36H_2O]$	8	18.0

Barium-bearing minerals

Name	Structural formula and chemical composition	Percentage barium	$\bar{Z}$ value
Witherite	$4[BaCO_3]$	70	41.3
Barite	$4[BaSO_4]$	59	37.3
Psilomelane	$2[(Ba,Mn^{II})Mn_4^{IV}O_8(OH)_2]$	13	19.2 ± 0.5

Beryllium-bearing minerals

Name	Structural formula and chemical composition	Percentage beryllium	$\bar{Z}$ value
Beryllonite	$12[NaBePO_4]$	7	10.0
Chrysoberyl	$4[BeAl_2O_4]$	7	9.8
Euclase	$4[BeAlSiO_4.OH]$	6	9.8
Beryl	$2[Be_3Al_2Si_6O_{15}]$	5	10.2
Helvine	$[(Mn,Fe,Zn)_8Be_6Si_6O_{24}S_2]$	5	16.1
Gadolinite	$2[Be_2FeY_2Si_2O_{10}]$	4	22.5

Bismuth-bearing minerals

Name	Structural formula and chemical composition	Percentage bismuth	$\bar{Z}$ value
Bismuth (native)	Bi	100	83.0
Bismutite	$2[Bi_2O_3.CO_2]$	82	69.4
Bismuthinite	$4[Bi_2S_3]$	81	70.5
Tetradymite	$[Bi_2Te_2S]$	59	68.7
Cosalite	$2[CuPb_7Bi_8S_{22}]$	43	69.6

Boron-bearing minerals

Name	Structural formula and chemical composition	Percentage boron	$\bar{Z}$ value
Boracite	$Mg_3B_7O_{13}Cl$	19	9.0
Colemanite	$2[Ca_2B_6O_{11}.5H_2O]$	16	9.7
Danburite	$4[CaB_2Si_2O_8]$	9	11.1
Tourmaline	Complex Be,Al,Fe silicate	6	17.8
Datolite	$CaBSiO_4(OH)$	6	12.3

Cadmium-bearing minerals

Name	Structural formula and chemical composition	Percentage cadmium	$\bar{Z}$ value
Greenockite	2[CdS]	78	40.9
Sphalerite	4[ZnS]	Small	25.4

Calcium-bearing minerals

Name	Structural formula and chemical composition	Percentage calcium	$\bar{Z}$ value
Fluorite	$4[CaF_2]$	51	14.6
Calcite	$2[CaCO_3]$	40	12.6
Apatite	$2[Ca_5(PO_4)_3F]$	40	14.1
Wollastonite	$CaSiO_3$	35	13.6
Perovskite	$8[CaTiO_3]$	30	16.5
Idocrase	Ca,Mg,Fe,Al,OH silicate	30	13.2
Anhydrite	$4[CaSO_4]$	29	13.4
Zeolite	hydrous aluminosilicate usually of Ca and Na	29	13.9
Datolite	$CaBSiO_4(OH)$	28	12.3
Garnet	Complex silicate	27	14.3 ± 1.3
Gypsum	$CaSO_4.2H_2O$	23	12.1
Dolomite	$[(Ca,Mg)(CO_3)_2]$	22	10.9
Sphene	$4[CaTiSiO_5]$	20	14.7
Ankerite	$[Ca(Mg,Fe)(CO_3)_2]$	20	12.7 ± 0.2
Colemanite	$2[Ca_2B_6O_{11}.5H_2O]$	20	9.7
Powellite	$8[CaMoO_4]$	20	26.7
Epidote	$2[Ca_2(Al,Fe)_3Si_3O_{12}OH]$	18	14.2 ± 2.1
Danburite	$4[CaB_2Si_2O_8]$	16	11.1
Microlite	$8[(CaNa)_2(TaNb)_2(O,OH,F)_7]$	15	50.3 ± 1.9
Malayite	$CaSn(SiO_4)O$	15	29.1
Scheelite	$8[CaWO_4]$	14	51.8
Glauberite	$Na_2Ca(SO_4)_2$	14	12.1
Anorthite	$CaAl_2Si_2O_8$	14	12.0
Ilvaite	$4[CaFe_2^{II}Fe^{III}Si_2O_8OH]$	10	17.4
Amphibole	Complex Mg,Fe silicate (R,SiO_3) where R = Ca,Mg,Fe	various	11.4 ± 0.8

Cerium-bearing minerals

Name	Structural formula and chemical composition	Percentage cerium	$\bar{Z}$ value
Allanite	$(Ca,Fe^{II})_2(RE,Al,Fe^{III})_3Si_3O_{12}(OH)$	small amounts	28.9 ± 0.1
Pyrochlore	$8[(Ca,Na,Ce)(Nb,Ti,Ta)_2(O,OH,F)_7]$	few per cent	52.9 ± 1.7
Euxenite	$(Y,Er,Ce,La,U)(Nb,Ti,Ta)_2(O,OH)_6$	$\simeq 5\%$	57.1 ± 2.2
Monazite	$4[(La,Ce)PO_4]$	$<5\%$	37.3

Chlorine-bearing minerals

Name	Structural formula and chemical composition	Percentage chlorine	$\bar{Z}$ value
Halite	$4[NaCl]$	61	14.6
Sylvine	$4[KCl]$	48	18.0
Sal-ammoniac	$4[NH_4Cl]$	38	10.8
Carnallite	$12[KMgCl_3.6H_2O]$	38	13.0
Chlorargyrite (cerargyrite)	$4[AgCl]$	25	39.6
Atacamite	$4[Cu_2Cl(OH)_3]$	17	21.9
Calomel	$2[Hg_2Cl_2]$	15	70.5
Boracite	$Mg_3B_7O_{13}Cl$	9	9.0
Sodalite	$Na_8(AlSiO_4)_6Cl_2$	7	11.1
Pyromorphite	$2[Pb_5(PO_4)_3Cl]$	3	65.2
Vanadinite	$2[Pb_5(VO_4)_3Cl]$	3	64.0
Mimetite	$2[Pb_5(AsO_4)_3Cl]$	2	63.5

Chromium-bearing minerals

Name	Structural formula and chemical composition	Percentage chromium	$\bar{Z}$ value
Chromite	$8[FeCr_2O_4]$	47	19.9
Crocoite	$4[PbCrO_4]$	16	58.0

Cobalt-bearing minerals

Name	Structural formula and chemical composition	Percentage cobalt	$\bar{Z}$ value
Linnaeite	$8[Co_3S_4]$	58	22.4
Cobaltite	$4[CoAsS]$	36	27.6
Erythrite	$2[Co_3(AsO_4)_2.8H_2O]$	30	19.7
Smaltite	$8[CoAs_2]$	28	31.3
Skutterudite	$[CoAs_3]$		
Carrollite	$CuCo_2S_4$		

Copper-bearing minerals

Name	Structural formula and chemical composition	Percentage copper	$\bar{Z}$ value
Copper	Cu	100	29
Cuprite	$2[Cu_2O]$	89	26.7
Chalcocite	$96[Cu_2S]$	80	26.4
Tenorite	$4[CuO]$	80	24.8
Covelline	$6[CuS]$	67	24.6
Bornite	$8[Cu_5FeS_4]$	63	25.3
Atacamite	$4[Cu_2Cl(OH)_3]$	60	21.9
Malachite	$4[Cu_2CO_3(OH)_2]$	58	19.9
Brochantite	$4[Cu_4SO_4(OH)_6]$	56	20.3
Azurite	$2[Cu_3(CO_3)_2(OH)_2]$	55	19.4
Tennantite	$8[Cu_3AsS_3]$	53	26.4
Enargite	$2[Cu_3AsS_4]$	48	25.5
Tetrahedrite	$8[Cu_3SbS_3]$	47	32.5
Dioptase	$6[CuSiO_2(OH)_2]$	40	17.4
Chalcophyllite	$Cu_{18}Al_2(AsO_4)_3(SO_4)_3(OH)_{27}.36H_2O$	38	18.0
Chrysocolla	$CuSiO_3.2H_2O$ (approx.)	36 (approx.)	16.4
Chalcopyrite	$4[CuFeS_2]$	35	23.5
Stannite	$2[Cu_2FeSnS]$	30	30.5
Olivenite	$Cu_2(AsO_4)_2.5H_2O$	26	20.8
Chalcanthite	$CuSO_4.5H_2O$	25	14.1
Bournonite	$4[CuPbSbS_3]$	13	54.4
Torbernite	$Cu_2(UO_2)_2(PO_4)_2.8H_2O$	6	52.4

Erbium-bearing mineral

Name	Structural formula and chemical composition	Percentage erbium	$\bar{Z}$ value
Fergusonite	$8[(Y,Er)(Nb,Ta)O_4]$	about 29	44.1 ± 0.9

Fluorine-bearing minerals

Name	Structural formula and chemical composition	Percentage fluorine	$\bar{Z}$ value
Cryolite	$2[Na_3AlF_6]$	54	10.2
Fluorite	$4[CaF_2]$	49	14.6
Topaz	$4[Al_2SiO_4(F,OH)_2]$	9	10.5
Amblygonite	$2[(Li,Na)AlPO_4(F,OH)]$	6	10.3
Apatite	$2[Ca_5(PO_4)_3F]$	4	14.1
Lepidolite	$4[K(Li,Al)_3(Si,Al)_4O_{10}(F,OH)_2]$	4	9.8 ± 0.3
Microlite	$8[(Ca,Na)_2(Ta,Nb)_2(O,OH,F)_7]$	2	50.3 ± 1.9
Pyrochlore	$8[(Ca,Na,Ce)(Nb,Ti,Ta)_2(O,OH,F)_7]$	1	52.9 ± 1.7
Idocrase	Ca,Mg,Fe,Al,OH silicate	1	13.2

Germanium-bearing mineral

Name	Structural formula and chemical composition	Percentage germanium	$\bar{Z}$ value
Argyrodite	$32[Ag_8GeS_6]$	6	40.7

Gold-bearing minerals

Name	Structural formula and chemical composition	Percentage gold	$\bar{Z}$ value
Gold (native)	Au	100	79.0
Electrum	Au,Ag alloy	80 (approx.)	Variable 72.6 (approx.)
Calaverite	$2[AuTe_2]$	44	63.8
Sylvanite	$2[AuAgTe_4]$	24	57.9

Iron-bearing minerals

Name	Structural formula and chemical composition	Percentage iron	$\bar{Z}$ value
Iron (native)	Fe	100	26.0
Magnetite	$8[Fe_3O_4]$	72	21.0
Hematite	$2[Fe_2O_3]$	70	20.1
Pyrrhotine	$2[FeS]$	64	22.4
Goethite	$4[FeO.OH]$	63	19.2
Limonite	mixed hydrated oxides of iron	60	18.7
Fayalite	Fe_2SiO_4	55	18.7
Olivine	$4[(Mg,Fe)_2SiO_4]$	up to 55	18.7
Tourmaline	Complex B,Fe,Al silicate	49	17.8
Siderite	$2[FeCO_3]$	48	16.5
Marcasite	$2[FeS_2]$	47	20.7
Pyrite	$4[FeS_2]$	47	20.7
Bravoite	$(Fe,Ni)S_2$	47	20.8 ± 1
Ilvaite	$4[CaFe_2^{II}Fe^{III}Si_2O_8OH]$	41	17.4
Pentlandite	$4[Fe,Ni)_9S_8]$	39	23.4 ± 0.1
Ilmenite	$2[FeTiO_3]$	37	19.0
Franklinite	$8[(Zn,Mn,Fe^{II})(Fe^{III},Mn^{III})_2O_4]$	36	20.8 ± 0.1
Triphylite	$4[LiFePO_4]$	35	15.5
Jarosite	$KFe_3(SO_4)_3(OH)_6$	34	15.8
Arsenopyrite	$4[FeAsS]$	34	27.3
Garnet	complex silicate	up to 34	14.3 ± 1.3
Epidote	$2[Ca_2(Al,Fe)_3Si_3O_{12}OH]$	31	14.2 ± 2.1
Biotite	complex silicate	31	13.6 ± 2.2
Chalcopyrite	$4[CuFeS_2]$	30	23.5
Löllingite	$2[FeAs_2]$	27	31.1
Scorodite	$8[FeAsO_4.2H_2O]$	26	21.3
Chromite	$8[FeCr_2O_4]$	25	19.9
Triplite	$8[(Mn,Fe)_2PO_4F]$	25	17.7 ± 0.1
Helvine	$[(Mn,Fe,Zn)_8Be_6Si_6O_{24}S_2]$	16–24	16.1
Melanterite	$FeSO_4.7H_2O$	20	12.2
Hypersthene	$(Mg,Fe)_2Si_2O_6$	11–20	12.9 ± 0.7
Staurolite	$[(Fe,Mg)_4Al_{18}Si_8O_{46}(OH)_2]$	12–13	12.5 ± 0.1
Columbite	$4[(Fe,Mn)(Nb,Ta)_2O_6]$	7–8	34.7 ± 1.7
Ankerite	$[Ca(Mg,Fe)(CO_3)_2]$	11–14	12.7 ± 0.2
Pyroxene	$RSiO_3$ (R = Ca, Mg, Fe, Al, etc.)	0–12	13.1 ± 0.8
Bornite	$8[Cu_5FeS_4]$	11	25.3
Wolframite	$2[(Fe,Mn)WO_4]$	9	51.2
Tantalite	$4[(Fe,Mn)(Ta,Nb)_2O_6]$	5	54.8 ± 0.9
Serpentine	hydrated Mg silicate	up to 6	10.6 ± 0.4
Jamesonite	$2[Pb_4FeSb_6S_{14}]$	3	55.2
Brannerite	$(U,Y,Ca,Fe,Th)_3Ti_5O_{16}$ (?)	up to 2	56.3 ± 4.1

Lanthanum-bearing mineral

Name	Structural formula and chemical composition	Percentage lanthanum	$\bar{Z}$ value
Monazite	$4[(La,Ce)PO_4]$	up to 56	37.3

Lead-bearing minerals

Name	Structural formula and chemical composition	Percentage lead	$\bar{Z}$ value
Lead (native)	Pb	100	82.0
Galena	$4[PbS]$	87	73.2
Cerussite	$4[PbCO_3]$	78	65.3
Pyromorphite	$2[Pb_5(PO_4)_3Cl]$	76	65.2
Vanadinite	$2[Pb_5(VO_4)_3Cl]$	73	64.0
Clausthalite	$4[PbSe]$	72	68.8
Mimetite	$2[Pb_5(AsO_4)_3Cl]$	70	63.5
Anglesite	$4[PbSO_4]$	68	59.4
Crocoite	$4[PbCrO_4]$	64	58.0
Altaite	$4[PbTe]$	62	70.6
Wulfenite	$8[PbMoO_4]$	56	58.6
Boulangerite	$Pb_5Sb_4S_{11}$	55	61.5
Descloizite	$4[PbZnVO_4OH]$	51	51.3
Stolzite	$8[PbWO_4]$	46	68.4
Bournonite	$4[CuPbSbS_3]$	42	54.4
Jamesonite	$2[Pb_4FeSb_6S_{14}]$	40	55.2
Cosalite	$2[CuPb_7Bi_8S_{22}]$	37	69.6
Zincenite	$PbSb_2S_4$	36	54.3

Magnesium-bearing minerals

Name	Structural formula and chemical composition	Percentage magnesium	$\bar{Z}$ value
Periclase	4[MgO]	60	10.4
Brucite	$Mg(OH)_2$	42	9.4
Olivine	$4[(Mg,Fe)_2SiO_4]$	up to 35	14.6 ± 4.1
Forsterite	Mg_2SiO_4	35	10.6
Magnesite	$2[MgCO_3]$	29	8.9
Serpentine (antigorite-chrysotile)	hydrated Mg silicate	26	10.6
Enstatite	$MgSiO_3$	24	10.6
Amphibole	Complex Mg,Fe silicate: $RSiO_3$ (R = Ca, Mg, Fe)	up to 24	11.4 ± 0.8
Talc	$4[Mg_3Si_4O_{10}(OH)_2]$	19	10.5
Boracite	$Mg_3B_7O_{13}Cl$	19	9.0
Hypersthene	$(Mg,Fe)_2Si_2O_6$	18	12.9 ± 0.7
Spinel	$8[MgAl_2O_4]$	17	10.6
Biotite	Complex K,Mg,Fe aluminosilicate	16	13.6 ± 2.2
Garnierite	$(Ni,Mg)Si_2O_5(OH)_4$	14	14.7 ± 0.8
Dolomite	$[CaMg(CO_3)_2]$	13	10.9
Pyroxene	Complex metasilicate: $RSiO_3$ (R = Ca, Mg, Fe, Al)	11	13.1 ± 0.8
Epsomite*	$MgSO_4.7H_2O$	10	9.0
Carnallite*	$12[KMgCl_3.6H_2O]$	9	13.0
Ankerite	$Ca(Mg,Fe)(CO_3)_2$	7	12.7 ± 0.2

*Readily dehydrated by high vacuum pressure and high energy electron beams, therefore $\bar{Z}$ values vary accordingly.

Manganese-bearing minerals

Name	Structural formula and chemical composition	Percentage manganese	$\bar{Z}$ value
Hausmannite	$8[Mn_3O_4]$	72	20.2
Alabandine	$4[MnS]$	63	21.7
Pyrolusite	$2[MnO_2]$	63	18.7
Manganite	$MnO.OH$	62	18.5
Psilomelane	$2[(Ba,Mn^{II})Mn_4^NO_8(OH_2)]$	50	19.2 ± 0.5
Braunite	$8[Mn_3^{II}Mn_2^{III}SiO_3O_{12}]$	50	17.4
Rhodochrosite	$2[MnCO_3]$	48	15.9
Rhodonite	$10[MnSiO_3]$	42	16.4
Franklinite	$8[(Zn,Mn,Fe^{II})(Fe^{III},Mn^{III})_2O_4]$	up to 31	20.8 ± 0.1
Triplite	$8[(Mn,Fe)_2PO_4F]$	24	17.7 ± 0.1
Helvine	$[(Mn,Fe,Zn)_8Be_6Si_6O_{24}S_2]$	24	16.1
Wolframite	$2[(Fe,Mn)WO_4]$	9	51.2
Columbite	$4[(Fe,Mn)(Ta,Nb)_2O_6]$	8	34.7 ± 1.7
Tantalite	$4[(Fe,Mn)(Ta,Nb)_2O_6]$	6	54.8 ± 0.9

Mercury-bearing minerals

Name	Structural formula and chemical composition	Percentage mercury	$\bar{Z}$ value
Mercury (native)	Hg	100	80
Cinnabar	$3[HgS]$	86	71.2
Calomel	$2[Hg_2Cl_2]$	85	70.5
Amalgam (mercury-rich)	Hg with Ag	74	71.3
Coloradoite	$HgTe$	61	69.1
Livingstonite	$4[HgSb_4S_7]$	22	48.8
Amalgam (silver-rich)	Ag with Hg	5	48.6

Molybdenum-bearing minerals

Name	Structural formula and chemical composition	Percentage molybdenum	$\bar{Z}$ value
Molybdenite	$2[MoS_2]$	60	31.6
Powellite	$8[CaMoO_4]$	48	26.7
Wulfenite	$8[PbMoO_4]$	26	58.6

Nickel-bearing minerals

Name	Structural formula and chemical composition	Percentage nickel	$\bar{Z}$ value
Millerite	$3[NiS]$	65	23.8
Zaratite	$[Ni_3CO_3(OH)_4.4H_2O]$	47	17.1
Niccolite	$2[NiAs]$	44	30.8
Gersdorffite	$4[NiAsS]$	35	27.9
Breithauptite	$2[NiSb]$	33	43.5
Rammelsbergite	$2[NiAs_2]$	28	31.6
Pentlandite	$4[(Fe,Ni)S]$	28 (max)	23.4 ± 0.1
Ullmannite	$4[NiSbS]$	28	39.4
Garnierite	$[(NiMg)_3Si_2O_5(OH)_4]$	less than 20	14.7 ± 0.8
Bravoite	$(Fe,Ni)S_2$	about 10	20.8 ± 0.1

Niobium-bearing minerals

Name	Structural formula and chemical composition	Percentage niobium	$\bar{Z}$ value
Columbite	$4[(Fe,Mn)(Nb,Ta)_2O_6]$	24	34.7 ± 1.7
Fergusonite	$8[(Y,Er)(Nb,Ta)O_4]$	29	44.1 ± 0.9
Microlite	$8[(Ca,Na)_2(Ta,Nb)_2(O,OH,F)_7]$	7	50.3 ± 1.9
Tantalite	$4[(Fe,Mn)(Nb,Ta)_2O_6]$	2	54.8 ± 0.9

Nitrogen-bearing minerals

Name	Structural formula and chemical composition	Percentage nitrogen	$\bar{Z}$ value
Nitratine	$2[NaNO_3]$	17	8.6
Nitre	$4[KNO_3]$	14	12.1

Phosphate minerals

Name	Structural formula and chemical composition	Percentage phosphate	$\bar{Z}$ value
Beryllonite	$12[NaBePO_4]$	24	10.0
Amblygonite	$2[(Li,Na)AlPO_4(F,OH)]$	21	10.3
Triphylite	$4[LiFePO_4]$	20	15.5
Apatite	$2[Ca_5(PO_4)_3F]$	18	14.1
Xenotime	$8[YPO_4]$	17	24.2
Triplite	$8[(Mn,Fe)_2PO_4F]$	14	17.7
Monazite	$4[(La,Ce)PO_4]$	14	37.3
Pyromorphite	$2[Pb_5(PO_4)_3Cl]$	7	65.2
Torbernite	$Cu(UO_2)_2(PO_4)_2.8H_2O$	7	52.4
Autunite	$Ca(UO_2)_2(PO_4)_2.10H_2O$	7	50.9

Platinum-bearing minerals

Name	Structural formula and chemical composition	Percentage platinum	$\bar{Z}$ value
Platinum (native)	Pt	100	78.0
Sperrylite	$4[PtAs_2]$	57	58.5

Potassium-bearing minerals

Name	Structural formula and chemical composition	Percentage potassium	$\bar{Z}$ value
Sylvine	$4[KCl]$	52	18.0
Nitre	$4[KNO_3]$	39	12.1
Orthoclase	$KAlSi_3O_8$	14	11.8
Carnallite	$12[KMgCl_3.6H_2O]$	14	13.0
Muscovite	$4[KAl_3Si_3O_{10}(OH)_2]$	10	11.3
Carnotite	$[KUO_2VO_4.\frac{1}{3}H_2O]$	9	57.3
Biotite	Complex K,Mg,Fe aluminosilicate	9	13.6 ± 2.2
Alunite	$K_2Al_6(OH)_{12}(SO_4)_4$	8	9.9
Jarosite	$KFe_3(SO_4)_2(OH)_6$	8	15.8
Alum	$4[KAl(SO_4)_2.12H_2O]$	8	9.9

Selenium-bearing minerals

Name	Structural formula and chemical composition	Percentage selenium	$\bar{Z}$ value
Clausthalite	$4[PbSe]$	28	68.8
Naumannite	Ag_2Se	27	43.5

Silicate minerals

Name	Structural formula and chemical composition	Percentage silicon	$\bar{Z}$ value
Quartz	$3[SiO_2]$	47	10.8
Zeolite	hydrous aluminosilicate	41	13.9
Petalite	$4[LiAlSi_4O_{10}]$	37	10.5
Albite	$NaAlSi_3O_8$	32	10.7
Beryl	$2[Be_3Al_2Si_6O_{18}]$	31	10.2
Pyrophyllite	$Al_2Si_4O_{10}(OH)_2$	31	10.6
Anorthite	$CaAlSi_3O_8$	30	12.0
Orthoclase	$KAlSi_3O_8$	30	11.8
Spodumene	$4[LiAlSi_2O_6]$	30	10.3
Talc	$4[Mg_3Si_4O_{10}(OH)_2]$	30	10.5
Enstatite	$MgSiO_3$	28	10.6
Amphibole	complex Mg, Fe silicate: $RSiO_3$ (R = Mg, Ca, Fe)	28	11.4
Hypersthene	$(Mg,Fe)_2SiO_6$	26	12.9 ± 0.7
Phenakite	$6[Be_2SiO_4]$	26	8.9
Pyroxene	Complex metasilicate: $RSiO_3$ (R = Ca, Mg, Fe, Al)	26	13.1 ± 0.8
Analcime	$Na_2Al_2Si_4O_{12}.6H_2O$	26	10.4
Montmorillonite	$Al_2(Si_4O_{10})(OH)_2.nH_2O$	25	10.4 ± 0.1
Biotite	Complex K, Mg, Fe aluminosilicate	25	13.6 ± 2.2
Wollastonite	$CaSiO_3$	24	13.6
Danburite	$4[CaB_2Si_2O_8]$	23	11.1
Natrolite	$Na_2Al_2Si_3O_{10}.2H_2O$	22	10.3
Kaolinite	$Al_4(Si_4O_{10})(OH)_3$	22	10.2
Rhodonite	$10[MnSiO_3]$	21	16.4
Muscovite	$4[KAl_3Si_3O_{10}(OH)_2]$	21	11.3
Datolite	$CaBSiO_4(OH)$	20	12.3
Serpentine	hydrated Mg silicate	20	10.6 ± 0.4
Forsterite	Mg_2SiO_4	20	10.6
Euclase	$4[BeAlSiO_4OH]$	19	9.8
Olivine	$4[(Mg,Fe)_2SiO_4]$	20	14.6 ± 0.1
Garnet	Complex silicate	19	14.3 ± 1.3
Epidote	$2[Ca_2(Al,Fe)_3Si_3O_{12}OH]$	19	14.2
Dioptase	$6[CuSiO_2(OH)_2]$	18	17.4
Idocrase	Ca, Mg, Fe, Al, OH silicate	18	13.2
Garnierite	$(Ni,Mg)_3Si_2O_5(OH)_4$	18	14.7 ± 0.8
Sodalite	$Na_8(AlSiO_4)_6SO_4$	17	11.1
Kyanite	$4[Al_2SiO_5]$	17	10.7
Andalusite	$4[Al_2SiO_5]$	17	10.7
Chrysocolla	$CuSiO_3.2H_2O$	16	16.4
Topaz	$4[Al_2SiO_4(F,OH)_2]$	16	10.5
Zircon	$4[ZrSiO_4]$	15	24.8
Helvine	$(Mn,Fe,Zn)_8Be_6Si_6O_{24}S_2$	15	16.1
Braunite	$8[Mn_3^{II}Mn_2^{III}Si_3O_{12}]$	15	17.4
Sphene	$4[CaTiSiO_5]$	14	14.7
Ilvaite	$4[CaFe_2^{II}Fe^{III}Si_2O_8OH]$	14	17.4
Fayalite	Fe_2SiO_4	14	18.7

continued over

Silicate minerals (*continued*)

Name	Structural formula and chemical composition	Percentage silicon	$\bar{Z}$ value
Lepidolite	$4[K(Li,Al)_3(Si,Al)O_{10}(F,OH)_2]$	13	9.8
Staurolite	$[(Fe,Mg)_4Al_{18}Si_8O_{46}(OH)_2]$	13	12.5
Willemite	$6[Zn_2SiO_4]$	13	21.7
Calamine	$2[Zn_4Si_2O_7(OH)_2.H_2O]$	12	20.6
Malayite		11	29.1
Tourmaline	Complex boro-silicate of iron and aluminium – of variable composition	11	17.8
Thorite	$ThSiO_4$	9	67.2
Uranophane	$Ca(UO_2)_2Si_2O_7.6H_2O$	7	55.6

Silver-bearing minerals

Name	Structural formula and chemical composition	Percentage silver	$\bar{Z}$ value
Silver (native)	Ag	100	47.0
Amalgam (silver-rich)	Ag with Hg	up to 95	48.6
Argentite	$2[Ag_2S]$	87	43.0
Argyrodite	$32[Ag_8GeS_6]$	77	40.7
Chlorargyrite (cerargyrite)	$4[AgCl]$	75	39.6
Polybasite	$16[Ag_{16}Sb_2S_{11}]$	74	42.7
Dyscrasite	Ag_3Sb	73	48.1
Naumannite	Ag_2Se	73	43.5
Stephanite	$4[Ag_5SbS_4]$	68	42.6
Proustite	$2[Ag_3AsS_3]$	65	38.9
Hessite	Ag_2Te	63	48.9
Pyrargyrite	$2[Ag_3SbS_3]$	60	42.4
Miargyrite	$8[AgSbS_2]$	37	41.9
Amalgam (mercury-rich)	Hg with Ag	26	71.3
Electrum	Ag,Au alloy	20 (approx.)	72.6 (approx.)
Sylvanite	$2[Au,AgTe_4]$	13	57.9

Sodium-bearing minerals

Name	Structural formula and chemical composition	Percentage sodium	$\bar{Z}$ value
Halite	$4[NaCl]$	39	14.6
Cryolite	$2[Na_3AlF_6]$	33	10.2
Nitratine	$2[NaNO_3]$	27	8.6
Sodalite	$Na_8(AlSiO_4)_6Cl_2$	19	11.1
Beryllonite	$12[NaBePO_4]$	18	10.0
Glauberite	$Na_2Ca(SO_4)_2$	17	12.1
Natron*	$Na_2CO_3.10H_2O$*	16	7.9
Mirabilite*	? $Na_2SO_4.10H_2O$*	14	8.8
Natrolite*	$Na_2Al_2Si_3O_{10}.2H_2O$*	12	10.3
Analcime*	$Na_2Al_2Si_4O_{12}.2H_2O$*	10	10.4
Albite	$NaAlSi_3O_8$	9	10.7
Amblygonite	$2[(Li,Na)AlPO_4(F,OH)]$	3	10.3

*These minerals readily dehydrated by high vacuum pressure and high energy electron beams, therefore $\bar{Z}$ values vary accordingly.

Strontium-bearing minerals

Name	Structural formula and chemical composition	Percentage strontium	$\bar{Z}$ value
Strontianite	$4[SrCO_3]$	59	25.6
Celestine	$4[SrSO_4]$	48	23.7

Sulphur-bearing minerals

Name	Structural formula and composition	Percentage sulphur	$\bar{Z}$ value
Sulphur (native)	S	100	16.0
Pyrite	$4[FeS_2]$	54	20.7
Marcasite	$2[FeS_2]$	54	20.7
Bravoite	$(Fe,Ni)S_2$	54	20.8
Linnaeite	$8[CoS_4]$	42	22.4
Molybdenite	$2[MoS_2]$	40	31.6
Orpiment	$4[As_2S_3]$	39	26.4
Alabandite	$4[MnS]$	37	21.7
Pyrrhotite	$2[FeS]$	37	22.4
Chalcopyrite	$4[CuFeS_2]$	35	23.5
Millerite	$3[NiS]$	35	23.8
Covelline	$6[CuS]$	34	24.6
Pentlandite	$4[(Fe,Ni)_9S_8]$	33	23.4 ± 0.1
Enargite	$2[Cu_3AsS_4]$	33	25.5
Sphalerite	$4[ZnS]$	33	25.4
Stannite	$2[Cu_2FeSnS_4]$	30	30.5
Realgar	$16[AsS]$	30	27.9
Stibnite	$4[Sb_2S_3]$	28	41.1
Tennantite	$8[Cu_3AsS_3]$	27	26.4
Bornite	$8[Cu_5FeS_4]$	26	25.3
Anhydrite	$4[CaSO_4]$	24	13.4
Glauberite	$Na_2Ca(SO_4)_2$	23	12.1
Tetrahedrite	$8[Cu_3SbS_3]$	23	32.5
Zincenite	$PbSb_2S_4$	22	24.8
Greenockite	$2[CdS]$	22	40.9
Jamesonite	$2[Pb_4FeSb_6S_{14}]$	22	55.2
Miargyrite	$8[AgSbS_2]$	22	41.9
Bournonite	$4[CuPbSbS_3]$	20	54.4
Kermesite	$8[Sb_2S_2O]$	20	41.9
Arsenopyrite	$8[FeAsS]$	20	27.3
Chalcocite	$96[Cu_2S]$	20	26.4
Bismuthinite	$4[Bi_2S_3]$	19	70.5
Boulangerite	$Pb_5Sb_4S_{11}$	19	61.5
Proustite	$2[Ag_3AsS_3]$	19	38.9
Gypsum	$8[CaSO_4.2H_2O]$	19	12.1
Cobaltite	$4[(Co)AsS]$	19	27.6
Gersdorffite	$4[NiAsS]$	19	27.9
Cosalite	$2[CuPb_7Bi_8S_{22}]$	18	69.6
Pyrargyrite	$2[Ag_3SbS_3]$	18	42.4
Celestine	$4[SrSO_4]$	18	23.7
Argyrodite	$32[Ag_8GeS_6]$	17	40.7
Stephanite	$4[Ag_5SbS_4]$	16	42.6
Alunite	$[K_2Al_6(SO_4)_2(OH)_6]$	16	11.2
Polybasite	$16[Ag_{16}Sb_2S_{11}]$	15	42.7
Ullmanite	$4[NiSbS]$	15	39.4
Cinnabar	$3[HgS]$	14	71.2
Barite	$4[BaSO_4]$	14	37.3
Alum	$4[KAl(SO_4)_2.12H_2O]$	14	9.9

Sulphur-bearing minerals (*continued*)

Name	Structural formula and chemical composition	Percentage sulphur	$\bar{Z}$ value
Argentite	$2[Ag_2S]$	13	43.0
Jarosite	$KFe_3(SO_4)_2(OH)_6$	13	15.8
Epsomite	$4[MgSO_4.7H_2O]$	13	9.0
Chalcanthite	$2[CuSO_4.5H_2O]$	13	14.1
Galena	$4[PbS]$	13	73.2
Melanterite	$[FeSO_4.7H_2O]$	12	12.2
Anglesite	$4[PbSO_4]$	11	59.4
Mirabilite	$Na_2SO_4.10H_2O$	10	8.8
Brochantite	$4[Cu_4SO_4(OH)_6]$	7	20.3
Helvine	$[(Mn,Fe,Zn)_8Be_6Si_6O_{24}S_2]$	6	16.1
Tetradymite	$[Bi_2Te_2S]$	5	68.7
Chalcophyllite	$[Cu_{18}Al_2(AsO_4)_3(OH)_{27}.36H_2O]$	3	18.0

Tantalum-bearing minerals

Name	Structural formula and chemical composition	Percentage tantalum	$\bar{Z}$ value
Microlite	$8[(Ca,Na)_2(Ta,Nb)_2(O,OH,F)_7]$	56	50.3 ± 1.9
Tantalite	$4[(Fe,Mn)(Ta,Nb)_2O_6]$	33	54.8 ± 0.9
Fergusonite	$8[(Y,Er)(Nb,Ta)O_4]$	6	44.1 ± 0.9
Columbite	$4[(Fe,Mn)(Nb,Ta)_2O_6]$	5	34.7 ± 1.7

Tellurium-bearing minerals

Name	Structural formula and chemical composition	Percentage tellurium	$\bar{Z}$ value
Sylvanite	$2[AgAuTe_4]$	63	57.9
Calaverite	$2[AuTe_3]$	56	63.8
Coloradoite	$HgTe$	39	69.1
Altaite	$4[PbTe]$	38	70.6
Hessite	Ag_2Te	37	48.9
Tetradymite	$[Bi_2Te_2S]$	36	68.7

Thorium-bearing minerals

Name	Structural formula and chemical composition	Percentage thorium	$\bar{Z}$ value
Thorianite	$4[ThO_2]$	88	80.1
Thorite	$ThSiO_4$	72	67.2
Brannerite	$(U,Y,Ca,Fe,Th)_3Ti_5O_{16}$ (?)	7	56.3 ± 4.1

Tin-bearing minerals

Name	Structural formula and chemical composition	Percentage tin	$\bar{Z}$ value
Tin (native)	Sn	100	50.0
Cassiterite	$2[SnO_2]$	79	41.1
Malayite	$CaSn(SiO_4)O$	45	29.1
Stannite	$2[Cu_2FeSnS_4]$	28	30.5

Titanium-bearing minerals

Name	Structural formula and chemical composition	Percentage titanium	$\bar{Z}$ value
Rutile	$2[TiO_2]$	60	16.4
Anatase	TiO_2	60	16.4
Brookite	TiO_2	60	16.4
Perovskite	$8[CaTiO_3]$	35	16.5
Ilmenite	$2[FeTiO_3]$	32	19.0
Sphene (titanite)	$4[CaTiSiO_5]$	24	14.7
Brannerite	$(U,Y,Ca,Fe,Th)_3Ti_5O_{16}$ (?)	23	56.3
Pyrochlore	$8[(Ca,Na,Ce)(Nb,Ti,Ta)_2(O,OH,F)]$	up to 10	52.9 ± 1.7
Euxenite	$(Y,Er,Ce,La,U)(Nb,Ti,Ta)_2(O,OH)_6$	up to 10	57.1 ± 2.2

Tungsten-bearing minerals

Name	Structural formula and chemical composition	Percentage tungsten	$\bar{Z}$ value
Scheelite	$8[CaWO_4]$	64	51.8
Wolframite	$2[(Fe,Mn)WO_4]$	61	51.2
Stoltzite	$8[PbWO_4]$	40	68.4

Uranium-bearing minerals

Name	Structural formula and chemical composition	Percentage uranium	$\bar{Z}$ value
Uraninite	UO_2	88	82.0
Uranophane	$Ca(UO_2)_2Si_2O_7.6H_2O$	56	55.6
Carnotite	$KUO_2VO_4.\frac{1}{2}H_2O$	55	57.3
Torbernite	$Cu(UO_2)_2(PO_4)_2.8H_2O$	51	52.4
Autunite	$Ca(UO_2)_2(PO_4)_2.10H_2O$	50	50.9
Brannerite	$(U,Y,Ca,Fe,Th)_3Ti_5O_{16}$ (?)	41	56.3 ± 4.1

Vanadium-bearing minerals

Name	Structural formula and chemical composition	Percentage vanadium	$\bar{Z}$ value
Descloizite	$4[Pb,Zn,VO_4OH]$	13	51.3
Carnotite	$[KUO_2VO_4.\frac{1}{2}H_2O]$	13	57.3
Vanadinite	$2[Pb_5(VO_4)_3Cl]$	11	64.0

Yttrium-bearing minerals

Name	Structural formula and chemical composition	Percentage yttrium	$\bar{Z}$ value
Xenotime	$8[YPO_4]$	48	24.2
Gadolinite	$2[Be_2FeY_2Si_2O_{10}]$	38	22.5
Fergusonite	$8[(Y,Er)(Nb,Ta)O_4]$	15	44.1 ± 0.9
Brannerite	$(U,Y,Ca,Fe,Th)_3Ti_5O_{16}$ (?)	3	56.3 ± 4.1

Zinc-bearing minerals

Name	Structural formula and chemical composition	Percentage zinc	$\bar{Z}$ value
Zincite	$2[ZnO]$	80	25.7
Sphalerite	$4[ZnS]$	67	25.4
Hydrozincite	$2[Zn_5(CO_3)_2(OH)_6]$	60	20.9
Willemite	$6[Zn_2SiO_4]$	59	21.7
Calamine	$2[Zn_4Si_2O_7(OH)_2.H_2O]$	54	20.6
Smithsonite	$2[ZnCO_3]$	52	19.3
Gahnite	$8[ZnAl_2O_4]$	36	17.3
Descloizite	$4[PbZnVO_4OH]$	16	51.3
Franklinite	$8[(Zn,Mn,Fe^{II})(Fe^{III},Mn^{III})_2O_4]$	up to 27	20.8

Zirconium-bearing minerals

Name	Structural formula and chemical composition	Percentage zirconium	$\bar{Z}$ value
Zircon	$4[ZrSiO_4]$	50	24.8

Selected Bibliography

Chapter 1. Introduction

Berry, L. G., Mason, B. and Dietrich, R. V. (1983). *Mineralogy* (2nd edn.), W. H. Freeman & Co., San Francisco, p. 561.

British Geological Survey (1984). *World Mineral Statistics, 1978–1982*, H.M.S.O., London.

Chayes, F. (1949). A simple point counter for thin section analysis. *Am. Mineral.*, (34) 1, pp. 600–601.

Correns, C. W. (1964). *Introduction to Mineralogy, Crystallography and Petrology* (2nd edn.), George Allen & Unwin, London, p. 484, (transl. Johns, W. D.).

De Lesse, A. (1848). Procédé mechanique pour déterminer la composition des roches. *Ann. Mines* 4, (13) p. 379.

Emburey, P. and Fuller, J. P. (Eds.) (1980). *A Manual of New Mineral Names*. 1892–1978. Oxford University Press, Oxford.

Glagolev, A. A. (1934). Quantitative analysis with the microscope by the 'point' method, *Eng. Min. J.*, (135) p. 399.

Govett. G. J. S. and Govett, M. H. (Eds.) (1976). *World Mineral Supplies – Assessment and Perspective*, Elsevier, Amsterdam, p. 472.

Herdan, G. (1953). *Small Particle Statistics*, Elsevier, Amsterdam.

Hurburt, C. S. and Klein, C. (1971). *Manual of Mineralogy (after James D. Dana)*, John Wiley & Sons, London, p. 532.

Jones, W. R. (1950). *Minerals in Industry*, Penguin Books, London, p. 224.

Kirsch, H. (1968). *Applied Mineralogy for Engineers, Technologists and Students*. Chapman & Hall, London, p. 233 (transl. K. A. Jones).

Knill, J. L. (Ed.) (1978). *Industrial Geology*, Oxford University Press, Oxford, p. 344.

Kostov, I. (1968). *Mineralogy*, Oliver & Boyd, Edinburgh, (transl. P. G. Emburey and J. Phemister), p. 587.

Putnis, A. and McConnell, J. D. C. (1980). *Principles of Mineral Behaviour*, Blackwell Scientific Publications, London, p. 257.

Rosival, A. (1898). Veber geometrische Gesteinanylsen USW., *Verhandl. K.K. Geol. Reich Wien*, 5–6, p. 143.

Saltykov, S. A. (1958). *Stereometric metallography* (2nd edn.), Metallurgizdat, Moscow, p. 446.

Shand, S. J. (1916). A recording micrometer for geometrical rock analysis. *J. Geol.*, (24) pp. 394–404.

Sinkankas, J. (1966). *Mineralogy, a First Course*, Van Nostrand Co. Inc., New York, p. 587.

Thomson, E. (1930). Quantitative microscopic analysis, *J. Geol.*, 38, p. 193.

Chapter 2. Sampling

Adler, H. L. and Roessler, E. B. (1977). *Introduction to Probability and Statistics* (6th edn.), W. H. Freeman & Co., San Francisco, p. 426.

Gy, P. M. (1982). *Sampling of Particulate Materials*, Elsevier, Amsterdam, p. 431.

Jones, M. P. and Beaven, C. H. J. (1971). Sampling of non-Gaussian mineralogical distributions. *Trans. Inst. Min. Metall.*, (80) B316.

Smith, R. and James, G. V. (1981). *The Sampling of Bulk Materials*. Royal Society of Chemistry, London, p. 191.

Chapter 3. Fractionation

British Standards Institution (1976). *Methods For Use of Test Sieves*, B.S.I. 1976.

Holmes, A. (1930). *Petrographic Methods and Calculations*, Thomas Murby & Co., London, p. 515.

Jones, M. P., Burley, J. R. J. and Simovic, M. (1974). Determining the amount and the size distribution of cassiterite in hard-rock tin ores. *Proc. 4th World Conf. on Tin*, Vol. 2. International Tin Council, London, p. 274.

Zussman, J. (Ed.) (1977). *Physical Methods in Determinative Mineralogy*, Academic Press, London (Chap. 1), pp. 1–34.

Chapter 4. Identification

Jones, M. P. and Fleming, M. G. (1965). *Identification of Mineral Grains*, Elsevier, Amsterdam, p. 102 (extensive Bibliography).

Schouten, C. (1962). *Determination Tables for Ore Microscopy*, Elsevier, Amsterdam, p. 242.

Uytenbogaardt, W. and Burke, E. A. J. (1971). *Tables for Microscopic Identification of Ore Minerals*, Elsevier, Amsterdam, p. 430.

Zussman, J. (Ed.) (1977). *Physical Methods in Determinative Mineralogy*, Academic Press, London.

Chapter 5. The polarising microscope

Kostov, I. (1968). *Mineralogy*, Oliver and Boyd, Edinburgh (transl. P. G. Emburey and J. Phemister).

Chapter 6. Image analysis – theory

Image analysis

Anon. (1973). *Bibliography of Articles on Automatic Quantitative Microscopy*, Imanco, Cambridge.

Chayes, F. (1956). *Petrographic Modal Analysis*, John Wiley & Sons, Inc., New York.

DeHoff, R. T. and Rhines, F. N. (1968). *Quantitative Microscopy*, McGraw-Hill Book Co., New York, p. 422.

Stereology

Jones, M. P. (1975). Automatic stereological analysis by electron probe X-ray microanalyser. *In* Holt, D. *et al.* (Eds.): "*Quantitative Scanning Electron Microscopy*", Academic Press, London, pp. 531–549.

Saltykov, S. A. (1958). *Stereometric Metallography* (2nd edn.), Metallurgizdet, Moscow, p. 466.

Underwood, E. E. (1972). *The Mathematical Foundations of Quantitative Stereology*. A.S.T.M. Philadelphia, STP 504, pp. 3–38.

Underwood, E. E. (Ed.) (1976). *Proc. 4th International Congress for Stereology*, Gaithersburg, Maryland, September 1975. National Bureau of Standards Spec. Publ. 431, p. 540.

Chapter 7. Modern image analysers

Fisher, C. (1971). Analysing images by computer. *New Scientist* (**51**) p. 676.

Jesse, A. (1974). Bibliography of automatic image analysis, *Microscope* (**22**) p. 89.

Seidel, K. (1969). The Quantimet as an integrating microphotometer, *Pract. Metallography* (**6**) p. 723.

Sturgess, G. L. and Braggins, D. W. (1972). Performance criteria for image analysis equipment, *Microscope* (**20**) p. 275.

Chapter 8. Stereology

Elias, H. (Ed.) (1967). *Proc. 2nd International Congress for Stereology*, Chicago, 1967. Springer-Verlag OHG, New York.

Underwood, E. E. (1970). *Quantitative Stereology*, Addison-Wesley, Cambridge, Mass.

Chapter 9. X-rays and electron beams

Adler, I. (1966). *X-ray Emission Spectrography in Geology*, Elsevier, Amsterdam, p. 258.

Birks, L. S. (1963). *Electron Probe Microanalysis*, Interscience, Publishers, London, p. 253.

Holt, D. B. *et al.* (1974). *Quantitative Scanning Electron Microscopy*, Academic Press, London, p. 570.

Jenkins, R. and De Vries, J. L. (1967). *Practical X-ray Spectrometry*, Phillips Technical Library, Eindhoven, p. 181.

Reed, S. J. B. (1975). *Electron Probe Microanalysis*, Cambridge University Press, Cambridge, p. 400.

Rogers, A. W. (1979). *Techniques of Autoradiography*, Elsevier, Amsterdam, p. 429.

Tousimis, A. J. and Morton, L. (Eds.) (1969). *Electron Probe Microanalysis*, Academic Press, London, p. 450.

Wenk, H. R. (Ed.) (1976). *Electron Microscopy in Mineralogy*, Springer-Verlag, Berlin, p. 564.

Woldseth, R. (1973). *X-ray Energy Spectrometry*, Kevex Corporation, Burlingame, Calif.

Zussman, J. (Ed.) (1977). *Physical Methods in Determinative Mineralogy*, Academic Press, London.

Chapter 10. Mineralogy in mineral processing

Burt, R. O. (Ed.) (1984). *Gravity Concentration Technology*, Elsevier, Amsterdam, p. 605.

Fander, H. W. (1985). *Mineralogy for Metallurgists: An Illustrated Guide*, Institute of Mining & Metallurgy, London, p. 77.

Gaudin, A. M. (1939). *Principles of Mineral Dressing*, McGraw-Hill Book Co., New York, p. 554.

Jones, M. P. (1979). *Mineral Processing*, Open University, Milton Keynes, Unit 3, T352, p. 80.

Laskowski, J. (Ed.) (1981). *Proc. 13th International Mineral Processing Congress*, Warsaw, 1979. Elsevier, Amsterdam, p. 2096.

Wills, B. A. (1985). *Mineral Processing Technology*, Pergamon Press, Oxford, p. 629.

Appendix 1. Practicals

Blyth, H. N. and Eldridge, A. (1953). Purpose in fine sizing and comparison of methods. In: *Recent Developments in Mineral Dressing*. Institute Mining and Metallurgy, London, pp. 11–30.

Appendix 2. Determinative scheme

Bowie, S. H. U. and Taylor, K. (1958). *A system of ore mineral identification*, Proc. UN Intern. Conf. Peaceful Uses At. Energy, 2nd, Geneva, 3, pp. 527–540.

Dana, E. S. (1949). *The System of Mineralogy*, Wiley, New York, 1:834 pp.; 2:1124 pp.

Feigl, F. (1985). *Spot Tests in Inorganic Analysis*, Elsevier, Amsterdam, 600 pp.

Jedwab, J. (1957). *Coloration de surface du beryl* (extract), Bull. Soc. Belg. Geol., Paléontol., Hydrol., 66, pp. 133–136.

Winchell, A. N. (1942). *Elements of Mineralogy*, Prentice-Hall, New York, 535 pp.

Glossary

Absorb: to take in and make part of a porous substance.

Adsorb: to take up and hold molecules of a foreign substance (as an extremely thin layer) on the surface of a solid substance.

Alluvial deposit: a deposit formed by stream action: modern alluvial deposits are found in stream beds; older deposits may be found away from existing streams.

Alpha-particle: positively charged nuclear particle that consists of two protons and two neutrons and which is produced in radioactive transformations.

Amorphous: having no definite form: having no readily-discernible crystalline structure: non-crystalline.

Analyser: a polarising device used in the petrological microscope.

Anion: a negatively charged ion.

Anisotropic: exhibiting properties that have different values when measured in different directions through a substance.

Anneal: to maintain at elevated temperature to encourage recrystallisation.

Atomic number: a number that is characteristic of a chemical element: it represents the number of protons in its nucleus.

Atomic number (mean): the calculated mean atomic number of a mineral; it is obtained by adding together the mass ratios multiplied by the atomic numbers of each elemental component of that mineral, e.g. mean atomic number ($\bar{z}$) of galena, PbS, is $(0.866 \times 82) + (0.134 \times 16) = 73.2$.

Beach sand: a granular mineral deposit formed on a beach: fossil beach sands may be found away from modern beaches.

Beta-particle: an electron or positron ejected from the atomic nucleus of an element during radioactive decay.

Bias: deviation of the expected value of a statistical estimate from the quantity it estimates.

Biaxial: having two optic axes.

Birefringence: the refraction by a crystal of a single ray of light into two directions so as to form two rays.

Cathodoluminescence: fluorescence in the visible range produced by bombarding material with high energy electrons.

Centrifuge: a device that uses centrifugal force to separate particles of different densities.

Classification: the operation of sorting mineral particles according to their settling speeds in a fluid (usually water).

Cleavage: the property of a crystallised mineral to split along definite planes.

Collimator: a device for producing a parallel beam of light, or other radiation.

Composite particle: a particle consisting of more than one mineral.

Composite probe: a probe which has passed through more than one mineral within a single particle.

Concentration criterion: a criterion that indicates whether two minerals of differing densities are likely to be separable by a separation process that relies upon gravitational forces.

Coning and quartering: a sampling method.

Contiguous: in contact with/along a boundary or at a point.

Covalent bond: a non-ionic chemical bond formed by shared electrons.

Crystalline: having a regular, well-defined atomic structure.

Cyclo-sizer : a device for fractionating mineral particles according to their settling velocities in a fluid (usually water).

Dense medium: a mixture of finely-divided solid particles and water used industrially for separating minerals on the basis of particle density.
Density: mass of a substance per unit volume.
Desorption: to remove a sorbed substance.
Diamagnetic: having a magnetic permeability less than that of a vacuum: diamagnetic minerals are slightly repelled by a magnet.
Dielectric constant: a measure of the ability of a substance to conduct direct electric current.
Drilling: the act of making holes in a solid rock.
Drilling (core): drilling with a hollow bit so as to retain a core (or cylinder) of largely undisturbed rock.
Drilling (percussion): drilling a hole by using a percussive (or hammering) action.

Endothermic: characterised or formed with the absorption of heat.
Exothermic: characterised or formed with the evolution of heat.
Ex-solution: the precipitation of a dissolved component from a solid substance.

Ferromagnetic: abnormally-high magnetism associated with a definite magnetic saturation point and appreciable residual magnetism.
"Float" product: the fraction that floats during a separation that uses heavy liquids.
Flotation: a process for separating minerals according to their natural (or induced) hydrophobicity – or water repellance.
Fluorescence: emission of light resulting from, and occurring only during, the absorption of radiation from an external source.
Frame (television): a complete image being formed by a television camera.
Fuller's earth: an earthy substance consisting chiefly of absorbent, non-plastic clay minerals.

Gel: a solid colloid.
Grade (of an ore or mineral concentrate): the proportion that consists of a selected valuable mineral or element.
Grain: a feature consisting of only one material; a mineral grain may occur as a free particle or it may be embedded in a rocky matrix.

Habit: the usual form adopted by a mineral.
Halogen elements: the elements chlorine, bromine, iodine, fluorine.
Heavy liquid: a liquid which is denser than water (density 1000 kgm^{-3}).
Histogram: a pictorial representation of a frequency distribution by means of rectangles whose widths represent class intervals and whose heights represent the corresponding frequencies.
Host (mineral): a large mineral grain that is older than, and encloses the minerals within it.
Hydrophilic: having a strong affinity for water.
Hydrophobic: lacking affinity for water (water-hating).

Image analyser: a device for measuring the geometric features seen on two-dimensional images.
Infrared: lying outside the visible spectrum at its red end; thermal radiation of wavelengths greater than those of visible light.
Isometric: a crystallographic system characterised by three equal axes at right angles to each other.
Isomorphous: similarity of crystalline form between substances of similar chemical composition.
Isotropic: exhibiting properties that have the same values, irrespective of the crystallographic direction in which they are measured.

Lamellae: thin, flat plates.
Liberated particle: a particle consisting of only one (liberated) mineral.
Liberation (of minerals): the freeing of a specified mineral from its associated matrix.

Magnetic permeability: the property of a magnetic material that determines the amount by which it modifies the magnetic flux in the region it occupies within a magnetic field.
Magnetic remanence: the magnetism remaining in a magnetised material when the magnetising force becomes zero.
Magnetic susceptibility: the ratio of the magnetisation in a material to the magnetising force.
Mean (value): average value (computed according to set procedure).
Median (value): that value (in an ordered set) that has the middle position.
Metamict (metamictisation): partial (or complete) destruction of the crystalline structure of a mineral by x-radiation derived from within that mineral.
Mineral: a naturally-occurring inorganic substance of characteristic (but not necessarily fixed) chemical composition and atomic structure.
Mineral deposit: an unusually rich, naturally-occurring concentration of a mineral(s) or of an element(s).
Mineral engineer: an engineer who is primarily concerned with the production of minerals.
Mineralogical feature: any feature (e.g. grain, particle, texture, porosity, etc) of mineralogical origin.
Mineralogy: the study of minerals – their origin, location, classification, nomenclature, etc.
Mineralogy (applied): mineralogy used in finding, winning and exploiting mineral resources.
Mining: the operations involved in the severance of mineral deposits from the earth's crust.
Mining engineer: an engineer who is primarily concerned with the extraction of mineral-rich rocks from the earth's crust.
Modal analysis: determination of the volumetric proportions of the various minerals in a specimen.
Mode (of a distribution): the most frequent value in a set of data.
Mode (of operation): the way in which a system is employed.
Monochromatic radiation:: radiation of a single wavelength.

Nomogram: a pictorial representation of lines marked off to scale and arranged so that a straight-edge joining known values on two lines will also show the value of an unknown point on a third line.

Optic axis: the line in a double refracting mineral that is parallel to the direction in which all the components of plane-polarised light travel with the same speed.
Ore: mineral deposit from which a metal can (ultimately) be produced by technological processes of mining and mineral processing that exist (or can reasonably be inferred).

Paramagnetic: having a small, positive magnetic susceptibility which varies only slightly with the magnetising force.
Parameter: a quantity, such as a mean or variance, that describes a statistical population.
Particle: a discrete (usually small) fragment of rock: it can contain any number of minerals. A particle that consists of only one mineral is said to be "liberated".
Partitioning: sharing of an element within the various minerals of a rock or an ore.
Peristaltic pump: a pump that uses a peristaltic action.
Pegmatite: a coarse, granitic rock that occurs in dykes or veins.
Petrological microscope: a microscope used especially in the study of the origin, history, occurrence, structure, chemical composition and classification of rocks.
Petrologist: one who specialises in the study of the origin and composition of rocks.
Pixel: picture "point"; small analysed area on an image.
Pleochroism: the property of a mineral of showing different colours when viewed by light vibrating parallel to different axes.
Polarise: to cause light waves to vibrate in a single direction.

Polariser: a device (usually, nowadays, a polaroid sheet) that is used to polarise a light beam in a specified direction.

Polychromatic: showing a mixture of colours.

Polymorph: any of the crystalline forms of a polymorphic substance.

Polymorphism: occurring in various forms.

Population: mass of material from which a sample is drawn.

Pseudomorph: a mineral having the characteristic outer form of another mineral.

Pulp, mineral: a mixture of water and fine-grained, solid particles.

Pycnometer: a special bottle used during the measurement of mineral density.

Quantum (quanta): a very small increment (or parcel) into which many forms of energy are subdivided.

Radioactive: the property of some elements to emit spontaneously alpha, beta and gamma rays by the disintegration of the nuclei of their atoms (e.g. uranium).

Random: lacking a definite plan or pattern.

Random (points): points chosen such that each has an equal probability of occurrence.

Raster: the area covered by a television "frame".

Refractive index: ratio between velocity of light in air and its velocity in a mineral.

Riffler: a device used for sampling granular materials.

Rock: a naturally-occurring, coherent aggregate of mineral grains.

Sample: a representative part of a larger whole: a finite part of a statistical population whose properties are studied to gain information about the whole.

Sample (representative): a small portion which, on average, has the same values for certain properties as the bulk material from which it was derived.

Sampled population: the population from which a sample is drawn.

Sampling frame: the set of instructions for collecting the samples.

Sampling (simple random): all possible values of the property being investigated have equal chances of being included in a sample.

Sampling (stratified): sampling of a total population that has been divided into convenient sub-populations.

Sampling (systematic): samples taken by a fixed, cyclic procedure.

Scalar property: a property that has a magnitude that is not affected by the direction in which it is measured.

Screen: a device for sizing mineral particles.

Screening: the action of passing (mineral) particles through a screen.

"Sink" product: the material that does not float in a heavy liquid.

Solid-solution: a solid containing a dissolved substance.

Specific gravity: mass per unit volume of a substance, or the ratio of the density of a substance to the density of water.

Spectrum: an array of components (as of x-radiation) separated and arranged in order of wavelength.

Standard deviation: a measure of the dispersion (or spread) of a frequency distribution.

Stereology: the branch of science concerned with developing and testing inferences about the three-dimensional characteristics of materials that, ordinarily, can only be observed from two-dimensional sections.

Stereoscopic microscope: a microscope that allows objects to be viewed in three dimensions.

Stokesian diameter: the diameter of a sphere having a known terminal velocity in a fluid.

Sub-sampling: further sampling of a sample.

Target population: population about which information is required.

Teetering: vacillating; in a heavy liquid, a teetering particle wavers between floating and sinking.

Traverse: linear route taken by, for instance, an electron beam across a specimen.

Twins: a compound crystal composed of two or more parts of the same crystallographic kind which have grown together in a specific manner.

Uniaxial: having only one optic axis.

Vector property: a property that has both magnitude and direction.
Vein: a narrow band of minerals (often orientated at a steep, nearly vertical angle).
Velocity, terminal: the maximum velocity achieved by a particle (or body) falling through a fluid.

X-ray: electromagnetic radiation with wavelength less than 10 nm.

Index

bold page numbers refer to figures; italic page numbers refer to tables
Note: Many minerals that appear in the Appendices are not included in the index.